Advances in Industrial Control

Advances in Industrial Control is a series of monographs and contributed titles focusing on the applications of advanced and novel control methods within applied settings. This series has worldwide distribution to engineers, researchers and libraries.

The series promotes the exchange of information between academia and industry, to which end the books all demonstrate some theoretical aspect of an advanced or new control method and show how it can be applied either in a pilot plant or in some real industrial situation. The books are distinguished by the combination of the type of theory used and the type of application exemplified. Note that "industrial" here has a very broad interpretation; it applies not merely to the processes employed in industrial plants but to systems such as avionics and automotive brakes and drive-train. This series complements the theoretical and more mathematical approach of Communications and Control Engineering.

Indexed by SCOPUS and Engineering Index.

Proposals for this series, composed of a proposal form (please ask the in-house editor below), a draft Contents, at least two sample chapters and an author cv (with a synopsis of the whole project, if possible) can be submitted to either of the:

Series Editor

Professor **Michael J. Grimble**: Department of Electronic and Electrical Engineering, Royal College Building, 204 George Street, Glasgow G1 1XW, United Kingdom; e-mail: m.j.grimble@strath.ac.uk

or the

In-house Editor

Mr. **Oliver Jackson**: Springer London, 4 Crinan Street, London, N1 9XW, United Kingdom; e-mail: oliver.jackson@springer.com

Proposals are peer-reviewed.

Publishing Ethics

Researchers should conduct their research from research proposal to publication in line with best practices and codes of conduct of relevant professional bodies and/or national and international regulatory bodies. For more details on individual ethics matters please see:
https://www.springer.com/gp/authors-editors/journal-author/journal-author-helpdesk/publishing-ethics/14214

Hai-An Zhu

Model-Based Control of Mass–Stiffness–Damping Systems

Springer

Hai-An Zhu
Department of Engineering
Omni Technologies
Singapore, Singapore

ISSN 1430-9491 ISSN 2193-1577 (electronic)
Advances in Industrial Control
ISBN 978-3-031-97591-2 ISBN 978-3-031-97592-9 (eBook)
https://doi.org/10.1007/978-3-031-97592-9

This Springer imprint is published by the registered company Springer Nature Switzerland AG
The registered company address is: Gewerbestrasse 11, 6330 Cham, Switzerland

Dedicated to Prof. Nan-Ning Zheng, my role model and source of inspiration.

Series Editor's Foreword

Researchers in control engineering often produce general theories while the engineers who must design, implement, and maintain control systems are focused on the peculiarities of their applications. In most cases, researchers develop algorithms for general control problems where there is a well-defined mathematical description, while engineers often have more immediate concerns over the limitations of equipment, quality of control, safety, security, and plant downtime.

The monograph series *Advances in Industrial Control* (AIC) attempts to bridge the divide by encouraging the development of advanced control techniques that can be applied in practice. It encourages the study of challenging application problems that need more advanced control design methods, that are model-based, or data-driven.

The development of new control theory, techniques, and technology has an impact on all areas of engineering applications. The study of application problems stimulates interest in engineering design issues and can provoke the creation of new control design methods. The series enables researchers to introduce new ideas that may be motivated by the challenging problems they meet in the applications of interest. It raises awareness of the various advances in control methods while explaining the challenges that can arise.

This AIC monograph, *Model-Based Control of Mass–Stiffness–Damping Systems*, covers the modeling of systems and both basic and advanced control methods. Chapter 1 provides a brief introduction and overview of application areas. Chapter 2, on mathematical modeling, is relevant to many applications but is particularly useful for vehicle-control problems that are of great current interest. Other applications, like wind-turbine controls, are also subject to intense research activity. When good physical system models are not easily available, attention turns to the use of system identification methods as described in Chap. 3. Batch and recursive-least-squares estimation methods are described.

To obtain a design model on which control algorithms can be based often requires some simplification of the physical system equations which can be time consuming and requires a good degree of expertise. Chapter 4 involves methods for model reduction, which is often a faster and easier way of obtaining a simplified model although not without its own challenges. The "modal" reduction method and ideas are interesting because they are not so well known despite the propensity of modes in dynamic systems like aircraft systems to have severe effects on achievable performance. Chapter 5, on the controllability and observability of linear systems, is necessary for completeness but it also has value when trying to understand the behavior of a linearized nonlinear system.

Chapter 6 introduces a basic control approach for nonlinear physical systems, namely model-based feedback linearization. This is often a practical way of obtaining a model for control but it does of course involve approximations that may not be valid in cases with difficult nonlinear dynamics. The main differences between parametric and structural nonlinearities are explained, since these often determine the type of control to be applied, and the topic of decoupling is also introduced.

Multi-loop control system structures are often used in applications. The synthesis of outer-loop controllers is covered in Chap. 7. These employ state-feedback controllers which act upon serially pre-compensated mass–stiffness–damping systems after the use of decoupling and linearization. Chapter 8 considers model-based decoupling methods and considers robustness problems which often arise in decoupling approaches.

Chapter 9, on model-based disturbance rejection and uncertainty, introduces feedforward and internal-model principal compensation methods. Chapter 10 concerns the use of model-following control to improve the robustness of model-based controllers in the presence of different types of uncertainty. Various ideas are proposed tailored to particular robustness problems. Chapter 11 considers structural properties of model-following methods and the topics of robustness for both stability and performance. Chapter 12 involves enhancing feedforward control methods, which is a very desirable ambition since feedforward is widely used in industry but often involves rather dubious assumptions.

Chapter 13 is on enhanced model-reference adaptive control. It is the case that real applications can involve adaptive controllers, but they must often be protected by the use of jacking software and only have a limited authority; however, it is useful to explore the opportunities that adaptation can provide in specific types of dynamic system. Chapter 14 is on improving internal-model control methods. Arising quite naturally from different optimal control philosophies, these methods have been proposed in various practical forms and are linked to the structure of so-called Smith predictors. They are sensitive to model uncertainties and improving robustness is an important goal.

Chapter 15, on machine learning in modeling and control, is very topical and the introductory material and comparison of recursive least squares and modern machine learning is helpful to those getting into the challenging world of artificial intelligence and machine learning.

This text on model-based control for mass–stiffness–damping systems includes a range of examples and useful lists of references. It should be valuable for control design engineers working on such systems and to researchers in advanced control. The text covers several useful topics in the control of dynamical systems and is a welcome addition to the series.

Glasgow, UK
May 2025

Michael J. Grimble

Preface

The field of control systems has seen significant advancements over the decades, particularly with the increasing complexity of dynamic systems that require robust and adaptive control strategies. Among these, mass–stiffness–damping systems—those governed by second-order differential equations—stand out due to their prevalence across various engineering disciplines, from aerospace and robotics to automotive and civil infrastructure.

This book was conceived to bridge the gap between theoretical advancements and practical applications in the control of such systems. The objective is twofold: to equip industry practitioners with actionable insights for improving the reliability and performance of their systems, and to provide postgraduate students with a comprehensive understanding of advanced control strategies that go beyond the classical approaches.

Divided into three main parts, the book begins with an exploration of the modeling techniques essential for understanding mass–stiffness–damping systems. This foundation is critical as adequate modeling serves as the bedrock upon which effective control strategies are built. The second and third parts delve into the basics of control and then into high-performance, model-based control techniques, offering readers a structured approach to tackle the inherent complexities and uncertainties that characterize real-world mass–stiffness–damping systems.

In writing this book, I have drawn from both the latest research and practical experiences. The aim has been to present these concepts in a way that is both rigorous and accessible, ensuring that the material is applicable to current industrial challenges while also paving the way for future developments in control systems.

Whether you are an engineer seeking to enhance the performance of a complex system, or a student aiming to deepen your understanding of advanced control techniques, I hope this book serves as a valuable resource in your journey.

Singapore | Hai-An Zhu

Acknowledgements This work has been shaped and strengthened through the guidance, support, and collaboration of many individuals to whom I am deeply grateful.

I would like to express my sincere thanks to Profs. Nan-Ning Zheng, You-Bai Xie, Zi-Yu Liu, Ren-Chu Zhu, and Wen-Hua Huang at Xi'an Jiaotong University; Prof. James C. Hung at the University of Tennessee; Profs. Chee-Leong Teo, Geok-Soon Hong, and Aun-Neow Poo at the National University of Singapore; Prof. Masayoshi Tomizuka at the University of California, Berkeley; Prof. Robert N. K. Loh at Oakland University; Prof. Charng-Ning Chen at Nanyang Technological University; and Prof. Peter B. Luh at the University of Connecticut. Their mentorship and academic influence have been instrumental in shaping my professional growth and research direction.

My deepest appreciation also goes to Robert Kirchgaessner at FESTO; Hans Falkenberg, C. T. Khoo, and Jos Kea at PHILIPS; Indrajit Purkayastha, Larry A. Sollecito, Darryl L. Wilson, and Jody A. Markopoulos at General Electric; and Yvonne F. Yu at Omni Technologies. Their support provided critical opportunities to translate theoretical concepts into real-world applications. Working with them has been intellectually stimulating and professionally rewarding.

I am equally grateful to Qing Zhao, Ji-Fu Yan, Lie Yu, You-Yun Zhang, Jin-Quan Ren, Xian-Zhong Xu, Shen Zhang, Yu-Ying Wang, Qing Li, Zhao Yang, Wen-Jian Cai, and many other colleagues whose feedback, insights, and discussions have significantly contributed to the ideas and developments presented in this book.

Finally, I would like to thank Oliver Jackson and Jayanthi Krishnamoorthi for their dedicated editorial guidance throughout the publication process. Their contributions have been invaluable to the realization of this work.

Singapore
May 2025

Hai-An Zhu

Competing Interests The author has no competing interests to declare that are relevant to the content of this manuscript.

Contents

Chapter 1
Introduction

Advanced control of real-world dynamic systems governed by second-order differential equations has garnered extensive research and development attention due to their complexity. These systems not only demand high-performance controllers but also present challenging control problems of broad interest. From a control science and technology perspective, progress in this area holds significant potential for contributing to both current and future technological advancements.

1.1 Mass–Stiffness–Damping Systems (MKC Systems)

Dynamic systems governed by second-order differential equations are foundational in engineering and physics, as they describe the dynamics of physical systems where inertia, stiffness (or elasticity), and damping are critical factors. These systems are typically modeled using the mass–stiffness–damping paradigm or the mass–spring–damper model. The general form of a second-order differential equation for such systems can be expressed as:

$$m\ddot{x}(t) + c\dot{x}(t) + kx(t) = f(t),$$

where

- m represents the mass of the system (inertia),
- c is the damping coefficient,
- k is the stiffness coefficient (spring constant),
- $x(t)$ is the displacement of the system over time,
- $\dot{x}(t)$ and $\ddot{x}(t)$ are the first and second derivatives of $x(t)$, representing velocity and acceleration, respectively,
- $f(t)$ is the external force applied to the system.

H. Zhu, *Model-Based Control of Mass–Stiffness–Damping Systems*, Advances in Industrial Control, https://doi.org/10.1007/978-3-031-97592-9_1

The equation described above pertains to a single-degree-of-freedom (1-DoF), or single-variable, system. For multivariable systems, the above equation becomes a second-order matrix differential equation, with the respective quantities represented by matrices and vectors. The term "MKC systems" is also derived from the primary physical parameters—mass (M), stiffness (K), and damping (C)—that characterize the second-order differential equations governing such dynamics. Henceforth, the abbreviation "MKC system" will be employed, where appropriate, to address space constraints within the text. These equations are ubiquitous in mechanical, civil, aerospace, and electrical engineering, among other fields. Typical examples of real-world mass–stiffness–damping systems are:

1. Mechanical (Mass–Spring–Damper) Systems:

   - *Automotive Suspension Systems:* These systems are designed to absorb shocks from road irregularities, improving ride comfort and handling. The mass corresponds to the car body, the spring to the suspension, and the damper to the shock absorber.
   - *Building and Bridge Oscillations:* In civil engineering, buildings and bridges are designed to withstand vibrations caused by earthquakes, wind, and other environmental forces. Tuned mass dampers (TMDs) are often employed to mitigate these vibrations.
   - *Seismic Isolation Systems:* These systems protect structures from seismic waves by introducing isolation bearings, which act like mass–spring–damper systems to absorb and dissipate seismic energy.

2. Electromechanical Systems:

   - *DC Motors:* The motion of the rotor in a DC motor is governed by second-order differential equations, where inertia (due to armature inductance, analogous to mass), stiffness (arising from magnetic forces and load stiffness, analogous to a spring), and damping (from armature resistance and mechanical friction) play critical roles.
   - *Piezoelectric Actuators:* These devices convert electrical energy into mechanical displacement, and their dynamics are often modeled as a second-order system due to the inherent stiffness and damping of the piezoelectric material.

3. Aerospace Systems:

   - *Flight Control Systems:* The dynamics of the flight control surfaces on an aircraft—such as ailerons, elevators, and rudders—are modeled using second-order differential equations. The control surfaces must counteract aerodynamic forces, where the inertia, damping (from air resistance), and stiffness (from control linkages) are critical.
   - *Satellite Attitude Control:* The orientation of satellites is controlled using reaction wheels or control moment gyroscopes, which can be modeled as second-order systems due to their inertial properties and the torques applied to maintain or change orientation.

4. Electrical Systems:

   - *RLC Circuits:* Electrical systems involving inductance, capacitance, and resistance can be naturally structured within a mass–stiffness–damping framework. In electrical engineering, second-order differential equations are used to model the behavior of series or parallel RLC (resistor–inductor–capacitor) circuits. The inductor L behaves like mass, storing kinetic energy in the form of a magnetic field; the capacitor C acts like a spring (specifically $1/C$ corresponds to stiffness), storing potential energy in the form of an electric field; and the resistor R functions like a damper, dissipating energy as heat. Thus, mass–spring–damper models can be directly mapped onto electrical systems through classical force–voltage or force–current analogies. These circuits can resonate at specific frequencies, with the inductance, capacitance, and resistance governing the system's effective inertia, stiffness, and damping characteristics.
   - *Power Transmission Lines:* Long transmission lines are often modeled with distributed parameters (high-frequency models): series inductance L, shunt capacitance C, and resistance R, forming an infinite sequence of mass–spring–damper analogs. The resulting partial differential equations (PDEs) (Telegrapher's equations) have second-order wave-like solutions, similar to mechanical vibration.

The current hot spots of mass–stiffness–damping systems in industry and research reflect the increasing demands for precision, reliability, and adaptability across various sectors. Key areas of focus include:

1. Robotics and Automation:

   - *Collaborative Robots (Cobots):* The development of cobots, which are designed to work alongside humans in industrial environments, is a major focus. These robots require advanced control systems to manage interactions with their environment, including dynamic coupling and nonlinearity, ensuring safety and precision.
   - *Robotic Manipulators:* Research and development are focused on improving the precision and adaptability of robotic arms, particularly in applications like assembly, surgery, and space exploration. Advanced control methods, including robust and adaptive controls, are being developed to handle the complex dynamics and varying loads these systems encounter.
   - *Autonomous Systems:* Autonomous vehicles and drones are a hot research and development area, where dynamic mass–stiffness–damping models are crucial for motion control, stability, and navigation. The focus is on real-time control systems that can adapt to rapidly changing environments.

2. Aerospace and Defense:

   - *Unmanned Aerial Vehicles (UAVs):* UAVs, including drones and military aircraft, are heavily reliant on mass–stiffness–damping dynamics for flight stability and control. Research and development are ongoing to improve

robustness and adaptability under various conditions, including extreme weather and combat scenarios.
- *Spacecraft Attitude Control:* The precise control of spacecraft orientation and stability in space, where environmental forces are minimal, is a critical area of research and development. Advanced control techniques are being developed to manage the complex dynamics of these systems, particularly for deep-space missions.
- *Hypersonic Vehicles:* Hypersonic flight introduces unique challenges in terms of stability and control due to the extreme speeds and dynamic pressures involved. Research and development are focused on developing control systems that can handle the highly nonlinear and time-varying dynamics of these vehicles.

3. Automotive and Vehicle Systems:

   - *Electric and Autonomous Vehicles:* The transition to electric vehicles (EVs) and the development of autonomous driving technologies are major industrial focuses. Dynamic mass–stiffness–damping systems are critical for vehicle stability control, suspension systems, and driver assistance systems. Research and development are ongoing to improve the precision and adaptability of these systems, particularly in varying driving conditions.
   - *Active Suspension Systems:* Active suspensions, which dynamically adjust the damping and stiffness of a vehicle suspension system in response to road conditions, represent an area of ongoing research and development. These systems are often modeled as mass–stiffness–damping dynamics and require advanced control methods for optimal performance.

4. Renewable Energy and Power Systems:

   - *Wind Turbines:* The control of wind turbine blades and the overall stability of the turbine structure are governed by mass–stiffness–damping dynamics. Research and development are focused on improving the robustness and efficiency of these control systems, particularly in offshore environments where conditions are more variable.
   - *Grid Stability and Control:* As renewable energy sources like wind and solar power become more prevalent, ensuring the stability of power grids, which are dynamic systems with second-order characteristics, is a critical research and development area. Advanced control methods are being developed to manage the variability and unpredictability of these energy sources.

5. Civil Engineering and Infrastructure:

   - *Earthquake-Resistant Structures:* The design of buildings and bridges to withstand earthquakes is a key area of research and development. These structures are normally modeled using mass–stiffness–damping dynamics, and research and development are focused on developing control systems, such as base isolation and TMDs, to enhance stability and safety during seismic events.

- *Smart Structures:* The development of smart materials and structures that can adapt to changing loads and environmental conditions is a growing research and development area. These systems often involve mass–stiffness–damping dynamics, and research and development are focused on integrating sensors and actuators for real-time control.

6. Biomedical Engineering:

- *Prosthetics and Exoskeletons:* The control of robotic prosthetics and exoskeletons, which assist individuals with mobility impairments, is an area of active research and development. These devices are usually modeled as mass–stiffness–damping systems, and research and development are focused on improving their adaptability and precision to match the user's movements.
- *Surgical Robotics:* The development of robotic systems for minimally invasive surgery involves mass–stiffness–damping dynamics for precise control of the surgical instruments. Research and development are focused on improving the robustness and accuracy of these systems to ensure patient safety and successful outcomes.

7. Manufacturing and Material Processing:

- *Precision Manufacturing:* In manufacturing processes that require high precision, such as semiconductor fabrication and 3D printing, dynamic mass–stiffness–damping models are used to control the motion and stability of the equipment. Research and development are focused on improving control systems to achieve higher precision and reliability in these processes.
- *Material Processing:* Processes like rolling, extrusion, and cutting involve dynamic systems where the interaction between the material and the machinery is modeled using second-order differential equations. Research and development are focused on optimizing these processes to improve product quality and reduce waste.

8. Smart Grids and Energy Management:

- *Demand Response and Load Balancing:* With the integration of renewable energy sources into the grid, managing the dynamic behavior of the grid, which can be modeled as a second-order system, is crucial. Research and development are focused on developing control systems that can balance load demand with supply in real time.
- *Energy Storage Systems:* The control of energy storage systems, such as batteries and supercapacitors, involves second-order dynamics. Research and development are focused on improving the stability and efficiency of these systems to support grid reliability and integration of renewable energy.

1.2 State-of-the-Art and Challenges in MKC System Control

Mass–stiffness–damping systems are integral to many industrial applications due to their complex dynamics. The most straightforward and widely used compensation technique for mass–stiffness–damping systems is the Proportional–Integral–Derivative (PID) controller. However, PID controllers are typically designed under the assumption of linear time-invariant (LTI) dynamics, which limits their effectiveness. Real-world mass–stiffness–damping systems, being inherently multi-degrees-of-freedom (m-DoF) and nonlinear, often challenge the assumptions of LTI models. Consequently, while PID controllers can deliver acceptable performance in some scenarios—particularly when system parameters are well-identified—their overall performance may fall short in more complex situations. This limitation underscores the need for more advanced control strategies to handle the dynamic complexities of mass–stiffness–damping systems effectively.

While classical control methods like PID remain popular, there is growing interest in more advanced techniques like robust control, adaptive control, and model predictive control to meet the challenges posed by modern applications. Generally, the control of mass–stiffness–damping systems involves three intertwined categories of methods and techniques:

1. Classical Control Methods:

   - *Proportional–Integral–Derivative Control:* Widely used in industrial applications, PID controllers are designed to regulate the output of a system by adjusting the control input based on the error signal [1]. For mass–stiffness–damping systems, PID controllers are effective in achieving desired transient and steady-state performance. However, tuning the PID gains for systems with high nonlinearity or time-varying parameters remains a challenge.
   - *Lead–Lag Compensation:* This method is used to shape the frequency response of a system, improving stability and performance [2]. It is particularly effective for mass–stiffness–damping systems with specific damping requirements.

2. Modern Control Techniques:

   - *Optimal Control (LQR/LQG):* Linear Quadratic Regulator (LQR) and Linear Quadratic Gaussian (LQG) methods are used for designing controllers that minimize a cost function, which typically includes terms for both the state and control effort of the system [3]. These methods are powerful for mass–stiffness–damping systems but require accurate system modeling and can be computationally intensive.
   - *State-Space Control (Observer Design):* In systems where all states are not directly measurable, state observers (e.g., Luenberger observers, Kalman filters) are employed to estimate the unmeasured states [4]. This method is

essential for the control of systems like robotic arms or aircraft, where not all states (e.g., velocities) can be directly measured.

- *Robust Control ($\mathcal{H}^\infty$, μ-Synthesis):* Robust control techniques are designed to handle system uncertainties and ensure stability and performance even in the presence of modeling errors or external disturbances [5]. For mass–stiffness–damping systems with high uncertainty and nonlinearity, robust control methods are critical in applications like aerospace and robotics, where operating conditions can vary widely.
- *Adaptive Control:* Adaptive control systems adjust their parameters in real time to cope with changes in system dynamics or external disturbances. This approach is being increasingly applied to mass–stiffness–damping systems [6] with time-varying parameters, high uncertainty, and nonlinearity, such as robotic manipulators handling variable payloads.

3. Advanced Control Techniques:
   - *Model Predictive Control (MPC):* MPC is a control strategy that optimizes the control input over a finite time horizon, considering system constraints [7]. It is gaining traction in industrial applications due to its ability to handle multivariable systems and constraints effectively, but requires significant computational resources, making real-time implementation challenging.
   - *Nonlinear Control:* For inherently nonlinear dynamic systems, techniques such as feedback linearization, sliding mode control, and backstepping are used [8]. These methods transform the nonlinear dynamics into a linear form or directly control the nonlinear dynamics, providing better performance than linear controllers in many cases.
   - *Data-Driven Control:* With the advent of machine learning, data-driven control approaches are gaining traction [9]. These methods do not require an explicit model of the system but instead rely on data to design controllers. The integration of artificial intelligence (AI) and machine learning with traditional control systems is advancing, particularly in applications requiring adaptability and real-time decision-making. While promising, the integration of data-driven techniques with traditional control methods remains an open research and development area.

It is widely recognized that most controllers are designed based on specific dynamic models of the plant. To maintain acceptable performance under changing parameters or varying operating conditions, two primary philosophies are employed in the control of uncertain systems: the robust control approach and the adaptive control approach.

The robust control approach involves designing controllers with fixed parameters and structures, intended to handle a predefined range of uncertainties without requiring real-time adjustments. This makes robust controllers relatively simple to implement, as they do not need continuous tuning in response to variations in the plant or environment. These controllers are specifically designed to ensure stability and performance despite the presence of uncertainties, delivering "acceptable" performance across a specified range of conditions.

On the other hand, the adaptive control approach focuses on designing controllers that "learn" and adapt to the uncertain parameters or structure of the system over time. Adaptive controllers continuously modify their gains or adjust the control structure in real time to better respond to the dynamic changes within the system. While this approach can address a broader range of uncertainties, it typically requires more complex implementation and real-time tuning.

Recently, there has been a trend toward combining these two approaches by "robustifying" adaptive controllers. This hybrid strategy aims to leverage the adaptability of the adaptive approach while incorporating the stability guarantees of robust control, thus providing a more comprehensive solution for managing uncertainties in dynamic systems [10, 11].

Despite significant advancements, issues of practical interest such as parameter variability, nonlinearity, and the need for real-time performance continue to drive research and innovation in this field. Typical challenges in control of mass–stiffness–damping systems are as follows:

- *Parameter Variability:* Many mass–stiffness–damping systems experience significant changes in their parameters due to environmental conditions, aging, or operational wear. This variability poses a challenge for maintaining consistent control performance.
- *Nonlinearity and Uncertainty:* Real-world mass–stiffness–damping systems usually exhibit nonlinear behavior, especially at the extremes of their operating ranges. Controlling such systems requires advanced techniques that go beyond the linear assumptions of classical methods. Dealing with inherent nonlinearity and uncertainty in dynamic mass–stiffness–damping systems continues to be a major research and development focus, with ongoing efforts to develop more robust and adaptive control methods.
- *High Precision Requirements:* In applications like robotics and aerospace, very high precision is required. This necessitates sophisticated control strategies that can handle small disturbances and uncertainties.
- *Computational Complexity:* As control algorithms become more sophisticated, their computational requirements increase, posing challenges for their implementation in real-time systems, especially in embedded environments.
- *Real-Time Constraints:* Many industrial applications require real-time control, which limits the complexity of the algorithms that can be implemented. Implementing advanced control strategies in real time, particularly in systems with fast dynamics, remains a significant challenge.
- *Robustness:* Ensuring that control systems remain stable and effective in the presence of uncertainties, disturbances, and modeling errors is a critical challenge, particularly for safety–critical applications like automotive and aerospace systems.

- *Interdisciplinary Integration with Other Technologies:* The increasing use of sensors, Internet of Things (IoT) devices, and AI, and big data in industrial applications requires control systems that can integrate with and leverage these modern technologies. However, ensuring reliability and robustness in such integrated systems presents challenges in terms of system complexity and cybersecurity.

To wrap up, the current hot spots in both industry and research for dynamic mass–stiffness–damping systems focus on advanced control methods, integration with emerging technologies, and tackling the challenges of nonlinearity, uncertainty, and real-time implementation.

1.3 Book Outline

This book is intended for industry practitioners and postgraduate students who are interested in applying advanced control strategies to their products or projects. In the realm of real-world applications, control practitioners now face the dual challenge of expanding their expertise to accommodate both complex models and systems with incomplete or inaccurate representations. On one hand, achieving optimal performance often necessitates the use of sophisticated models for control design, requiring the adaptation of control methods to these more intricate frameworks. On the other hand, because real-world system models are inherently imperfect, unmodeled dynamics and uncertainties present ongoing challenges for control engineers. This book offers a practical and distinctive approach to the modeling and control of mass–stiffness–damping systems, effectively bridging the gap between research and application.

The contents of this book are divided into three main parts. The first part focuses on the modeling of mass–stiffness–damping systems, while the second and third parts first describe algorithms and procedures relevant for basic control and go on to address model-based techniques for advanced control of these systems. The scope of model-based control of mass–stiffness–damping systems covered in this book is summarized in Fig. 1.1.

A solid foundation is key to success. In control practices, the process typically begins with modeling the plant to be controlled. A well-constructed model simplifies control design and enhances its effectiveness. Chapter 2 delves into the mathematical modeling of mass–stiffness–damping systems based on their physical principles. The chapter includes examples of mass–stiffness–damping systems that are of significant industrial interest today, such as active suspension systems for vehicles, robotic manipulators, large rotating machinery, and large-scale structures.

Occasionally, models cannot be easily derived due to a lack of data or the structural complexity of the plant systems. In such cases, techniques like parameter estimation and system identification become invaluable for system modeling. Chapter 3 introduces a systematic approach to parameter estimation for systems with incomplete information. This approach enables the straightforward determination

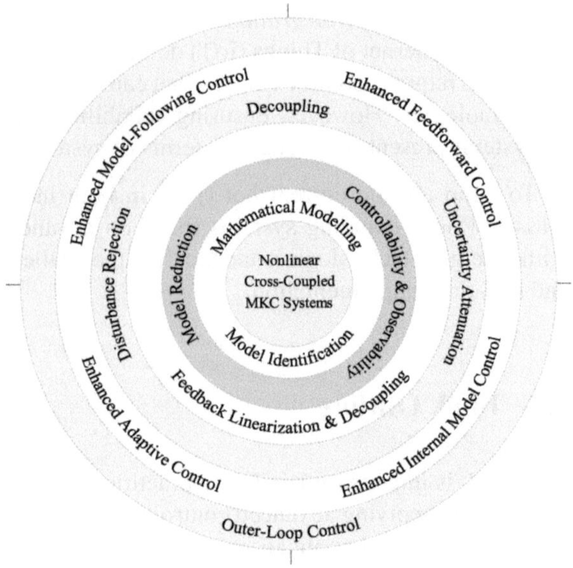

Fig. 1.1 Model-based control of mass–stiffness–damping systems

of parameters for linear, discrete- and/or continuous-time, multivariable dynamic models. The systematic technique offers several advantages over many existing methods, including simplicity in engineering implementation and flexibility for real-world applications. Beyond modeling, parameter estimation is also applicable to model reduction, decoupling control, and adaptive control, which are discussed in subsequent chapters.

High-order system models can sometimes be overly complex for control design and implementation. In Chap. 4, we explore model reduction techniques for large-scale systems. The methods presented in this chapter enable the creation of lower-order equivalent models that preserve the essential dynamics necessary for effective control. By applying these techniques, engineers can simplify control designs without compromising performance, making the control process more efficient and manageable.

Industrial mass–stiffness–damping systems are often complex, multivariable systems that present unique challenges in control design. Chapter 5 delves into the concepts of controllability and observability specific to these systems. By leveraging the structural characteristics of mass–stiffness–damping systems, this chapter introduces simplified criteria and algorithms that allow for real-time assessment of these critical properties. This is particularly important for mass–stiffness–damping systems like robotic manipulators, large-scale structures, and power systems, where changes in parameters or structural configurations during operation can significantly impact their controllability and observability. Ensuring real-time assessment and adaptation of these properties is crucial for maintaining system stability and performance under varying operational conditions.

Chapter 6 explores pre-compensation methods for model-based feedback linearization and dynamic decoupling of mass–stiffness–damping systems, using robotic manipulators characterized by structural nonlinearity and significant actuator dynamics as a primary example. This chapter presents two algorithms: one that is rigorous yet straightforward, and another that is simplified but highly effective. These model-based algorithms are designed to be intuitive and easy to implement, requiring minimal derivation and transformation. While the examples provided focus on robotic manipulators with specific dynamics, the underlying principles of these algorithms are broadly applicable to other mass–stiffness–damping systems as well.

After a bare mass–stiffness–damping system has been linearized and decoupled using the inner-loop pre-compensation techniques introduced in Chap. 6, the resulting pre-compensated system essentially consists of a series of isolated integrators. The logical next step is to equip this pre-compensated system with appropriate outer-loop controllers to ensure it operates as intended. Chapter 7 delves into the design of outer-loop controllers for closed-loop control of these inner-loop pre-compensated (linearized and decoupled) mass–stiffness–damping systems. The chapter also examines the steady-state performance of the closed-loop systems. To enhance steady-state performance, additional feedforward integral compensation is introduced as an extra degree-of-freedom in the control strategy. It is demonstrated that such a two-degrees-of-freedom control structure, combining feedback control and feedforward compensation, enables arbitrary pole/zero placement, offering significant flexibility in control design.

A critical aspect of multivariable mass–stiffness–damping systems is the dynamic coupling among their variables and subsystems. Conventional decoupling control schemes may not always be practical, either because precise models are unavailable under changing operating conditions or because the models are too complex for real-time control. In Chap. 8, a model-based decoupling strategy is introduced, leveraging the structural characteristics of complex mass–stiffness–damping systems to develop a decoupling control scheme suited for industrial applications. This scheme effectively mitigates cross-couplings among variables and subsystems, demonstrating robust performance even in the presence of nonlinear perturbations in system parameters. Moreover, it is designed with practical real-time control in mind, featuring a straightforward structure and ease of implementation.

In Chap. 9, model-based algorithms designed for disturbance rejection and uncertainty attenuation are discussed, leveraging the inherent physical properties of mass–stiffness–damping systems. The chapter explores disturbance observation and feedforward compensation techniques, which simplify control while delivering advanced performance. These enhanced algorithms offer high performance and are significantly simpler compared to conventional methods that lack the ability to reject disturbances and attenuate uncertainties.

Chapter 10 focuses on enhancing the robustness of model-following controllers. A straightforward uncertainty reduction technique, applicable across a wide range of scenarios, is introduced. This technique strengthens the robustness of conventional model-following controllers by effectively minimizing potential uncertainties.

Case studies involving mass–stiffness–damping systems demonstrate that significant performance improvements can be achieved with this uncertainty reducer. Importantly, the presented scheme is designed for easy implementation in practical controller design, making it accessible for real-world applications.

Chapter 11 further elaborates on the structural properties of the model-based control schemes discussed in Chap. 10. The enhanced uncertainty reduction scheme builds upon the conventional model-following structure. It is demonstrated that different configurations of these enhanced model-following schemes can achieve versatile performances in reducing uncertainties. For example, some of the new schemes can deliver excellent steady-state performance without requiring the addition of integrators in the forward paths.

Chapter 12 delves into model-based strategies for enhancing the robustness of feedforward control. Feedforward schemes have been frequently employed throughout this book. The chapter demonstrates that by incorporating the uncertainty reduction techniques presented earlier, the performance of feedforward control can be significantly improved, even when the underlying models are not perfectly accurate.

Chapter 13 addresses the challenge of stabilizing large-scale mass–stiffness–damping systems through decentralized adaptive pole-placement. The chapter introduces a simple and straightforward procedure for adaptive stabilization. Additionally, an enhanced model-reference adaptive scheme is presented to mitigate the effects of nonlinearity and cross-coupling while also suppressing external disturbances affecting the system.

A discussion on model-based control would be incomplete without addressing Internal Model Control (IMC). Chapter 14 introduces an enhanced IMC scheme that does not rely on the precise models typically required by conventional IMC. Designed with industrial applications in mind, the necessary structural modifications for robustness enhancement are both simple and effective. Moreover, the synthesis of additional components is straightforward and independent of the original IMC controller design.

In light of the growing interest in applying artificial intelligence to real-world control systems, Chap. 15 introduces the integration of machine learning into modeling and control. It highlights a lesser-known but significant insight: the recursive least-squares (RLS) algorithm, long utilized in system identification and adaptive control, is fundamentally a machine learning method. Due to its real-time adaptability, computational efficiency, and strong physical interpretability, RLS remains one of the most effective machine learning techniques for online estimation and control applications.

Chapter 16 offers concluding remarks and provides a perspective on future developments in the control of mass–stiffness–damping systems.

Appendix A provides a brief introduction to the method used in this book as a foundational tool for model-based linearization and decoupling of nonlinear, cross-coupled mass–stiffness–damping systems, serving as a quick reference.

Appendix B summarizes the pole-placement methods for second- and third-order linear dynamic systems discussed in this book.

Appendix C compiles additional materials related to the motion tasks of robotic manipulators featured in the examples of this book.

Appendix D details the experimental setup for readers interested in real-life implementation.

Finally, a comprehensive index is provided at the end of the book to facilitate cross-referencing of commonly used technical terms, enabling readers to quickly and easily locate specific subjects or concepts discussed throughout the text.

The contents of this book are structured progressively, beginning with modeling and advancing to control, covering both basic and advanced strategies. However, each chapter is designed to be self-contained, allowing readers to focus on specific topics of interest without needing to read through previous chapters. While some examples are cross-referenced for comparative purposes, readers can directly access any chapter without needing to follow the entire sequence.

References

1. Johnson MA, Moradi MH (2005) PID control: new identification and design methods. Springer, New York
2. Nise NS (2015) Control systems engineering. Wiley
3. Halperin I, Agranovich G, Ribakov Y (2021) Design of optimal feedback for structural control. CRC Press
4. Espinoza-Fraire AT, López AED, Morado RPP, Esqueda JAS (2023) Design of control laws and state observers for fixed-wing uavs: simulation and experimental approaches. Elsevier
5. Grimble MJ (2006) Robust industrial control systems: optimal design approach for polynomial systems. Wiley
6. Rigatos GG (2011) Modelling and control for intelligent industrial systems: adaptive algorithms in robotics and industrial engineering. Springer, Berlin
7. Kouvaritakis B, Cannon M (2016) Model predictive control: classical, robust and stochastic. Springer, Berlin
8. Grimble MJ, Majecki P (2020) Nonlinear industrial control systems: optimal polynomial systems and state-space approach. Springer, Berlin
9. Novara C, Formentin S (2019) Data-driven modeling, filtering and control: methods and applications. The Institution of Engineering and Technology
10. Landau ID, Airimitoaie TB, Castellanos-Silva A, Constantinescu A (2017) Adaptive and robust active vibration control: methodology and tests. Springer, Berlin
11. Lavretsky E, Wise KA (2024) Robust and adaptive control: with aerospace applications. Springer, Berlin

Part I
Modeling

The Road Map: Modeling

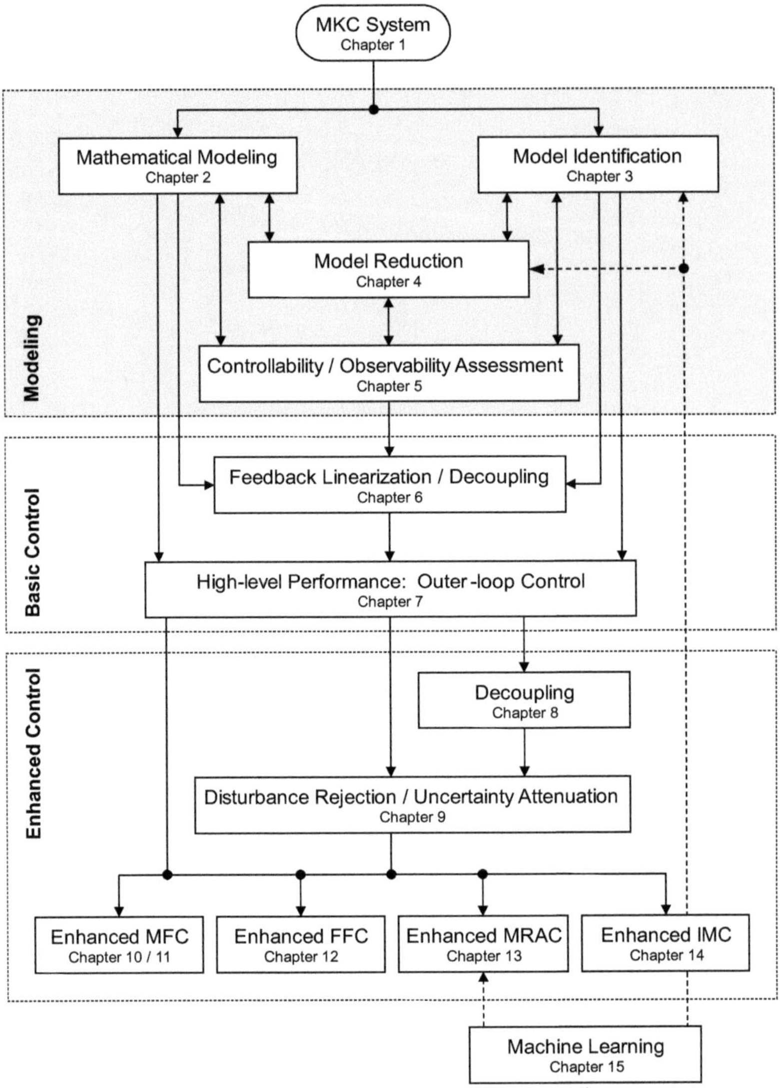

Chapter 2
Mathematical Models

To represent real-world systems in the context of control, various modeling methods are employed, each with distinct advantages and limitations. These methods are designed to address specific questions and focus areas, rather than offering a one-size-fits-all solution. This book explores the diverse range of modeling approaches from a systems and control perspective, with a particular focus on mass–stiffness–damping (mass, stiffness, damping) systems. The emphasis here is on understanding the dynamic behavior and the effects of applied forces within these systems.

Models are generally classified based on whether system parameters are distributed (continuum) or concentrated (rigid bodies). For systems with distributed parameters, the behavior is typically described by partial differential equations (PDEs), as it varies with both location and time. In contrast, systems with concentrated parameters are modeled using ordinary differential equations (ODEs), where the behavior is dependent solely on time. For the purposes of control, this book focuses on mathematical models represented by ordinary differential equations, which are well-suited for analyzing and controlling the dynamic behavior of mass–stiffness–damping systems.

2.1 Introduction

The dynamic behavior of many engineering systems can be effectively modeled using the following second-order matrix ordinary differential equation:

$$\mathbf{M}\ddot{\boldsymbol{x}} + (\mathbf{C}_D + \mathbf{C}_C)\dot{\boldsymbol{x}} + \mathbf{K}\boldsymbol{x} = \boldsymbol{f}. \tag{2.1}$$

In this equation, $\boldsymbol{x}$ and $\boldsymbol{f}$ are $n \times 1$ vectors, representing the state variables and external forces, respectively, while $\mathbf{M}$, $\mathbf{C}_D$, $\mathbf{C}_C$, and $\mathbf{K}$ are $n \times n$ matrices that represent the mass, damping, and stiffness properties of the system.

H. Zhu, *Model-Based Control of Mass–Stiffness–Damping Systems*, Advances in Industrial Control, https://doi.org/10.1007/978-3-031-97592-9_2

The model described by Eq. (2.1) is highly versatile, enabling the accurate representation of a wide range of physical systems through the appropriate selection of the "through" variables $\boldsymbol{f}$ (such as force in mechanical systems or current in electrical systems) and the "across" variables $\boldsymbol{x}$ (such as displacement in mechanical systems or voltage in electrical systems). This equation highlights the analogies that exist among various types of dynamic systems, making it broadly applicable to mechanical structures, electrical circuits, thermal systems, and beyond. It is a fundamental tool in diverse applications, providing a reliable framework for analyzing the dynamic behavior of these systems, which is crucial for accurate modeling and control.

Table 2.1 presents a structured mapping of the two classical analogies between mechanical systems (mass–spring–damper) and electrical systems (RLC circuits).

The force–voltage analogy constitutes the predominant framework in control engineering, particularly in the modeling and analysis of electromechanical systems such as motors and actuators. In contrast, the force–current analogy finds application in specialized domains, including magnetic circuit representation and network synthesis. Although the two analogies yield distinct equivalent circuit topologies, both faithfully preserve the dynamic structure of second-order systems and accurately replicate the associated differential equations. Consequently, the mass–spring–damper behavior is maintained across mappings. These structured correspondences enable systematic translation between mechanical and electrical models, thereby facilitating rigorous control design, dynamic simulation, and system identification across domains.

In describing physical systems, certain assumptions can often be made without losing generality, particularly when considering energy storage and dissipation:

- $\mathbf{M} = \mathbf{M}^{\mathrm{T}} > 0$: Represents generalized capacitive storage. This symmetric and positive-definite matrix $\mathbf{M}$ typically corresponds to the inertia or mass in mechanical systems, or capacitance in electrical systems, ensuring that energy storage is positive and well-defined.
- $\mathbf{K} = \mathbf{K}^{\mathrm{T}} \geq 0$: Represents generalized inductive storage. This symmetric and semi-definite matrix $\mathbf{K}$ relates to the stiffness in mechanical systems or inductance in electrical systems, ensuring non-negative potential energy storage.

Table 2.1 Mechanical–electrical analogies

Mechanical quantity	Force–voltage analogy	Force–current analogy
Force F	Voltage V	Current I
Velocity $v = \dot{x}$	Current I	Voltage V
Displacement x	Electric charge q	Magnetic flux ϕ
Mass m	Inductance L	Capacitance C
Stiffness k (Spring constant)	Inverse capacitance $1/C$	Inverse inductance $1/L$
Damping c	Resistance R	Conductance $G = 1/R$
External mechanical force	Voltage source	Current source

- $\mathbf{C}_D = \mathbf{C}_D^{\mathrm{T}} \geq 0$: Represents generalized energy dissipators. This symmetric and semi-definite matrix $\mathbf{C}_D$ corresponds to dissipative elements like damping in mechanical systems or resistance in electrical systems, ensuring that energy dissipation is non-negative and contributes to system stability.
- $\mathbf{C}_C = -\mathbf{C}_C^{\mathrm{T}}$: Represents generalized conservative elements. This skew-symmetric matrix $\mathbf{C}_C$ corresponds to elements that conserve energy, such as gyroscopic forces in mechanical systems, which do not contribute to energy dissipation but influence the dynamics of the system.

These assumptions help simplify the modeling of physical systems by ensuring that the mathematical descriptions align with physical principles like energy conservation, dissipation, and storage.

The model described by Eq. (2.1) can arise directly from lumped-parameter models or from finite approximations of distributed parameter systems typically governed by PDEs.

In the lumped-parameter approach, we often assume that the mass of a body is concentrated at a single point, usually at its center of gravity, under the assumption of a homogeneous and isotropic mass distribution. This simplifies the analysis by treating the body as a point-mass, which allows all forces acting on the body to be considered as acting at this central point. Line and area loads are similarly converted into equivalent forces applied at the center of gravity.

The concept of point-mass multibody systems, where bodies are idealized as rigid entities with mass concentrated at their center of gravity, is intuitive for most engineers. This book adopts the multibody system method as a fundamental approach. In this framework, mechanisms are viewed as assemblies of rigid bodies connected through ideal joints. These bodies are assumed to have no deformation, and their mass properties are concentrated at their respective centers of gravity. Additionally, massless force elements like springs and dampers, as well as constraints, can act between these bodies.

This method allows for a simplified yet accurate representation of complex mechanical systems, making it a powerful tool in modeling and control applications for engineering systems. By focusing on the essential dynamic properties of the system—such as inertia, damping, and stiffness—this approach facilitates the development of effective control strategies while maintaining physical realism.

One significant class of mass–stiffness–damping systems that is currently of great interest is large space structures (LSSs). These systems are large-scale distributed parameter systems that are typically discretized using the finite-element method (FEM) to transform them into a form represented by Eq. (2.1). In the context of LSS, the vector $\boldsymbol{x}$ represents displacements, the vector $\boldsymbol{f}$ denotes external forces, $\mathbf{M}$ is the mass matrix, $\mathbf{K}$ is the stiffness matrix, $\mathbf{C}_{\mathrm{D}}$ is the damping matrix, and $\mathbf{C}_{\mathrm{C}}$ represents the matrix associated with gyroscopic forces. Typically, the dimension n can be very large, leading to high-order models, and the resulting matrices are often sparse.

In the discretization process, FEM divides the continuous structure into a finite number of elements, each represented by a set of nodal points. The physical properties—such as mass, stiffness, and damping—are then mapped onto these elements, resulting in large but sparse matrices. These properties ensure that the dynamic behavior of the structure, including vibrations, can be accurately modeled and analyzed.

This approach is particularly important in the study of LSS because it allows engineers to manage the complexities associated with the large scale and distributed nature of these systems. By converting a continuous system into a discrete one, FEM facilitates the use of numerical methods to solve the equations of motion, providing insights into the structural dynamics that are crucial for control design and stability analysis. This makes the finite-element discretization of LSS a cornerstone in the modeling and control of large-scale mass–stiffness–damping systems.

2.2 The Simplest MKC Model

A typical mass–spring–damper model is composed of discrete mass nodes distributed throughout an object, interconnected by a network of springs and dampers. The most basic configuration of this type of system is illustrated in Fig. 2.1.

The dynamic behavior of this single-degree-of-freedom (1-DoF) model is governed by the following second-order ODE:

$$m\ddot{x} + c\dot{x} + kx = f. \tag{2.2}$$

In this equation, m, k, and c represent the mass, stiffness of the spring, and damping coefficient of the damper, respectively. The variables x, $\dot{x}$, and $\ddot{x}$ denote the displacement, velocity, and acceleration of the mass. The force f is the net force acting on the system, which includes both the actuating force and any external disturbances driving the motion of the mass. A block diagram representing this 1-DoF, second-order dynamic mass–stiffness–damping system is shown in Fig. 2.2.

The state-space representation of the mathematical model given in Eq. (2.2) is:

$$\begin{bmatrix} \dot{x} \\ \ddot{x} \end{bmatrix} = \begin{bmatrix} 0 & 0 \\ -\frac{k}{m} & -\frac{c}{m} \end{bmatrix} \begin{bmatrix} x \\ \dot{x} \end{bmatrix} + \begin{bmatrix} 0 \\ \frac{1}{m} \end{bmatrix} f. \tag{2.3}$$

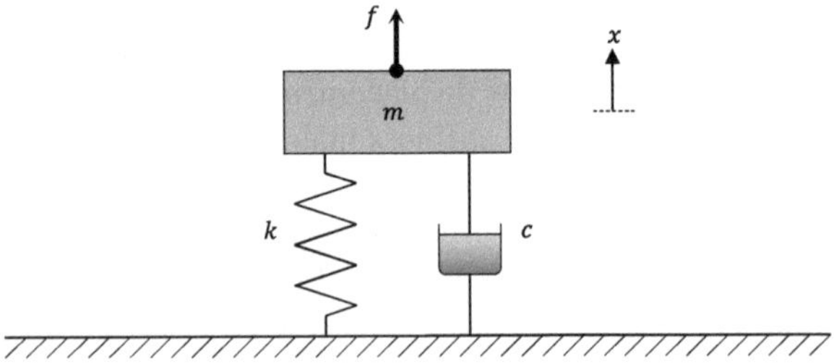

Fig. 2.1 Simplest mass–stiffness–damping system

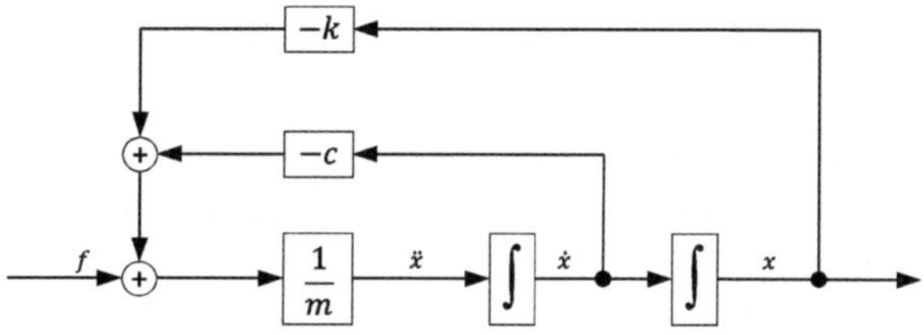

Fig. 2.2 Block diagram of a 1-DoF mass–stiffness–damping system

The transfer function of the system, from the input force f to the output displacement x, is given by:

$$\mathrm{P}(s) = \frac{\mathrm{X}(s)}{\mathrm{F}(s)} = \frac{1}{ms^2 + cs + k}. \tag{2.4}$$

This model represents the simplest yet most fundamental form of a mass–stiffness–damping system. As we will explore throughout this book, with appropriate transformations, linearization, and decoupling, many complex mass–stiffness–damping systems can be reduced to or approximated by such basic building blocks. This reduction is key in simplifying the analysis and control design for more intricate systems.

2.3 Active Suspension Systems

Conventional passive suspensions rely on springs and dampers positioned between the body of the vehicle and its wheel assemblies. The characteristics of these spring-damper systems must be carefully tuned to prioritize among several competing objectives, such as passenger comfort, road handling, and minimizing suspension deflection. Passenger comfort typically requires minimizing the acceleration experienced by the body of the vehicle due to road disturbances, while road handling and suspension deflection are concerned with maintaining tire contact with the road and preventing excessive movement within the suspension system [1].

Active suspension systems, on the other hand, offer a more sophisticated solution by using controlled actuators between the chassis and the wheel assemblies. These actuators can dynamically adjust the suspension characteristics in real time, allowing for a better balance between the conflicting objectives of comfort, handling, and deflection. Moreover, active suspensions can adapt to changing passenger preferences and varying road conditions, offering a significant improvement over the traditional passive systems. This adaptability is key in modern automotive design, where enhancing both safety and ride quality is increasingly important [2, 3].

2.3.1 Quarter-Vehicle Active Suspension Model

To analyze the dynamics and design controls for vehicles, quarter-vehicle models are frequently employed due to their simplicity and effectiveness in capturing essential dynamics. An example of a quarter-vehicle model for an active suspension system is depicted in Fig. 2.3. This model represents a two-degrees-of-freedom (2-DoF) multibody system where the mass m_b corresponds to the vehicle chassis (body), and m_w represents the wheel assembly. The spring k_s and damper c_s model the passive components of the suspension system, specifically the spring and shock absorber between the vehicle body and the wheel assembly. The spring k_t accounts for the compressibility of the pneumatic tire. The variables x_1, x_2, and r represent body travel, wheel travel, and road disturbance, respectively, while the force f_s, applied between the body and wheel assembly, is generated by the active suspension system.

As seen in Fig. 2.3, the model is essentially constructed by stacking two simple mass–stiffness–damping modules, one on top of the other. The resulting system can be represented by the following linearized fourth-order state-space equation:

$$\begin{bmatrix} \dot{x}_1 \\ \ddot{x}_1 \\ \dot{x}_2 \\ \ddot{x}_2 \end{bmatrix} = \begin{bmatrix} 0 & 1 & 0 & 0 \\ -\frac{k_s}{m_b} & -\frac{c_s}{m_b} & \frac{k_s}{m_b} & \frac{c_s}{m_b} \\ 0 & 0 & 0 & 1 \\ \frac{k_s}{m_w} & \frac{c_s}{m_w} & -\frac{(k_s+k_t)}{m_w} & -\frac{c_s}{m_w} \end{bmatrix} \begin{bmatrix} x_1 \\ \dot{x}_1 \\ x_2 \\ \dot{x}_2 \end{bmatrix} + \begin{bmatrix} 0 & 0 \\ 0 & \frac{1}{m_b} \\ 0 & 0 \\ \frac{k_t}{m_w} & -\frac{1}{m_w} \end{bmatrix} \begin{bmatrix} r \\ f_s \end{bmatrix}. \quad (2.5)$$

The block diagram in Fig. 2.4 illustrates that the system consists of two subsystems, which are cross-coupled by the stiffness k_s and damping c_s. This coupling is critical in the design of control strategies for such systems.

It is important to note that the spring constant k_t of the tire is highly uncertain, as it varies with tire pressure and wear conditions. For simplicity, the tire damping c_t is not included in the model, since it is often negligible, combined with the overall system damping, or treated as part of the system uncertainties.

Active suspension systems typically use actuators—whether electric, hydraulic, or another type—to achieve desired performance. These actuators have their own dynamic characteristics, which must be accounted for in the control design. For

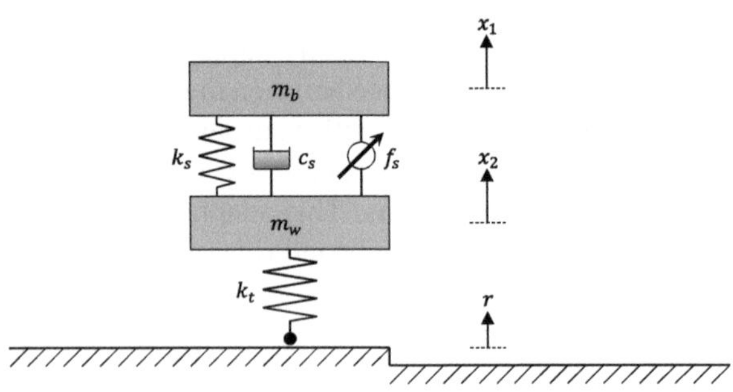

Fig. 2.3 Quarter-vehicle suspension model

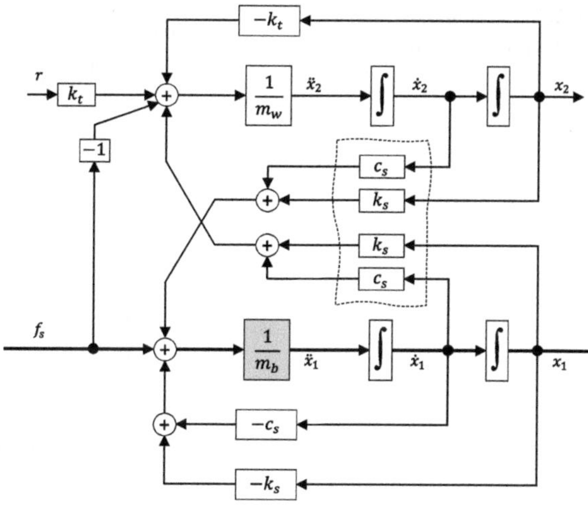

Fig. 2.4 Block diagram of a quarter-vehicle active suspension system

instance, a hydraulic actuator used in active suspension systems is modeled by the first-order transfer function:

$$\mathrm{P}_a(s) = \frac{g_a}{T_a s + 1}, \tag{2.6}$$

where g_a and T_a represent the gain and time constant of the actuator, respectively. This model provides an approximation of the actual actuator dynamics, which are frequency-dependent and therefore inherently uncertain. Additionally, the allowable travel of the active suspension system is limited by the displacement constraints of the actuator, typically allowing only a few millimeters of movement.

Electric actuators exhibit dynamic behaviors similar to those of hydraulic actuators, with distinct characteristics that are essential for their effective use in active suspension systems and other industrial applications. These dynamics, including the response time, torque production, and sensitivity to control signals of the actuator, will be thoroughly explored in later chapters of this book. Understanding these dynamics is crucial for the precise control and integration of electric actuators into advanced control systems, ensuring optimal performance and reliability across various operating conditions.

2.3.2 *Half-Vehicle Active Suspension Model*

To analyze the interaction between adjacent vehicle modules more accurately for improved vehicle performance, half-vehicle suspension models are essential. Traditionally, a half-vehicle model is created by combining two quarter-vehicle models with a common body mass representing half of the chassis. This configuration results in a model with four degrees-of-freedom: two for the vertical motion of the wheel

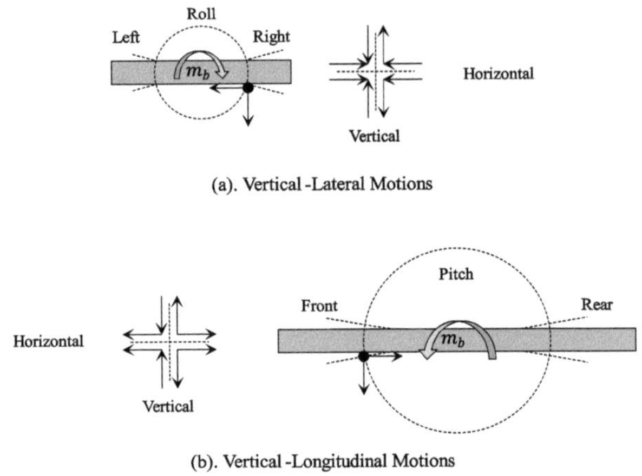

Fig. 2.5 Decomposition of roll and pitch motion

masses, one for the vertical motion of the body mass, and one for the roll or pitch angle of the vehicle body. These models are often referred to as roll models or pitch models.

Roll and pitch motions can be effectively decomposed into vertical and horizontal components, as illustrated in Fig. 2.5. The vertical component can be directly compensated by active suspension systems, which are designed to manage the response of the vehicle to vertical forces such as those from road irregularities and load shifts. These systems work by adjusting the damping and stiffness characteristics of the suspension in real time, enhancing ride comfort and stability.

The horizontal component, however, is less directly influenced by vertical control mechanisms and requires alternative solutions for effective management. Technologies such as active steering, torque vectoring, and adaptive cruise control play a crucial role in addressing horizontal forces. Active steering adjusts the steering angle based on vehicle speed and road conditions, improving cornering and handling. Torque vectoring distributes torque between wheels to optimize traction and control, particularly during acceleration and cornering. Adaptive cruise control maintains a safe distance from vehicles ahead by automatically adjusting speed, which can influence both longitudinal and lateral dynamics.

These complementary systems, as explained in Table 2.2, provide a holistic approach to managing the complex dynamics of roll and pitch, ensuring that both vertical and horizontal components are effectively controlled for improved vehicle performance and safety.

Moreover, the chassis of a vehicle is not a perfectly rigid structure; it can flex and deflect under various load conditions, which significantly influences the overall dynamics of the vehicle. These deflections impact how forces are distributed across the vehicle, affecting its stability, handling, and ride comfort. The lumped-mass model depicted in Fig. 2.6 takes these factors into account by focusing on the distribution and coupling of mass, stiffness, and damping within the chassis, offering a more accurate representation of the vertical dynamics of the vehicle.

Table 2.2 Roll and pitch motion of the chassis

Motion	Vertical	Horizontal
Roll (left–right)	Active suspension	• Lateral (side-to-side) • Active steering, torque vectoring
Pitch (front–rear)	Active suspension	• Longitudinal (forward/backward) • Adaptive cruise control

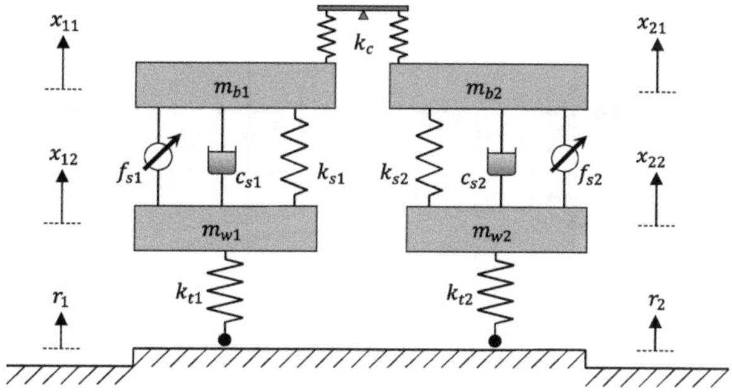

Fig. 2.6 Half-vehicle suspension model

In Fig. 2.6, additional springs with stiffness k_c represent the cross-coupling between the two lumped masses of the quarter-vehicle bodies. This 4-degrees-of-freedom multibody mass–stiffness–damping system allows for a more nuanced understanding of the actual dynamics of a half-vehicle by capturing how vertical forces are managed across the chassis. This model is particularly effective for simulating real-world conditions where the interplay between the different sections of the vehicle significantly affects overall performance. By accounting for these interactions, engineers can better design suspension systems that optimize vehicle stability, improve ride comfort, and extend the lifespan of various components, making the lumped-mass model a valuable tool in modern automotive engineering practices.

The block diagram of this half-vehicle model (as shown in Fig. 2.7) illustrates how the two quarter-vehicle systems are cross-coupled by the stiffness k_c. By using this model, engineers can better understand the dynamic behavior of the vehicle, enabling them to design more effective suspension controls that improve ride comfort, handling, and stability.

In comparison with conventional half-vehicle models, this lumped-mass half-vehicle model offers several distinct advantages. First, it more accurately captures the dynamic behavior of the vehicle by accounting for deflections in the chassis, which is not a perfectly rigid slab. This model also simplifies the analysis of roll and pitch motions by effectively decomposing them into vertical and horizontal components, making the complex interactions within the vehicle system more transparent and easier to manage. Additionally, the lumped-mass model enhances the clarity and effectiveness of control strategies by making the internal dynamics of the vehicle system more physically meaningful, facilitating decoupling between the quarter-vehicle systems once the cross-coupling stiffness k_c is determined.

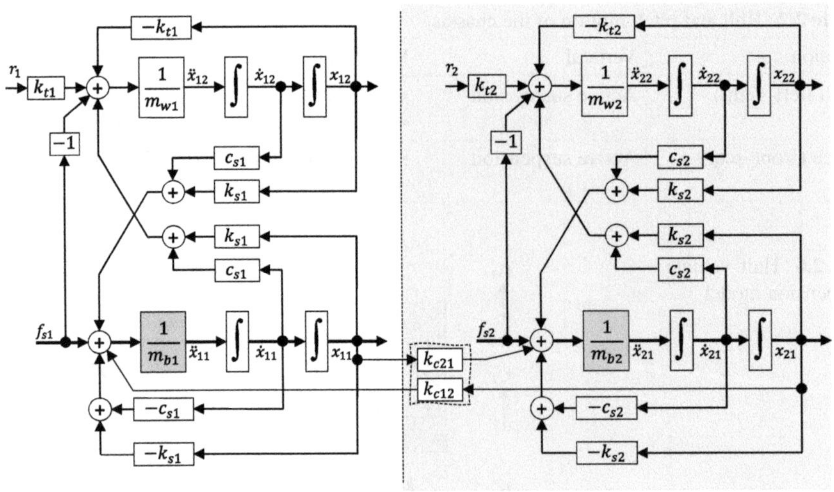

Fig. 2.7 Block diagram of a half-vehicle active suspension system

Overall, the lumped-mass model shown in Fig. 2.6 offers a more accurate and physically insightful framework for analyzing and controlling vehicle dynamics, particularly when considering the interactions between different vehicle modules. This model is particularly beneficial in industrial practices where precise control of vehicle behavior under varying conditions is critical.

2.3.3 Full-Vehicle Active Suspension Model

When investigating the dynamics and control of a complete vehicle, full-vehicle suspension models are essential. Typically, these models are constructed by combining four quarter-vehicle suspension models with a common chassis mass, resulting in a system with seven degrees-of-freedom: four for the vertical motion of the wheel masses, one for the vertical motion of the chassis mass, and two for the roll and pitch angles of the vehicle body. However, as discussed earlier, using a lumped-mass model is more advantageous for control design, where this model exhibits eight degrees-of-freedom, as shown in Table 2.3.

Table 2.3 Comparison of vehicle suspension models

Model	Roll-pitch model	Lumped-mass model
Half-vehicle	$(2 + 1 + 1)$ degrees-of-freedom	2×2 degrees-of-freedom
Full-vehicle	$(4 + 1 + 2)$ degrees-of-freedom	2×4 degrees-of-freedom

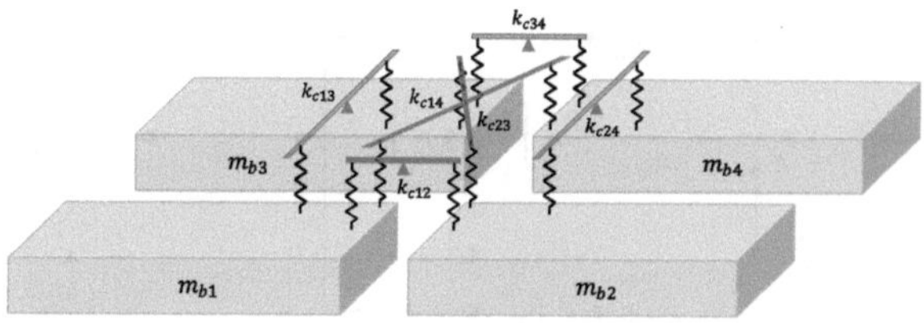

Fig. 2.8 Cross-coupling among lumped body masses

The cross-couplings among the lumped masses of the four quarter-vehicle bodies are depicted in Fig. 2.8, illustrating how these interactions influence the overall dynamics of the vehicle.

The block diagram of the full-vehicle active suspension system, as shown in Fig. 2.9, demonstrates how the four subsystems are interconnected and influence one another through cross-coupling. This interconnectedness makes it essential to consider the lumped-mass model when designing control strategies.

In comparison with conventional full-vehicle models, this lumped-mass full-vehicle model offers the following advantages:

- *Cross-Coupling Representation:* The inclusion of stiffness between the quarter-vehicle bodies accurately captures the interaction dynamics between them, providing a more realistic depiction of the behavior of the vehicle. This is crucial for understanding the complex interactions within the suspension system and developing appropriate control strategies.
- *Enhanced Control Transparency:* The lumped-mass model clarifies the internal dynamics of a full-vehicle system, facilitating the design of physically meaningful control strategies. Once the cross-coupling stiffness k_c is determined, decoupling the quarter-vehicle systems becomes more straightforward, enabling more effective control.

This model is particularly advantageous for implementing advanced control solutions, especially in modern vehicles where decentralized control strategies are increasingly adopted due to the availability of high-performance, low-cost microcontrollers. These controllers allow for more sophisticated, real-time control algorithms that can significantly enhance vehicle performance and safety.

2.4 Robotic Manipulators

A robotic manipulator is a quintessential example of a multibody mass–stiffness–damping (mass, stiffness, damping, and coupling) system, resembling a mechanical arm that mimics the functions of a human limb. These manipulators can either function as standalone mechanisms or as integral components of more complex robotic systems [4]. The links of the manipulator are interconnected by joints that allow for either rotational motion, such as in articulated robots, or translational (linear)

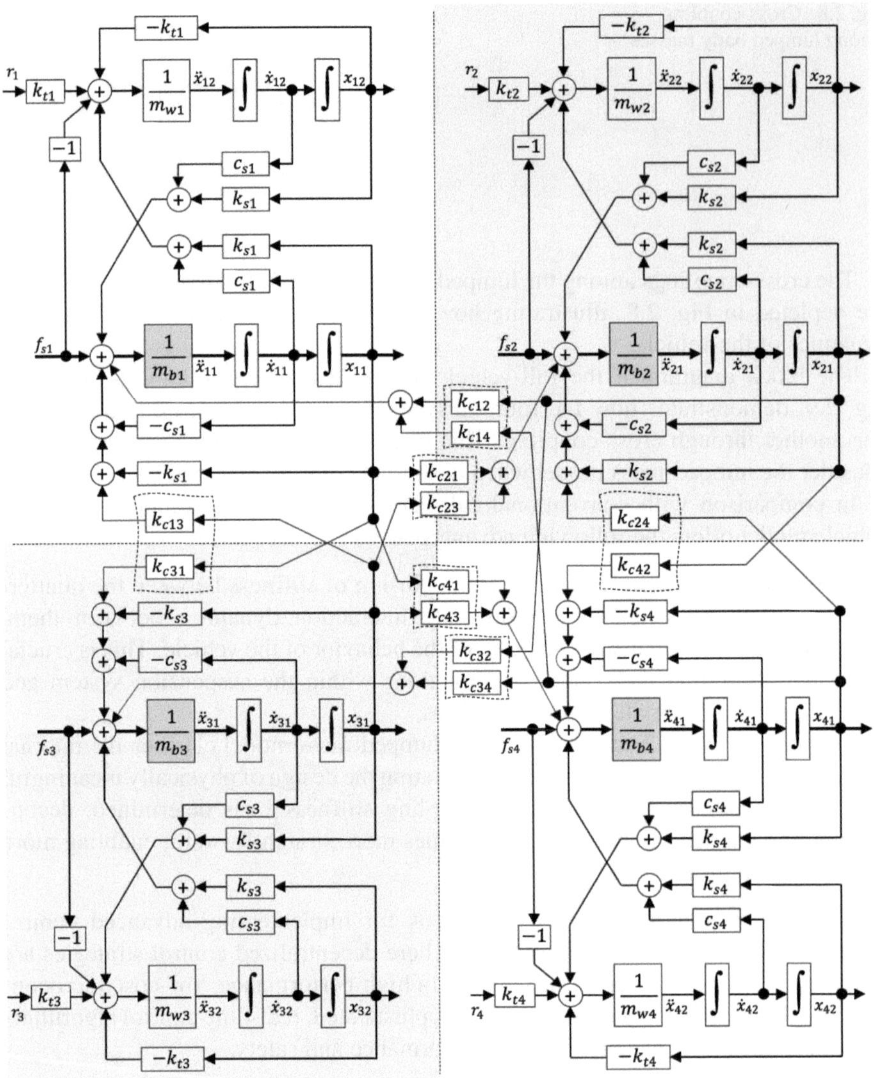

Fig. 2.9 Block diagram of a full-vehicle active suspension system

displacement. These links form a kinematic chain [5], the terminus of which is known as the *end effector*, analogous to the human hand. In industrial applications, end effectors are often simple tools like probes or grippers, but in the context of dynamic control, the end effector is typically considered part of the variable payload.

Robotic manipulators come in various configurations, each suited to specific tasks. The most common types include:

- *Articulated Manipulator:* Resembling the human arm, with multiple rotary joints.
- *SCARA Manipulator:* Ideal for pick-and-place tasks with rotational and linear motion.
- *Cartesian (Gantry) Manipulator:* Operates in linear motions along the x-, y-, and z-axes.
- *Cylindrical Manipulator:* Combines rotary motion with linear displacement along a cylindrical coordinate system.
- *Spherical (Polar) Manipulator:* Allows for rotational motion along multiple axes and linear movement along a radial arm.
- *Parallel (Delta) Manipulator:* Known for high-speed, high-precision tasks, utilizing a closed-loop kinematic chain.

For the purpose of this book, two commonly used robotic manipulators are examined. To simplify the analysis without losing generality, these manipulators are modeled in equivalent simplified configurations [6]. This approach facilitates a more straightforward understanding of their dynamic behavior while still capturing the essential characteristics needed for advanced control strategies.

2.4.1 Single-Link Manipulator

As a representative example of nonlinear mass–stiffness–damping systems, a single-link manipulator is used for both numerical and experimental analysis of the control schemes discussed in this book. As depicted in Fig. 2.10, this manipulator operates in a vertical plane, with its mass modeled as a point-mass concentrated at the free upper end of the link. This simplified configuration is particularly effective for demonstrating the performance characteristics of various control schemes, as it captures the essential nonlinear dynamics while remaining accessible for analysis.

The mechanical dynamics of this single-link manipulator can be formulated using the following equation:

$$ml^2\ddot{q} + mgl \cdot \cos q + \tau_D = \tau_C, \tag{2.7}$$

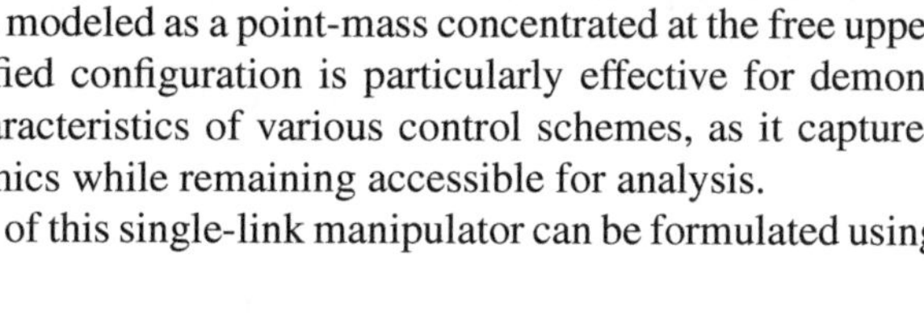

Fig. 2.10 Simplified configuration of a single-link manipulator

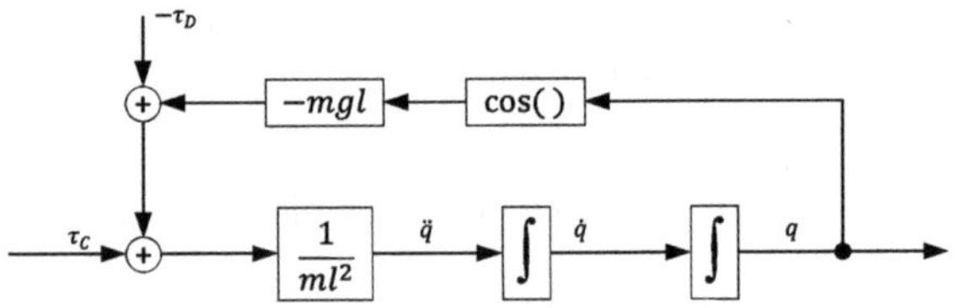

Fig. 2.11 Block diagram of the single-link manipulator

where

- l is the length of the link,
- q is the angular displacement of the link,
- m is the total mass, defined as $m := m_0 + m_p$ (where m_0 is the point-mass of the link and m_p is the payload at the free end),
- τ_C represents the control torque applied to the manipulator, and
- τ_D is the disturbance torque acting at the joint of the manipulator.

The block diagram of this system is shown in Fig. 2.11. The 1-DoF system illustrated here is structurally similar to the mass–spring–damper system depicted in Fig. 2.2, with one key difference: the linear stiffness element in the mass–stiffness–damping model is replaced by a nonlinear stiffness term, $\cos(q)$, and the system lacks inherent damping.

For convenience, polar coordinates have been adopted to model the dynamics of this rotating manipulator, which exhibits nonlinear motion. The relationship between Cartesian and polar coordinate systems is illustrated in Fig. 2.12.

As shown in Fig. 2.12, the conversion between polar and Cartesian coordinates is relatively straightforward. Given a pair of Cartesian coordinates (x, y), the corresponding polar coordinates can be determined using the following equations:

$$l = \sqrt{x^2 + y^2}, \tag{2.8}$$

and

$$q = \tan^{-1}\left(\frac{y}{x}\right). \tag{2.9}$$

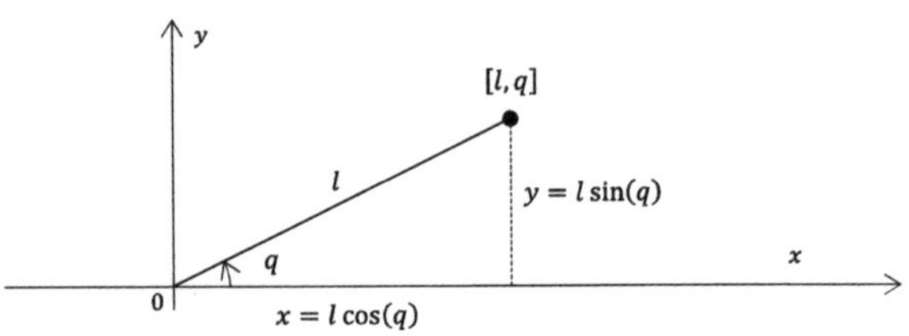

Fig. 2.12 Cartesian and polar coordinate systems

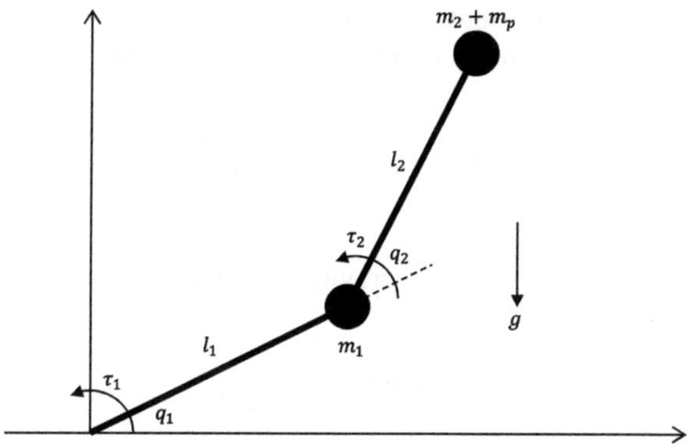

Fig. 2.13 Simplified configuration of a two-link manipulator

While converting to Cartesian coordinates may appear simple, it complicates the calculations in this context, particularly when dealing with the nonlinear dynamics inherent in this system. Therefore, using polar coordinates is more advantageous for analyzing and controlling the behavior of the system in practical applications.

2.4.2 Two-Link Manipulator

Robotic manipulators, as multibody mass–stiffness–damping systems, present significant challenges in control design due to their structural nonlinearities and complex dynamic behaviors. To address these challenges, this book frequently uses a two-link manipulator as a model for numerical studies and control performance analysis. As depicted in Fig. 2.13, the two links of this manipulator move in a vertical plane, with point-masses concentrated at the upper ends of each link. The lower end of the first link is anchored to a base, while the upper end of the second link is free to move, allowing for comprehensive demonstrations of control schemes.

The mechanical dynamics of this two-link manipulator are described by the following equations:

$$\mathbf{D}(\boldsymbol{q})\ddot{\boldsymbol{q}} + \boldsymbol{c}(\dot{\boldsymbol{q}}, \boldsymbol{q}) + \boldsymbol{g}(\boldsymbol{q}) + \boldsymbol{\tau}_D = \boldsymbol{\tau}_C, \tag{2.10}$$

where

- $\mathbf{D}(\boldsymbol{q})$ is the inertial matrix, which accounts for the combined mass and inertia of the links:

$$\mathbf{D}(\boldsymbol{q}) = \begin{bmatrix} M_2 l_2^2 + 2M_2 l_1 l_2 \mathrm{C}_2 + (m_1 + M_2) l_1^2 & M_2 l_2^2 + M_2 l_1 l_2 \mathrm{C}_2 \\ M_2 l_2^2 + M_2 l_1 l_2 \mathrm{C}_2 & M_2 l_2^2 \end{bmatrix}. \tag{2.11}$$

- $\boldsymbol{c}(\dot{\boldsymbol{q}}, \boldsymbol{q})$ represents the vector of centrifugal and Coriolis torques, highlighting the nonlinear cross-couplings between the links:

$$\boldsymbol{c}(\dot{\boldsymbol{q}}, \boldsymbol{q}) = \begin{bmatrix} -M_2 l_1 l_2 \mathrm{S}_2 \cdot \dot{q}_2^2 - 2M_2 l_1 l_2 \mathrm{S}_2 \cdot \dot{q}_1 \dot{q}_2 \\ M_2 l_1 l_2 \mathrm{S}_2 \cdot \dot{q}_1^2 \end{bmatrix}. \tag{2.12}$$

- $\boldsymbol{g}(\boldsymbol{q})$ is the vector of gravitational torques, showing how gravity affects the links:

$$\boldsymbol{g}(\boldsymbol{q}) = \begin{bmatrix} M_2 g l_2 \mathrm{C}_{12} + (m_1 + M_2) g l_1 \mathrm{C}_1 \\ M_2 g l_2 \mathrm{C}_{12} \end{bmatrix}. \tag{2.13}$$

Here, for $i = 1$ and 2, l_i and q_i are the length and angle of the i-th link, respectively. m_i is the point-mass of the i-th link, m_p the payload added on to the free upper end of the second link. τ_{Ci} is the control torque driving the i-th link, τ_{Di} the disturbance torque acting that the joint of the i-th link. For this manipulator,

- $M_2 := m_2 + m_p$, where m_2 is the point-mass of the second link and m_p is the payload at the upper end of the second link.
- $\boldsymbol{q} := \begin{bmatrix} q_1 \ q_2 \end{bmatrix}^{\mathrm{T}}$, the displacement vector of the links,
- $\boldsymbol{\tau}_C := \begin{bmatrix} \tau_{C1} \ \tau_{C2} \end{bmatrix}^{\mathrm{T}}$, the torques τ_{C1} and τ_{C2} are control inputs driving the links, while
- $\boldsymbol{\tau}_D := \begin{bmatrix} \tau_{D1} \ \tau_{D2} \end{bmatrix}^{\mathrm{T}}$, the torques τ_{D1} and τ_{D2} represent disturbance torques acting at the joints.

And $\mathrm{C}_1 := \cos(q_1)$, $\mathrm{C}_2 := \cos(q_2)$, $\mathrm{S}_2 := \sin(q_2)$, $\mathrm{C}_{12} := \cos(q_1 + q_2)$.

The block diagram of the two-link manipulator, shown in Fig. 2.14, reveals the complex interactions due to nonlinear cross-coupling. Unlike simpler mass–stiffness–damping systems, this model incorporates rotational motion, making the dynamics inherently nonlinear. When $l_2 = 0$, the two-link model simplifies to the single-link model discussed earlier.

The dynamic behavior of a robotic manipulator is heavily influenced by its parameters, such as mass, length, and actuator characteristics. These parameters are crucial in determining how the manipulator responds to various control strategies. In the examples provided in this book, a manipulator operating without any payload is

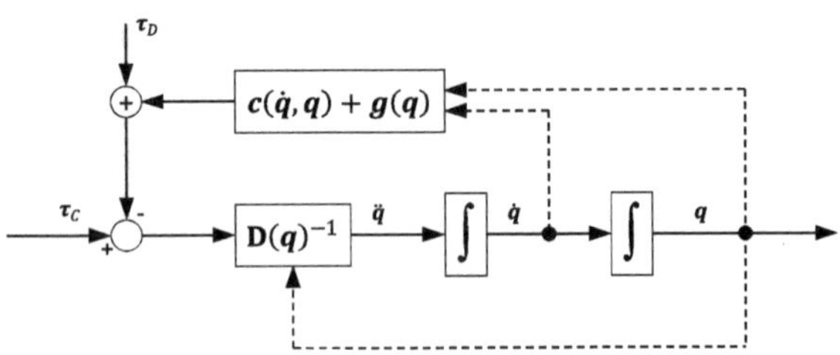

Fig. 2.14 Block diagram of the two-link manipulator

considered to be accurately modeled. This is because the parameters of the system remain constant, allowing for precise control based on the established model.

However, when a payload is introduced, particularly if the weight of the payload varies, the dynamics of the system change, leading to modeling uncertainties. These uncertainties make the manipulator more challenging to control accurately, as the original model no longer fully represents the behavior of the system. The added mass from the payload alters the inertia of the manipulator, potentially affecting its stability, response time, and overall performance.

This discrepancy between the model and the actual system behavior under varying payloads underscores the difficulties in maintaining precise control in real-world applications. In industrial practice, these challenges are often addressed by incorporating adaptive control strategies or robust control methods that can compensate for such uncertainties. These strategies adjust the controls in real-time to account for changes in the system dynamics, ensuring that the manipulator continues to operate effectively even when external conditions, such as payload variations, are present.

In essence, while a manipulator without payload can be accurately modeled and controlled, the introduction of a varying payload introduces complexities that require advanced control techniques to manage effectively. This highlights the importance of considering these factors in both the design and implementation of control systems in industrial robotic applications.

When modeling robotic manipulators, it is essential to consider the entire system, including actuators and mechanical transmission components. The main challenges in control design often arise within these mechanical transmission systems. However, with the advancement of drive technologies, particularly the introduction of direct-drive systems, many of the issues traditionally associated with mechanical gearing, such as backlash and excessive friction, can be effectively mitigated.

In a direct-drive system, the actuator is directly connected to the load without intermediary gears. This design eliminates backlash—a common problem in geared systems—and significantly reduces friction, which might only occur at the bearing interfaces. As a result, the mechanical structure becomes stiffer, leading to improved control precision and responsiveness. The absence of gears also reduces maintenance needs and increases the overall reliability of the system.

These improvements directly impact the performance of robotic manipulators. The enhanced stiffness and reduced friction allow for higher positioning accuracy and better repeatability, often by an order of magnitude compared to systems using traditional geared drives. Additionally, the simplified mechanical construction means that the system dynamics can be more easily identified and modeled with high accuracy, leading to more predictable dynamic responses.

These benefits make advanced control strategies, such as force control and compliant motion control, more practical and effective. By reducing uncertainties like friction and eliminating mechanical backlash, direct-drive systems pave the way for the successful implementation of high-performance control algorithms in industrial robotic applications.

2.5 Rotating Machineries

Rotating machinery encompasses a vast range of sizes, complexities, and configurations, found in applications as diverse as small electric motors in household appliances to large-scale turbo-generator sets. The basic rotor-bearing-foundation system is a fundamental element in countless systems, including centrifuges, vacuum pumps operating at speeds exceeding 100, 000 rpm, gyroscopes, machine spindles, helicopters, internal combustion engines (both gasoline and diesel), rotary and reciprocating compressors, and turbines of various types and sizes [7]. These machines play critical roles in many industries, from manufacturing to aerospace, highlighting the importance of understanding and controlling their dynamic behaviors.

The control of rotating machinery typically focuses on either the rotational speed or vibration management. In this book, the emphasis is primarily on vibration control, which is crucial for ensuring the longevity and reliability of rotating machines. Excessive vibration can lead to premature wear, component failure, and even catastrophic breakdowns, making effective vibration control a key aspect of maintenance and operational efficiency.

Vibration control in rotating machinery often necessitates indirect approaches, as direct access to the rotor is usually limited. Therefore, control strategies are frequently implemented through the bearings that support the rotors. The rotor-bearing system is thus a critical focus area for vibration control in rotating machinery [8, 9]. Bearings not only support the mechanical loads but also influence the dynamic behavior of the rotor. By adjusting the characteristics of the bearings—such as stiffness and damping—engineers can mitigate undesirable vibrations, enhancing the performance and stability of the entire system.

Given the complexities involved in the dynamics of rotor-bearing systems, advanced control techniques are often required. These may include active bearing systems, which use sensors and actuators to dynamically adjust bearing properties in response to changing operating conditions, thereby reducing vibration levels. This approach is particularly valuable in high-performance applications, such as in aerospace and precision manufacturing, where maintaining minimal vibration is critical for both safety and product quality.

Ultimately, while rotating machinery varies widely in application and design, the principles of vibration control through rotor-bearing systems remain universally applicable. This book delves into these principles, offering insights and techniques for effectively managing the dynamic challenges posed by rotating machinery across various industries.

As an example in this book, we analyze the rotor-bearing system of the low-pressure cylinder of a synthesis gas compressor used in a fertilizer plant. The rotor, as illustrated in Fig. 2.15, is supported at both ends by tilting-pad journal bearings, each consisting of five pads. These bearings are essential for maintaining stability and minimizing vibration in rotating machinery, playing a critical role in the overall performance and reliability of the system.

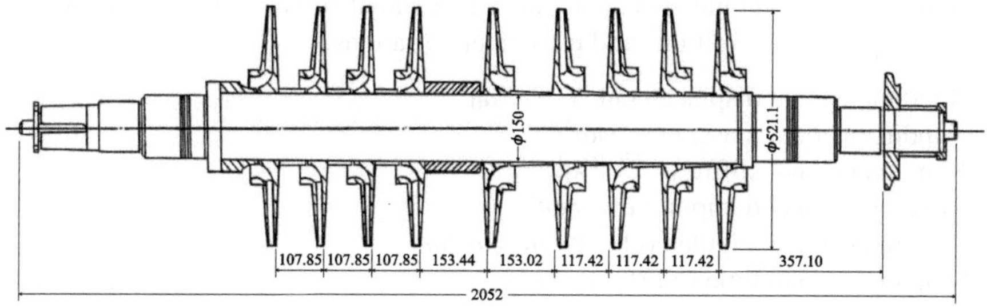

Fig. 2.15 Rotor of a compressor

The rotor is modeled as a composite system, consisting of a shaft with variable cross-sections and five discrete disks, as depicted in Fig. 2.16. For analytical purposes, this complex structure is further simplified into a model of a massless shaft with five point-masses, shown in Fig. 2.17. This approach, known as the lumped-parameter model, is commonly used in rotor dynamics to simplify the analysis while retaining the essential dynamic characteristics of the system.

When considering the dynamic effects of the tilting-pad bearings, the inertia of the pads is neglected.

The total mass of the rotor is approximately 516 kg, with the five discretized point-masses distributed as follows: $m_1 = 69.38$ kg, $m_2 = 130.41$ kg, $m_3 = 116.34$ kg, $m_4 = 144.37$ kg, and $m_5 = 55.47$ kg. This discretization allows for a more manageable analysis while maintaining the accuracy of the dynamic response.

For ease of analysis and calculation, it is standard practice to transform the equations of the system into a dimensionless form. The transformation formulas are provided, where variables with a "–" above the symbol represent nondimensional quantities. These include nondimensional displacement, mass, stiffness, damping,

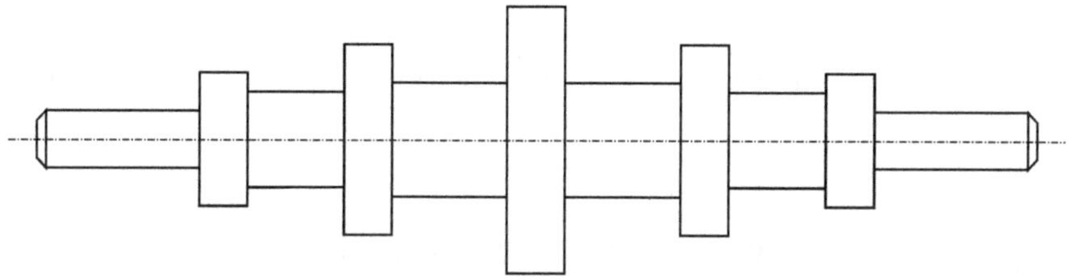

Fig. 2.16 Simplified rotor (illustrative only, not to scale)

Fig. 2.17 Point-mass model of the rotor

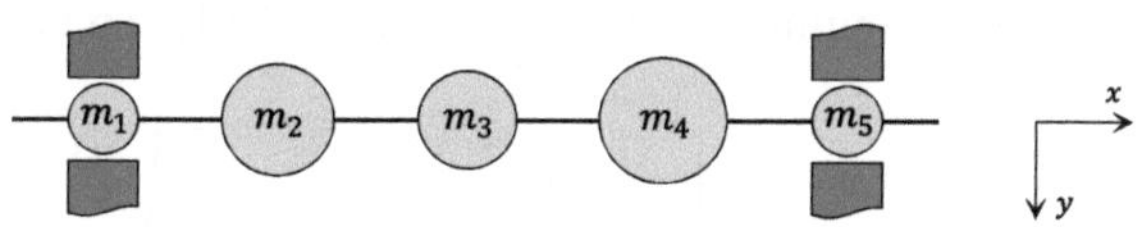

angular velocity, and time, which facilitate the comparison of system behavior under different operating conditions and parameter variations:

- Nondimensional displacement: $\overline{x} = x/d_0$
- Nondimensional mass: $\overline{m} = m\omega^2 f_m$
- Nondimensional stiffness: $\overline{k} = kf_m$
- Nondimensional damping: $\overline{c} = c\omega f_m$
- Nondimensional angular velocity: $\overline{\omega} = \omega/\omega_k$
- Nondimensional time (NDT): $\overline{t} = t/\omega$

where ω is the operating angular velocity, $\omega_k = 466.2/\text{s}$. is the first-order critical speed (angular velocity), $d_0 = 135$ mm is the average diameter, $f_m = 64/(\pi E d_0)$ is the conversion constant, $E = 2.06 \times 10^5$ N/mm^2 is the elastic modulus.

Vibration in rotating machinery, such as the whirling of a rotor around its equilibrium position, is a critical concern. The horizontal (x) and vertical (y) coordinates in the rotating plane of the shaft are used to describe this motion (the y-axis direction is 90° from the x-axis in the direction of shaft rotation). When using a second-order matrix ordinary differential equation to model the discretized and linearized dynamics of this multibody mass–stiffness–damping system, the equation of motion is as follows:

$$[\mathbf{M}][\ddot{\boldsymbol{x}}] + [\mathbf{C}]\begin{bmatrix} \dot{\boldsymbol{x}} \\ \dot{\boldsymbol{\phi}} \end{bmatrix} + [\mathbf{K}]\begin{bmatrix} \boldsymbol{x} \\ \boldsymbol{\phi} \end{bmatrix} = 0. \tag{2.14}$$

Here, **M**, **K**, and **C** represent the mass, stiffness, and damping matrices, respectively. The mass matrix **M** has a dimension of 10×10, while the stiffness and damping matrices, **K** and **C**, are 16×16. The discrepancy in matrix dimensions is due to the neglect of pad inertia in the tilting-pad bearings, which simplifies the model without compromising the accuracy required for control design.

The vectors $\boldsymbol{x}$ and $\boldsymbol{\phi}$ denote the 10×1 rotor-displacement vector and the 6×1npad tilt-angle vector, respectively:

$$\boldsymbol{x} = \begin{bmatrix} x_1 \ x_2 \ x_3 \ x_4 \ x_5 \ \ y_1 \ y_2 \ y_3 \ y_4 \ y_5 \end{bmatrix}^{\mathrm{T}}, \tag{2.15}$$

and

$$\boldsymbol{\phi} = \begin{bmatrix} \phi_{11} \ \phi_{12} \ \phi_{13} \ \phi_{51} \ \phi_{52} \ \phi_{53} \end{bmatrix}^{\mathrm{T}}. \tag{2.16}$$

These vectors are crucial for capturing the dynamic response of the rotor-bearing system under various operating conditions.

By re-arranging the state vector, Eq. (2.14) can be transformed into

$$\begin{bmatrix} 0 & \mathbf{I} \ 0 \\ \mathbf{M} & \\ 0 & \mathbf{C} \end{bmatrix}\begin{bmatrix} \ddot{\boldsymbol{x}} \\ \dot{\boldsymbol{x}} \\ \dot{\boldsymbol{\phi}} \end{bmatrix} = \begin{bmatrix} \mathbf{I} & 0 \ 0 \\ 0 & \\ 0 & -\mathbf{K} \end{bmatrix}\begin{bmatrix} \dot{\boldsymbol{x}} \\ \boldsymbol{x} \\ \boldsymbol{\phi} \end{bmatrix}. \tag{2.17}$$

After inverting the coefficient matrix on the left side of the above equation, the state-space representation in the form of

$$\begin{bmatrix} \ddot{x} \\ \dot{x} \\ \dot{\phi} \end{bmatrix} = [\tilde{\mathbf{A}}] \begin{bmatrix} \dot{x} \\ x \\ \phi \end{bmatrix}, \tag{2.18}$$

is obtained. Next, performing the following elementary matrix transformations

$$\begin{bmatrix} x \\ \dot{x} \\ \phi \end{bmatrix} = \begin{bmatrix} 0 & \mathbf{I} & 0 \\ \mathbf{I} & 0 & 0 \\ 0 & 0 & \mathbf{I} \end{bmatrix} \begin{bmatrix} \dot{x} \\ x \\ \phi \end{bmatrix}, \tag{2.19}$$

on the state variables in Eq. (2.18) yields the following state-space representation mathematical model

$$\begin{bmatrix} \dot{x} \\ \ddot{x} \\ \dot{\phi} \end{bmatrix} = [\mathbf{A}] \begin{bmatrix} x \\ \dot{x} \\ \phi \end{bmatrix}, \tag{2.20}$$

commonly used in control where $\mathbf{A}$ is 26×26 in this case.

The state-space representation of the system, a common format in control theory, is derived by transforming the equation into a first-order differential form. This representation is more suitable for designing and analyzing control systems, particularly for high-order, complex dynamics like those found in rotor-bearing systems.

In essence, this rotor-bearing model serves as a foundational example for the advanced control techniques discussed in this book. The simplifications made, such as using a point-mass model and neglecting pad inertia, are typical in industrial practice, where the goal is to balance accuracy with computational efficiency. This model forms the basis for exploring more complex control strategies aimed at improving the stability and performance of rotating machinery.

2.6 Large-Scale Structures

Vibration control of large-scale structures, such as wind turbine towers, long-span bridges, tall buildings, and large space structures, is crucial for ensuring their operational integrity, safety, and longevity. These structures are often subjected to dynamic loads, such as wind, seismic activity, and operational vibrations, which can lead to detrimental effects if not adequately controlled.

The modeling approach for these large-scale structures often mirrors the techniques used for multibody mass–stiffness–damping systems, as discussed in earlier sections. Specifically, these structures can be represented as systems of cross-coupled point-masses, where the relevant stiffness and damping properties define the interactions between these masses.

For instance, consider a wind turbine tower, depicted in Fig. 2.18a. Its lumped point-mass model, shown in Fig. 2.18b, resembles a vertically oriented rotor-bearing system, similar to the one in Fig. 2.17, with the upper end free and the lower end anchored to the ground. This analogy highlights the similarities in modeling these different types of structures, despite their distinct physical configurations.

The mathematical representation of such a multibody mass–stiffness–damping system, after discretization and linearization, can be described by the following second-order matrix ordinary differential equation:

$$\mathbf{M}\ddot{\boldsymbol{x}} + \mathbf{C}\dot{\boldsymbol{x}} + \mathbf{K}\boldsymbol{x} + \boldsymbol{f}_D = \boldsymbol{f}_C. \tag{2.21}$$

In this equation, **M**, **K**, and **C** are the mass, stiffness, and damping matrices, respectively. The vector $\boldsymbol{x}$ represents the displacements of the point-masses. The disturbance

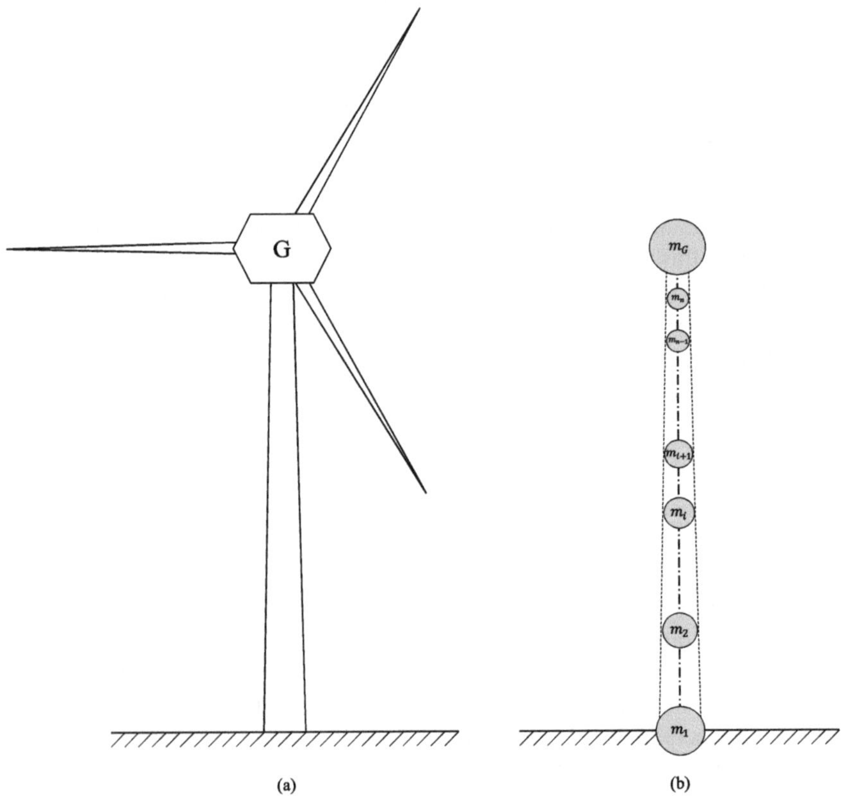

Fig. 2.18 Wind turbine tower

forces acting on the system are captured by the vector $\boldsymbol{f}_D$, while $\boldsymbol{f}_C$ represents the control forces that might be applied to mitigate vibrations. The dimensions of these matrices and vectors are determined by the degrees-of-freedom corresponding to the number of lumped masses.

Similarly, skyscrapers and other tall buildings can be modeled using vertical point-mass models akin to the one shown in Fig. 2.18b. For more complex structures, such as the example of a four-column building shown in Fig. 2.19, cross-coupled point-mass models can be used. These models account for interactions between different levels of the building and within each level, providing a more detailed representation of the dynamic behavior of the building.

Using this modeling approach, long-span bridges and large space structures [10] can also be represented as combinations of point-mass arrays. These arrays are based on the fundamental forms of cross-coupled shafts and columns, as seen in Figs. 2.17 and 2.18b. The resulting mathematical models of these mass–stiffness–damping systems adhere to the form presented in Eq. (2.21), allowing for consistent analysis and control design across various types of large-scale structures.

This systematic approach provides a unified framework for analyzing and controlling the vibrations of large-scale structures, enhancing their resilience and performance in real-world applications.

It is important to highlight that the actual locations of individual point-masses in many large-scale structures, such as wind turbines, skyscrapers, and bridges, are often physically accessible. This accessibility is critical for implementing control

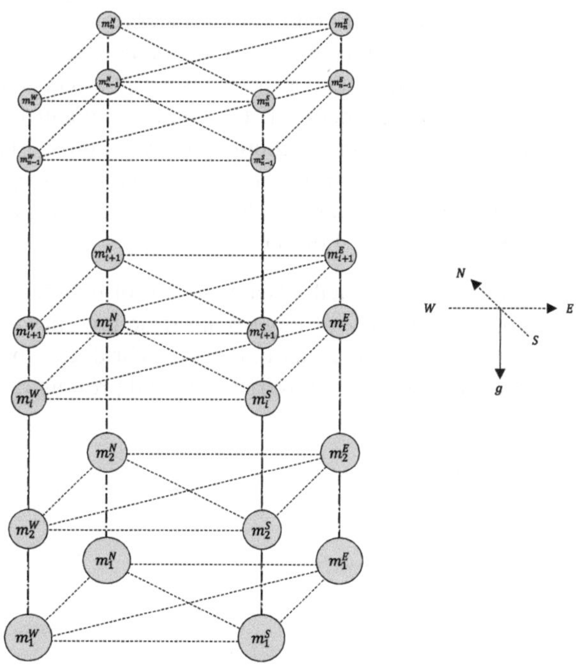

Fig. 2.19 Building of cross-coupled point-masses

strategies, as it allows for the installation of both passive compensation and active control elements, as shown in Fig. 2.20, directly at the desired points.

These elements, when properly installed, can effectively influence the dynamics and vibration behavior of the structure:

- *Passive Components* (k_{ci} and c_{ci})*:* The stiffness k_{ci} and damping c_{ci} components provide open-loop compensation. These elements locally adjust the dynamic properties at each point-mass, effectively tuning the response of the system [11]. For example, by altering the stiffness, one can modify the natural frequency of the structure, while changes in damping can affect how quickly vibrations dissipate.
- *Active Control Component* (f_{Ci})*:* The active component f_{Ci} provides closed-loop control, crucial for meeting the overall vibration control objectives. By actively adjusting the forces applied at specific points, this component can counteract disturbances in real-time, ensuring that the structure remains within safe operational limits.

A practical example of such solutions includes tuned mass dampers (TMDs) [12]. TMDs are widely used in skyscrapers and bridges to mitigate vibrations by introducing a counteracting force to the motion of the structure. The advantage of TMDs is that they can be tuned to target specific frequencies, making them highly effective in reducing resonant vibrations.

The following additional considerations should be noted when using compensation and control elements:

- *Mass* m_{ci} *Consideration:* When the mass m_{ci} of the compensation and control element is not negligible, it introduces an additional degree-of-freedom, thereby increasing the complexity and order of the system by two. This is an essential factor to consider during the design phase, as it affects the overall system dynamics.
- *Cross-Coupling Effects:* If the control element is installed between two point-masses, it creates cross-coupling between those masses. This cross-coupling can complicate the control strategy but also offers opportunities for more comprehensive control over the behavior of the system, particularly in managing multi-directional vibrations.

In practice, these methods are crucial for maintaining the structural integrity and operational efficiency of large-scale structures in various industries. For instance, skyscrapers in seismic zones or long-span bridges exposed to high winds rely heavily on these technologies to ensure safety and prolong their lifespan.

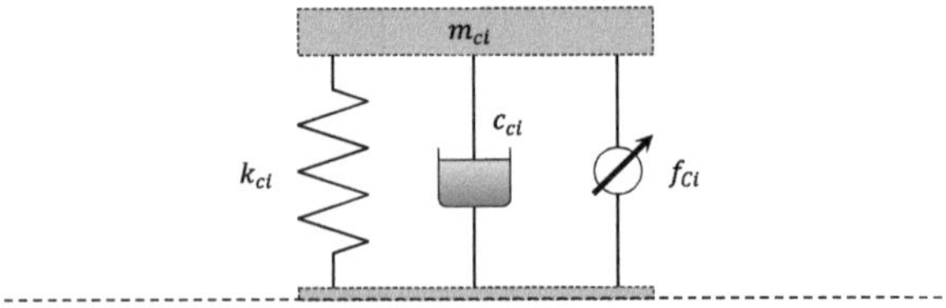

Fig. 2.20 Element of passive compensation and active control

2.7 Summary

Modeling typical mass–stiffness–damping systems has been extensively discussed in this chapter, illustrating that even the simplest 1-DoF model can serve as a foundational building block and subsystem for more complex, large-scale, multibody mass–stiffness–damping systems.

A crucial point in modeling multibody mass–stiffness–damping systems is a deep understanding of the physical characteristics of each subsystem and the cross-couplings between them. Gaining insights into the inner workings of the target system enables the creation of models that are effective for control purposes, without being burdened by unnecessary complexities.

The process of deriving a mathematical representation of the target system is only the beginning. Good modeling practice involves balancing the level of detail included in the final model and fine-tuning the parameters. This makes model validation an essential step to ensure that the model accurately reflects the behavior of the system under real-world conditions.

In practice, obtaining accurate parameters for lumped-parameter models can sometimes be challenging. In such situations, parameter estimation and system identification techniques, as discussed in the next chapter, can play a pivotal role in acquiring and validating the necessary parameters.

When dealing with high-order mathematical models that are too complex for effective system analysis and control design, model reduction techniques, covered in Chapter 4, offer valuable tools for simplifying these models while retaining the essential dynamics needed for control. By reducing the model order, engineers can focus on the most critical aspects of the system, facilitating more straightforward and effective control strategies in real-world applications.

In summary, a balanced approach to modeling—combining an understanding of physical principles, parameter estimation, model reduction, and model validation—leads to practical and efficient control solutions for complex multibody mass–stiffness–damping systems.

References

1. Goodarzi A, Lu Y, Khajepour A (2023) Vehicle suspension system technology and design. Springer, Berlin
2. Guglielmino E, Sireteanu T, Stammers CW, Ghita G, Giuclea M (2008) Semi-active suspension control: improved vehicle ride and road friendliness. Springer, Berlin
3. Savaresi SM, Poussot-Vassal C, Spelta C, Sename O, Dugard L (2010) Semi-active suspension control design for vehicles. Elsevier
4. Schilling RJ (1990) Fundamentals of robotics: analysis and control. Prentice-Hall
5. Craig JJ (2022) Introduction to robotics: mechanics and control. Pearson Education
6. Spong MW, Hutchinson S, Vidyasagar M (2020) Robot modeling and control. Wiley
7. Friswell MI, Penny JET, Garvey SD, Lees AW (2010) Dynamics of rotating machines. Cambridge University Press

8. Ren JQ, Xie YB, Zhu HA (1988) Investigation into the vibration behavior of a rotor-bearing system with non-linearized hydrodynamic bearing forces. In: Proceedings of the 6th international modal analysis conference, USA, vol 2, pp 1669–1674
9. Matsushita O, Tanaka M, Kobayashi M, Keogh P, Kanki H (2019) Vibrations of rotating machinery. In: Advanced rotordynamics: applications of analysis, troubleshooting and diagnosis, vol 2. Springer, Berlin
10. Nurre GS, Ryan RS, Scofield HN, Sims JL (1984) Dynamics and control of large space structures. J Guid Control Dyn 7(5):514–526
11. Yang F, Sedaghati R, Esmailzadeh E (2022) Vibration suppression of structures using tuned mass damper technology: a state-of-the-art review. J Vibr Control 28(7–8):812–836
12. Koutsoloukas L, Nikitas N, Aristidou P (2022) Passive, semi-active, active and hybrid mass dampers: a literature review with associated applications on building-like structures. Develop Built Environ 12:100094

Chapter 3
Model Identification

In industrial practices, the derivation of mathematical models using available parameters is a cornerstone of controller design. However, real-world systems often present complexities that challenge straightforward modeling. These complexities may result from the inherent complexity of the system, the presence of nonlinearities, or time-varying parameters. In such cases, relying solely on exact models may not be feasible. Instead, estimated lumped-parameter equivalent models can provide a practical solution for control design.

This chapter introduces a systematic approach to parameter estimation for dynamic systems, offering a unified framework for estimating parameters of both linear discrete-time and continuous-time multivariable systems. This approach stands out due to its simplicity in engineering implementation and its flexibility in accommodating real-world applications where systems may have time-varying or nonlinear characteristics.

3.1 Introduction

In modern control engineering and its practical applications, the analysis and design of control systems fundamentally rely on accurate quantitative models of the plants being controlled. Consequently, obtaining a precise model of the plant is a critical first step in developing effective control algorithms [1]. While conceptual models can often be derived from the physical characteristics of the plant, it is equally important to employ system identification techniques to develop both parametric [2, 3] and nonparametric [4] models, especially when the actual parameters of the plants are not readily available. These identification techniques are integral to advanced system analysis, modeling, and control. As such, the theory and practice of system identification and parameter estimation have continually evolved to meet the increasing complexity and demands of modern control systems [5, 6].

H. Zhu, *Model-Based Control of Mass–Stiffness–Damping Systems*, Advances in Industrial Control, https://doi.org/10.1007/978-3-031-97592-9_3

A wide array of algorithms has been developed for system identification, each offering unique strengths tailored to specific operating conditions [7, 8]. However, many of these algorithms encounter significant limitations in real-time applications, particularly in scenarios where system parameters are time-varying or exhibit nonlinear behavior—challenges frequently encountered in industrial practice. Among the various techniques, recursive least-squares estimation is widely recognized for its efficacy in the modeling and control of dynamic systems. The systematic approach detailed in this chapter, based on the least-squares method, has demonstrated significant advantages in real-world applications, offering not only simplicity in engineering implementation but also the flexibility needed to adapt to diverse practical conditions.

The main content of this chapter is structured in a progressively logical sequence as follows:

i. *Estimation Algorithm for Linear Difference Equations:* A modified algorithm is presented for the identification of linear single-input, single-output (SISO) dynamic systems represented by difference equations. This algorithm has the required accuracy and robustness of parameter estimation in linear SISO systems, making it more suitable for practical applications.
ii. *Estimation Algorithm for Discrete State Equations:* Building on the algorithm introduced in (i), a parameter estimation algorithm for linear multi-input, multi-output (MIMO) discrete-time state equations is detailed. This approach extends the capability of the SISO method to handle more complex MIMO systems. Real-world applications have demonstrated that this estimation technique is both feasible and effective, particularly in systems with multiple interacting variables.
iii. *Estimation Algorithm for Continuous State Equations:* Expanding on the methods outlined in (i) and (ii), a direct algorithm is introduced for the parameter estimation of linear MIMO continuous-time state equations [9]. This algorithm employs the trapezoidal rule for numerical integration, converting continuous-time differential equations into a difference format, thus enabling recursive computation. The use of an appropriate sampling interval is crucial, as it ensures that the presented identification technique yields accurate and reliable results in continuous-time systems.

In addition to these core algorithms, the chapter also addresses related issues, such as the identification of dynamic systems with time-varying and/or nonlinear parameters. These discussions provide insights into the challenges and potential solutions for dealing with more complex and realistic industrial systems.

3.2 Algorithm for Linear Difference Equation

Given that the system identification algorithms discussed in this chapter are based on the least-squares estimation method [8], it is essential to first introduce a modified version of the algorithm that is specifically tailored for practical, real-world applications.

We assume that the plant to be identified is a SISO linear time-invariant dynamic system, described by the following n-th order linear difference equation with constant coefficients:

$$\begin{aligned} y(k) + a_1 y(k-1) + \cdots + a_n y(k-n) \\ = b_0 u(k) + b_1 u(k-1) + \cdots + b_l u(k-l) \\ + e(k), \quad l \le n, \end{aligned} \tag{3.1}$$

where $u(k)$ and $y(k)$ represent the measured system input and output, respectively, and $e(k)$ is the residual error.

This system can also be compactly represented using the unit backward-shifting operator z^{-1} as:

$$\mathrm{A}(z^{-1})y(k) = \mathrm{B}(z^{-1})u(k) + e(k), \tag{3.2}$$

where

$$\begin{aligned} \mathrm{A}(z^{-1}) &= 1 + a_1 z^{-1} + a_2 z^{-2} + \cdots + a_n z^{-n}, \\ \mathrm{B}(z^{-1}) &= b_0 + b_1 z^{-1} + b_2 z^{-2} + \cdots + b_l z^{-l}. \end{aligned}$$

The objective of system identification here is to determine the coefficients in the polynomials $\mathrm{A}(z^{-1})$ and $\mathrm{B}(z^{-1})$, ensuring that the sum of squared errors,

$$\sum_{k=1}^{n+N} e^2(k),$$

is minimized.

Equation (3.1) can be further expanded as:

$$y(k) = -\sum_{i=1}^{n} a_i y(k-i) + \sum_{j=0}^{l} b_j u(k-j) + e(k). \tag{3.3}$$

Defining the $1 \times (n+l+1)$ parameter vector and $(n+l+1) \times 1$ state vector as:

$$\boldsymbol{\theta} = [a_1 \cdots a_n \mid b_0\ b_1 \cdots b_l],$$

$$\boldsymbol{\phi}(k) = \left[-y(k-1) \cdots - y(k-n) \mid u(k)\, u(k-1) \cdots u(k-l)\right]^{\mathrm{T}}.$$

Equation (3.3) becomes:

$$y(k) = \boldsymbol{\theta}\boldsymbol{\phi}(k) + e(k). \tag{3.4}$$

Assuming that the system has been sampled N times (where $N \geq n + l + 1$), and N sets of input–output data have been recorded, substituting this data into Eq. (3.4) yields a set of difference equations that can be expressed in matrix form as:

$$\boldsymbol{y}_N = \boldsymbol{\theta}\boldsymbol{\Phi}_N + \boldsymbol{e}_N, \quad N \geq n + l + 1, \tag{3.5}$$

where

$$\begin{aligned} \boldsymbol{y}_N &= \left[y(n+1)\ \ y(n+2)\ \cdots\ y(n+N)\right], \\ \boldsymbol{e}_N &= \left[e(n+1)\ \ e(n+2)\ \cdots\ e(n+N)\right], \end{aligned}$$

are $N \times 1$ vectors, and

$$\begin{aligned} \boldsymbol{\Phi}_N &= \left[\boldsymbol{\phi}(n+1)\ \ \boldsymbol{\phi}(n+2) \cdots \boldsymbol{\phi}(n+N)\right] \\ &= \begin{bmatrix} -y(n) & -y(n+1) & \cdots & -y(n+N-1) \\ \vdots & \vdots & \ddots & \vdots \\ -y(1) & -y(2) & \cdots & -y(N) \\ u(l+1) & u(l+2) & \cdots & u(l+N) \\ \vdots & \vdots & \ddots & \vdots \\ u(1) & u(2) & \cdots & u(N) \end{bmatrix}, \end{aligned}$$

is an $(n + l + 1) \times N$ matrix.

The least-squares estimation of the parameter vector $\boldsymbol{\theta}$ is obtained by minimizing the objective function:

$$\begin{aligned} J &= \sum_{k=n+1}^{n+N} e^2(k) = \boldsymbol{e}_N \boldsymbol{e}_N^{\mathrm{T}} \\ &= (y_N - \boldsymbol{\theta}\boldsymbol{\Phi}_N)(y_N - \boldsymbol{\theta}\boldsymbol{\Phi}_N)^{\mathrm{T}} \\ &= \sum_{k=n+1}^{n+N} \left[y(k) - \boldsymbol{\theta}\boldsymbol{\phi}(k)\right]^2. \end{aligned} \tag{3.6}$$

Setting the gradient to zero:

$$\left.\frac{\partial J}{\partial \boldsymbol{\theta}}\right|_{\theta = \hat{\theta}} = 0,$$

yields the least-squares estimation:

$$\hat{\boldsymbol{\theta}} = \mathbf{y}_N \boldsymbol{\Phi}_N^{\mathrm{T}} \left(\boldsymbol{\Phi}_N \boldsymbol{\Phi}_N^{\mathrm{T}} \right)^{-1}. \tag{3.7}$$

The existence of $\hat{\boldsymbol{\theta}}$ is guaranteed as long as the matrix $\left(\boldsymbol{\Phi}_N \boldsymbol{\Phi}_N^{\mathrm{T}} \right)$ is non-singular.

In industrial applications where fast, real-time parameter estimation is necessary—especially for systems with time-varying or nonlinear behavior—recursive algorithms are preferred. The recursive form of Eq. (3.7) is given by:

$$\hat{\boldsymbol{\theta}}(k+1) = \hat{\boldsymbol{\theta}}(k) + r(k+1)\left[y(k+1) - \hat{\boldsymbol{\theta}}(k)\boldsymbol{\phi}(k) \right] \boldsymbol{\phi}^{\mathrm{T}}(k)\mathbf{P}(k), \tag{3.8a}$$

$$\mathbf{P}(k+1) = \left[\mathbf{P}(k) - r(k+1)\mathbf{P}(k)\boldsymbol{\phi}(k)\boldsymbol{\phi}^{\mathrm{T}}(k)\mathbf{P}(k) \right] f^{-1}, \tag{3.8b}$$

$$r(k+1) = \left[f + \boldsymbol{\phi}^{\mathrm{T}}(k)\mathbf{P}(k)\boldsymbol{\phi}(k) \right]^{-1}, \tag{3.8c}$$

where $\mathbf{P}(k)$ is an $(n+l+1) \times (n+l+1)$ matrix. The initial values are set as:

$$\hat{\boldsymbol{\theta}}(0) = [0], \quad \mathbf{P}(0) = c\mathbf{I},$$

with c being a sufficiently large constant, typically in the range of 10^5 to 10^{10}.

The variable f, known as the forgetting factor, is introduced to handle time-varying and/or nonlinear systems by mitigating data saturation and enhancing the ability of the algorithm to track changes in system parameters. The error objective function now becomes:

$$J = \sum_{k=n+1}^{n+N} f^{n+N-k} \left[y(k) - \boldsymbol{\theta}\boldsymbol{\phi}(k) \right]^2. \tag{3.9}$$

This formulation ensures that older data is exponentially "forgotten", allowing the algorithm to focus on more recent data.

In practice, f typically ranges from 0.95 and 1.00, depending on the specific application. A value close to 1 is appropriate for nearly time-invariant systems, while a smaller f increases the tracking speed of changing parameters but also amplifies the impact of noise.

The trade-off between tracking speed and noise sensitivity highlights the importance of selecting an appropriate forgetting factor, as discussed in the literature [10].

3.3 Algorithm for Linear Discrete State Equation

Building on the algorithm presented in the previous section, a straightforward parameter estimation algorithm for the discrete-time state-space model of a linear MIMO dynamic system is introduced here.

Consider a linear time-invariant discrete-time dynamic system described by:

$$\boldsymbol{x}(kT+T) = \mathbf{F}\,\boldsymbol{x}(kT) + \mathbf{G}\,\boldsymbol{u}(kT) + \boldsymbol{e}(kT), \tag{3.10}$$

where $\boldsymbol{x}(kT)$, $\boldsymbol{u}(kT)$, and $\boldsymbol{e}(kT)$ are the $n \times 1$, $l \times 1$, and $n \times 1$ state, input, and noise (or error) vectors, respectively. $\mathbf{F}$ and $\mathbf{G}$ are real constant matrices with appropriate dimensions, and T is the sampling interval. For simplicity, the time index kT will be referred to as k in the following equations.

Let:

$$\boldsymbol{\theta}_o = \left[f_{o1}\ f_{o2} \cdots f_{on} \mid g_{o1}\ \ g_{o2} \cdots g_{ol}\right], \tag{3.11}$$

where f_{oi} and g_{oj}, for $i = 1, \ldots, n$, $j = 1, \ldots, l$, and $o = 1, \ldots, n$ are the elements of the o-th row of the matrices $\mathbf{F}$ and $\mathbf{G}$, respectively, and

$$\begin{aligned} \boldsymbol{\phi}(k) &= \left[x_1(k)\ \ x_2(k)\ \cdots\ x_n(k) \mid u_1(k)\ \ u_2(k)\ \cdots\ u_l(k)\right]^{\mathrm{T}} \\ &= \begin{bmatrix} \boldsymbol{x}(k) \\ \boldsymbol{u}(k) \end{bmatrix}. \end{aligned} \tag{3.12}$$

Thus, each individual equation in Eq. (3.10) can be expressed as:

$$x_i(k+1) = \boldsymbol{\theta}_i \boldsymbol{\phi}(k) + e_i(k), \quad i = 1, \ldots, n. \tag{3.13}$$

Using Eq. (3.13), the system Eq. (3.10) can be rewritten in matrix form as:

$$\boldsymbol{x}(k+1) = \Theta \boldsymbol{\phi}(k) + \boldsymbol{e}(k), \tag{3.14}$$

where

$$\Theta = [\boldsymbol{\theta}_1\ \ \boldsymbol{\theta}_2\ \ \cdots\ \ \boldsymbol{\theta}_n]^{\mathrm{T}} = [\,\mathbf{F} \mid \mathbf{G}\,]. \tag{3.15}$$

According to the least-squares estimation method outlined in the previous section, the estimate of Θ can be obtained as:

$$\hat{\Theta} = \boldsymbol{X}_N \boldsymbol{\Phi}_N^{\mathrm{T}} \left(\boldsymbol{\Phi}_N \boldsymbol{\Phi}_N^{\mathrm{T}}\right)^{-1}. \tag{3.16}$$

The recursive algorithm for Eq. (3.16) is then formulated as:

$$\hat{\Theta}(k+1) = \hat{\Theta}(k) + r(k+1)\left[\boldsymbol{x}(k+1) - \hat{\Theta}(k)\boldsymbol{\Phi}(k)\right]\boldsymbol{\Phi}^{\mathrm{T}}(k)\mathbf{P}(k), \tag{3.17a}$$

$$\mathbf{P}(k+1) = \left[\mathbf{P}(k) - r(k+1)\mathbf{P}(k)\mathbf{\Phi}(k)\mathbf{\Phi}^{\mathrm{T}}(k)\mathbf{P}(k)\right]f^{-1}, \tag{3.17b}$$

$$r(k+1) = \left[f + \mathbf{\Phi}^{\mathrm{T}}(k)\mathbf{P}(k)\mathbf{\Phi}(k)\right]^{-1}. \tag{3.17c}$$

This recursive algorithm is structurally similar to the one given in Eq. (3.8). The initial values for recursive estimation are:

$$\hat{\Theta}(0) = [0], \quad \mathbf{P}(0) = c\mathbf{I},$$

where c is a large constant, typically chosen in the range 10^5–10^{10}.

Applications have shown that this algorithm is not only effective but also easy to implement, making it highly suitable for real-world scenarios involving discrete-time MIMO dynamic systems. Its simplicity in execution, combined with robust performance, ensures reliable results in practical settings where efficiency and accuracy are paramount.

3.4 Algorithm for Linear Continuous State Equation

In most applications involving parameter estimation for physical systems, obtaining a continuous-time model is generally preferred due to its closer representation of the underlying physical processes.

For a MIMO linear time-invariant dynamic system described by the following continuous-time state-space representation:

$$\dot{\boldsymbol{x}}(t) = \mathbf{A}\boldsymbol{x}(t) + \mathbf{B}\boldsymbol{u}(t). \tag{3.18}$$

If its corresponding discrete-time state-space description is given by Eq. (3.10), and the input is held constant between sampling instants or the sampling interval T is sufficiently small, then the coefficient matrices of the two descriptions in Eqs. (3.10) and (3.18) are related by the following expressions [11]:

$$\mathbf{F} = \exp(\mathbf{A}T) = \mathbf{I} + \mathbf{A}T + \frac{(\mathbf{A}T)^2}{2!} + \cdots, \tag{3.19}$$

and

$$\mathbf{G} = \int_0^{\mathrm{T}} \exp(\mathbf{A}T)\mathbf{B}\mathrm{d}t = \left(\mathbf{I}T + \frac{\mathbf{A}T^2}{2!} + \frac{\mathbf{A}^2T^3}{3!} + \cdots\right)\mathbf{B}$$

$$= (\mathbf{F} - \mathbf{I})\mathbf{A}^{-1}\mathbf{B}. \tag{3.20}$$

The inverse relationship of Eq. (3.19) is:

$$(\mathbf{A})^{(k+1)} = (\mathbf{A})^{(k)} + \mathbf{F}^{-1}\frac{1}{T}\left(\mathbf{F} - \mathbf{F}^{(k)}\right), \tag{3.21}$$

where $(\mathbf{A})^{(k)}$ is the estimate of $\mathbf{A}$ at the k-th iteration, and:

$$\mathbf{F}^{(k)} = \exp\left[(\mathbf{A}T)^{(k)}\right] = \mathbf{I} + (\mathbf{A}T)^{(k)} + \frac{1}{2!}\left[(\mathbf{A}T)^{(k)}\right]^2 + \cdots. \tag{3.22}$$

The initial estimate of $\mathbf{A}$ is given by:

$$(\mathbf{A})^{(0)} = \frac{1}{2T}\left(\mathbf{F} - \mathbf{F}^{-1}\right). \tag{3.23}$$

The matrix $\mathbf{B}$ can be computed directly from Eq. (3.20) as:

$$\mathbf{B} = \mathbf{A}(\mathbf{F} - \mathbf{I})^{-1}\mathbf{G}, \tag{3.24}$$

provided that $(\mathbf{F} - \mathbf{I})$ is non-singular. This iterative algorithm is computationally efficient and converges rapidly, allowing for convenient conversion between continuous and discrete state-space descriptions. Consequently, the coefficient matrices of a linear continuous-time state equation can be estimated indirectly by first obtaining the discrete-time model and then converting it to the continuous-time form.

In addition to this indirect approach, a direct estimation algorithm for the parameter matrices of linear continuous-time state equations can be derived using the principles outlined in previous sections. The key idea is to convert the differential equation into a difference equation, enabling recursive computation [9]. This method yields a good approximation when the sampling interval T is sufficiently small.

Consider the continuous-time system described by Eq. (3.18). Integrating it from time kT to $(kT + T)$ gives:

$$\boldsymbol{x}(kT + T) - \boldsymbol{x}(kT) = \mathbf{A}\int_{kT}^{kT+T} \boldsymbol{x}(t)\mathrm{d}t + \mathbf{B}\int_{kT}^{kT+T} \boldsymbol{u}(t)\mathrm{d}t. \tag{3.25}$$

Applying the trapezoidal rule for numerical integration [9], this integral equation can be approximated as:

$$\begin{aligned}\boldsymbol{x}(k+1) - \boldsymbol{x}(k) = {} & \frac{1}{2}\mathbf{A}T[\boldsymbol{x}(k+1) + \boldsymbol{x}(k)] \\ & + \frac{1}{2}\mathbf{B}T[\boldsymbol{u}(k+1) + \boldsymbol{u}(k)] + \boldsymbol{e}(k),\end{aligned} \tag{3.26}$$

where $\boldsymbol{e}(k)$ represents modeling errors or random noise. Based on Eq. (3.26), we define the following state vectors:

$$\boldsymbol{w}(k) = \frac{2}{T}[\boldsymbol{x}(k+1) - \boldsymbol{x}(k)], \tag{3.27a}$$

$$\boldsymbol{z}(k) = \boldsymbol{x}(k+1) + \boldsymbol{x}(k), \tag{3.27b}$$

$$\boldsymbol{v}(k) = \boldsymbol{u}(k+1) + \boldsymbol{u}(k). \tag{3.27c}$$

Substituting these into Eq. (3.26) results in:

$$\boldsymbol{w}(k) = \mathbf{A}\boldsymbol{z}(k) + \mathbf{B}\boldsymbol{v}(k) + \boldsymbol{e}(k). \tag{3.28}$$

Defining the following vector similar to Eq. (3.12):

$$\boldsymbol{\phi}(k) = \begin{bmatrix} \boldsymbol{z}(k) \\ \boldsymbol{v}(k) \end{bmatrix}. \tag{3.29}$$

Equation (3.28) can be expressed as:

$$\boldsymbol{w}(k) = \Theta\boldsymbol{\phi}(k) + \boldsymbol{e}(k), \tag{3.30}$$

where the matrix:

$$\Theta = [\ \mathbf{A} \mid \mathbf{B}\ ], \tag{3.31}$$

is the parameter matrix to be estimated. Applying the least-squares estimation method again, the estimate of Θ is given by:

$$\hat{\Theta} = \boldsymbol{W}_N \boldsymbol{\Phi}_N^{\mathrm{T}} \left(\boldsymbol{\Phi}_N \boldsymbol{\Phi}_N^{\mathrm{T}}\right)^{-1}. \tag{3.32}$$

Its recursive form is:

$$\hat{\Theta}(k+1) = \hat{\Theta}(k) + r(k+1)\left[\boldsymbol{w}(k) - \hat{\Theta}(k)\boldsymbol{\phi}(k)\right]\boldsymbol{\phi}^{\mathrm{T}}(k)\mathbf{P}(k), \tag{3.33a}$$

$$\mathbf{P}(k+1) = \left[\mathbf{P}(k) - r(k+1)\mathbf{P}(k)\boldsymbol{\phi}(k)\boldsymbol{\phi}^{\mathrm{T}}(k)\mathbf{P}(k)\right]f^{-1}, \tag{3.33b}$$

$$r(k+1) = \left[f + \boldsymbol{\phi}^{\mathrm{T}}(k)\mathbf{P}(k)\boldsymbol{\phi}(k)\right]^{-1}. \tag{3.33c}$$

The initial values for recursive estimation are set as:

$$\hat{\Theta}(0) = [0], \quad \mathbf{P}(0) = c\mathbf{I},$$

where c is a large constant, typically in the range of 10^5–10^{10}.

Practical experience with this direct method has shown that it provides highly accurate results in parameter estimation, making it a reliable approach for identifying continuous-time models in real-world applications.

The identification algorithms discussed for discrete-time systems typically assume that input and output data are sampled simultaneously. However, in real-world data acquisition, there is often a time difference between the sampling of different signals. If this time difference is sufficiently small compared to the sampling interval and the dynamics of the plant, its impact may be negligible. However, when the time difference is more significant, it can introduce bias into the estimated models [12].

Research has shown that when the sampling delays are known, the data can be adjusted to produce models that correspond to synchronous sampling [13]. This adjustment is crucial for accurately modeling physical plants. The process is straightforward and can be applied after the initial parameter estimation. Essentially, the obtained model can be converted to one that reflects synchronous sampling, effectively eliminating the influence of time delays in the data sampling. This conversion acts as a post-processing step, ensuring the accuracy of the estimated model. It is important to use an appropriate model structure during the estimation process to facilitate this adjustment.

3.5 A Unified Framework

Based on the algorithms presented in the previous sections, a systematic approach to system identification is illustrated in Fig. 3.1.

The system identification process can be summarized into three main steps:

1. *Data Acquisition:* The first step involves obtaining or sampling the input and output data, either from computer simulations or from the actual operation of the plant system. This data forms the foundation for accurate model estimation.
2. *Model Estimation:* Using the sampled data, a state-space model—either in discrete or continuous form—is estimated by applying the algorithms presented in this chapter. This step is critical for developing a model that accurately reflects the dynamics of the plant.
3. *Model Refinement and Conversion:* If necessary, the estimated model can be further refined or converted, for instance, from a discrete-time to a continuous-time representation, to better suit specific application needs. Once this is done, the model is ready for use in control design or further analysis.

This systematic approach ensures that the process of system identification is both methodical and adaptable to various practical requirements.

In addition to obtaining a time-domain model, it may be necessary to derive the frequency-domain model of the plant. For a dynamic system where the output equation is given by:

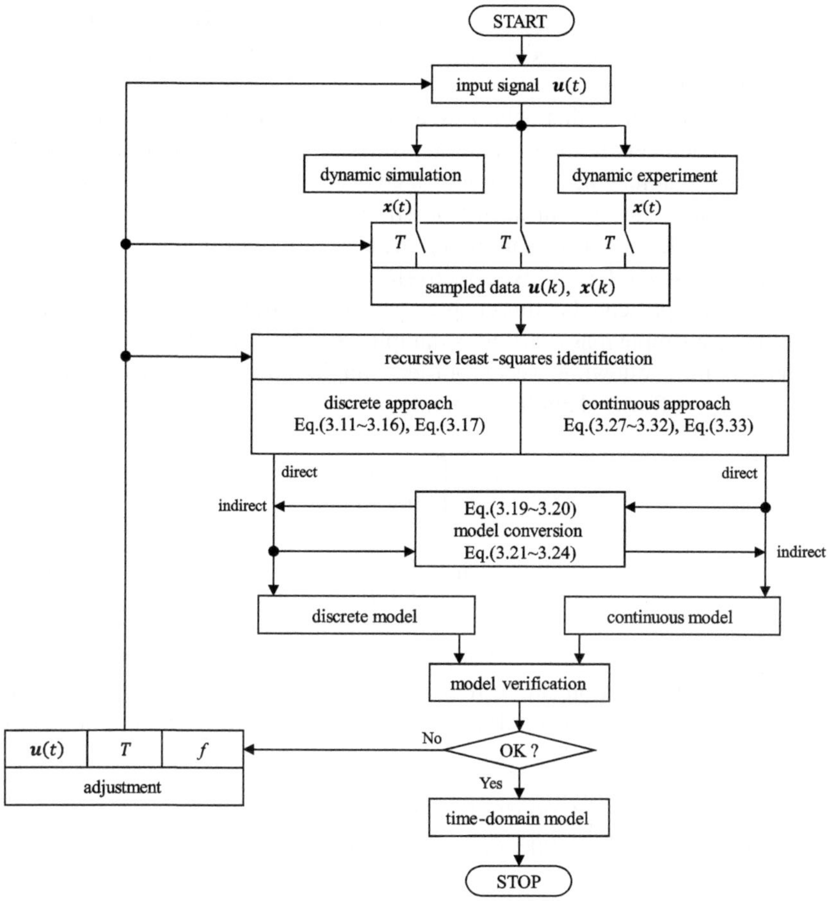

Fig. 3.1 Flowchart of the unified framework

$$y(k) = \mathbf{C}\boldsymbol{x}(k), \quad \text{or} \quad y(t) = \mathbf{C}\boldsymbol{x}(t),$$

the transfer function or transfer matrix of the discrete- or continuous-time system can be expressed as:

$$\mathbf{T}(z) = \mathbf{C}(z\mathbf{I} - \mathbf{F})^{-1}\mathbf{G},$$

for the discrete-time system, or

$$\mathbf{T}(s) = \mathbf{C}(s\mathbf{I} - \mathbf{A})^{-1}\mathbf{B},$$

for the continuous-time system. These relationships allow for the indirect attainment of frequency-domain models through the conversion of the state-space models derived in the time domain.

It is important to note that achieving a good model often requires multiple iterations and refinements in the estimation process, as system identification involves various factors that may necessitate informed decision-making. Prior knowledge or insights about the plant system, such as its structure and dynamics, can be invaluable in guiding these decisions and improving the accuracy of the model.

To demonstrate the feasibility, effectiveness, accuracy, and practical application of the algorithms presented in this chapter, we will illustrate the model identification process using a simple mass–stiffness–damping system as an example.

Consider the following second-order, single-input linear continuous-time dynamic mass–stiffness–damping system:

$$\mathbf{A} = \begin{bmatrix} -1 & 1 \\ 0 & -3 \end{bmatrix}, \tag{3.34a}$$

$$\mathbf{B} = \begin{bmatrix} 0 \\ 1 \end{bmatrix}. \tag{3.34b}$$

With a sampling interval $T = 0.01$, the corresponding coefficient matrices of its discrete-time state equation can be derived using the partial summations of the first 100 terms of the matrix series in Eqs. (3.19) and (3.20). The discrete-time matrices are obtained as:

$$\mathbf{F} = \begin{bmatrix} 9.900498\text{E}-01 & 9.802150\text{E}-03 \\ 0.0000000 & 9.704456\text{E}-01 \end{bmatrix}, \tag{3.35a}$$

$$\mathbf{G} = \begin{bmatrix} 4.933872\text{E}-05 \\ 9.851489\text{E}-03 \end{bmatrix}. \tag{3.35b}$$

Under the condition of zero initial states, if the system in Eq. (3.34) is excited by a unit step input

$$u(t) = 1, \quad t > 0,$$

and the sampling interval is $T = 0.01$, the algorithms presented in Eqs. (3.17) and (3.33) can be applied to the first 100 sets of sampled data from the step input and its state response. The estimated results are as follows:

$$\hat{\mathbf{A}} = \begin{bmatrix} -9.999942\text{E}-01 & 9.998927\text{E}-01 \\ 8.396009\text{E}-06 & -2.9997790 \end{bmatrix}, \tag{3.36a}$$

$$\hat{\mathbf{B}} = \begin{bmatrix} 3.321382\text{E}-05 \\ 9.999250\text{E}-01 \end{bmatrix}, \tag{3.36b}$$

and

$$\hat{\mathbf{F}} = \begin{bmatrix} 9.900498\text{E}-01 & 9.802110\text{E}-03 \\ 4.697536\text{E}-07 & 9.704453\text{E}-01 \end{bmatrix}, \tag{3.37a}$$

$$\hat{\mathbf{G}} = \begin{bmatrix} 4.934017\text{E}-05 \\ 9.851487\text{E}-03 \end{bmatrix}. \tag{3.37b}$$

These results demonstrate the high accuracy of the algorithms discussed in this chapter. Furthermore, model conversion between the continuous and discrete domains, as shown in Eqs. (3.36) and (3.37), can be easily performed using the relationships outlined in Eqs. (3.19)–(3.20) and (3.21)–(3.24).

Although this example is relatively simple, it is highly illustrative for practical applications, such as modeling a specific degree-of-freedom in a more complex mass–stiffness–damping system that changes over time, like the tire of a vehicle. It can also be valuable for analyzing cross-coupling between subsystems in larger mass–stiffness–damping systems. This process underscores the versatility and practical relevance of the identification algorithms presented in this chapter.

3.6 Summary

This chapter introduces a systematic approach for the recursive identification of state-space models for dynamic systems, mathematically unified through the least-squares method. Both discrete and continuous versions of the recursive algorithm in state-space form are presented.

The techniques discussed offer significant advantages, particularly in their ease of application in industrial scenarios. These methods enable engineers to develop advanced solutions for the real-time identification and control of complex systems with varying parameters.

The effectiveness and reliability of this parameter estimation approach have been validated through various applications, demonstrating its suitability for real-time modeling and control of dynamic systems with changing or nonlinear characteristics. A simulation example of a typical mass–stiffness–damping system further highlights the practical applicability of the algorithms discussed.

References

1. Isermann R (1989) Digital control systems: fundamentals, deterministic control, vol 1. Springer
2. Dasgupta S, Anderson BDO, Kaye RJ (1988) Identification of physical parameters in structured systems. Automatica 24(2):217–225
3. Xia L, Moore JB (1989) Recursive identification of over parametrized systems. IEEE Trans Autom Control 34(3):327–331

4. Kovanic P (1986) A new theoretical and algorithmical basis for estimation, identification and control. Automatica 22(6):657–674
5. Ljung L (2021) System identification: an overview, encyclopedia of systems and control. Springer, Berlin, pp 2302–2317
6. Kailath T, Sayed AH, Hassibi B (2000) Linear estimation. Prentice Hall
7. Goodwin GC, Payne RL (1977) Dynamic system identification: experiment design and data analysis. Academic Press
8. Ljung L (1999) System identification: theory for the user. Prentice Hall
9. Zhu HA, Hung JC (1990) Identification of dynamic systems: a systematic approach. In: Proceedings of the 16th IEEE international conference on industrial electronics (IECON'90), pp 131–136
10. Canetti RM, España MD (1989) Convergence analysis of the least-squares identification algorithm with a variable forgetting factor for time-varying linear systems. Automatica 25(4):609–612
11. Sinha NK, Lastman GJ (1982) Identification of continuous-time multivariable systems from sampled data. Int J Control 35(1):117–126
12. Sridharan G, Srisailam MC, Rao VS (1985) A note on the effect of asynchronous sampling on estimation accuracy. Automatica 21(4):491–493
13. Wittenmark B, Olsson PO (1988) Identification based on data with sampling delays. Automatica 24(2):271–273

Chapter 4
Model Reduction

There are instances when the original models of dynamic systems may be too complex or of too high an order to be practical for control purposes. In modern control engineering, when dealing with high-order dynamic systems, a commonly adopted strategy is to simplify the original system using model reduction techniques. These methods are crucial for achieving effective control without compromising the essential dynamics of the system. Unlike the modal reduction methods used in mechanics and modal analysis, model simplification in large-scale system theory has evolved significantly over the past fifty years to address these needs.

Model simplification methods for large-scale systems are generally categorized into two basic types: time-domain methods and frequency-domain methods. Time-domain methods typically focus on state-space representations, while frequency-domain methods are associated with transfer function matrix representations. Significant advancements have been made in frequency-domain model simplification, leading to the development of many effective approaches [1–4]. However, the development of time-domain methods has lagged behind, with few effective methods emerging since the early contributions by Davison [5, 6], Chidambara and Davison [7], Aoki [8, 9], Soong [10].

This chapter introduces a time-domain model simplification method based on the aggregation method proposed by Kymn [11], Hildenbrand [12], incorporating relevant techniques from system identification and modal analysis. This approach addresses the gap in time-domain model reduction, providing a practical tool for simplifying complex dynamic systems while preserving their essential characteristics for control purposes.

H. Zhu, *Model-Based Control of Mass–Stiffness–Damping Systems*, Advances in Industrial Control, https://doi.org/10.1007/978-3-031-97592-9_4

4.1 Introduction

The aggregation method, originally developed for analyzing static economic models, has a long history of application. The modern formulation of aggregation methods was proposed by Kymn [11], Hildenbrand [12] and was later extended by Aoki [8] to address control problems in large-scale dynamic systems. While the aggregation method is a powerful tool for model simplification, its significance extends beyond that. It serves as a foundational approach for other approximation techniques and provides valuable insights into the mechanisms underlying various model simplification methods.

Consider a high-order linear time-invariant (LTI) dynamic system described by the following continuous-time state-space representation:

$$\dot{\boldsymbol{x}}(t) = \mathbf{A}\boldsymbol{x}(t) + \mathbf{B}\boldsymbol{u}(t),\ \boldsymbol{x}(0) = \boldsymbol{x}_0, \tag{4.1a}$$

$$\boldsymbol{y}(t) = \mathbf{C}\boldsymbol{x}(t), \tag{4.1b}$$

where $\boldsymbol{x}(t)$, $\boldsymbol{u}(t)$, and $\boldsymbol{y}(t)$ are the $n \times 1$ state vector, $l \times 1$ input vector, and $p \times 1$ output vector, respectively. **A**, **B**, and **C** are constant matrices with appropriate dimensions. The goal of model reduction is to derive a simplified system that approximates the original high-order system, described as:

$$\dot{\boldsymbol{x}}_r(t) = \mathbf{A}_r\boldsymbol{x}_r(t) + \mathbf{B}_r\boldsymbol{u}(t),\ \ \boldsymbol{x}_r(0) = \boldsymbol{x}_{r0}, \tag{4.2a}$$

$$\boldsymbol{y}_r(t) = \mathbf{C}_r\boldsymbol{x}_r(t). \tag{4.2b}$$

Here, $\boldsymbol{x}_r(t)$ is the $r \times 1$ reduced state vector, and $\boldsymbol{y}_r(t)$ is the $p \times 1$ output vector of the reduced model. The matrices $\mathbf{A}_r$, $\mathbf{B}_r$, and $\mathbf{C}_r$ are the unknown constant matrices of the reduced system, which need to be determined through an appropriate model reduction method.

Using the concept of aggregation [1], we assume that the relationship between the original high-order system in Eq. (4.1) and the reduced system in Eq. (4.2) can be expressed by the following linear transformation:

$$\boldsymbol{x}_r(t) = \mathbf{R}\boldsymbol{x}(t), \tag{4.3}$$

where **R** is an $r \times n$ (with $r < n$) constant aggregation matrix, also known as the model reduction weighting matrix, and $\boldsymbol{x}_r(t)$ is the reduced state vector. Theoretically, the matrix **R** is arbitrary and can be chosen to suit various objectives, providing a high degree of flexibility in the model reduction process. However, in practice, **R** is typically determined by specific criteria, which will be discussed later in this chapter.

According to Eq. (4.3), the relationship between the original system in Eq. (4.1) and the reduced system in Eq. (4.2) must satisfy the following conditions for dynamic exactness (or perfect aggregation) [1]:

$$\mathbf{A}_r\mathbf{R} = \mathbf{R}\mathbf{A}, \tag{4.4a}$$

$$\mathbf{B}_r = \mathbf{R}\mathbf{B}, \tag{4.4b}$$

$$\mathbf{C}_r\mathbf{R} = \mathbf{C}. \tag{4.4c}$$

Given that $r < n$ and assuming the system is fully controllable and observable, these conditions for perfect aggregation generally cannot be exactly satisfied. Therefore, both the original high-order system and the reduced system should be viewed as approximations. The goal of the reduction process is to ensure that these approximations are close enough to serve the intended control or analysis purposes effectively.

4.2 A Time-Domain Approach to Model Reduction

The primary objective of model reduction for large systems using time-domain state-space models is to determine the reduced matrices $\mathbf{A}_r$, $\mathbf{B}_r$, and $\mathbf{C}_r$.

4.2.1 Determination of the Reduced Parameter Matrices

In principle, when an effective model reduction method is applied, the output of the reduced system, $\boldsymbol{y}_r(t)$, should closely approximate the output of the original system, $\boldsymbol{y}(t)$, as described by:

$$\boldsymbol{y}_r(t) \approx \boldsymbol{y}(t). \tag{4.5}$$

This condition directly leads to Eq. (4.4c). When the matrix $\mathbf{R}$ is non-singular, the unique solution for $\mathbf{C}_r$ can be obtained using the generalized inverse of $\mathbf{R}$ as follows:

$$\mathbf{C}_r = \mathbf{C}\mathbf{R}^{\mathrm{T}}\left(\mathbf{R}\mathbf{R}^{\mathrm{T}}\right)^{-1}. \tag{4.6}$$

However, determining the matrices $\mathbf{A}_r$ and $\mathbf{B}_r$ can be more challenging because their solutions are highly sensitive to the choice of $\mathbf{R}$. To address this, a special approach is presented in this chapter to derive $\mathbf{A}_r$ and $\mathbf{B}_r$.

The process begins by using the matrix $\mathbf{R}$ to pre-process the state-variable observations $\boldsymbol{x}(t)$, transforming them into the reduced state variables $\boldsymbol{x}_r(t)$ via Eq. (4.3). Subsequently, the parameter estimation methods described in Chap. 3 are applied to simultaneously and directly identify the reduced matrices $\hat{\mathbf{A}}_r$ and $\hat{\mathbf{B}}_r$, ensuring that the estimated state $\hat{\boldsymbol{x}}_r$ closely approximates the actual reduced state $\boldsymbol{x}_r$.

It is well established that the transient response of state variables captures the intrinsic dynamics of the system. Therefore, minimizing the following objective function for continuous systems:

$$J = \int_0^\infty [\mathbf{R}\boldsymbol{x}(t) - \hat{\boldsymbol{x}}_r(t)]^{\mathrm{T}} [\mathbf{R}\boldsymbol{x}(t) - \hat{\boldsymbol{x}}_r(t)], \tag{4.7}$$

or for discrete systems:

$$J = \sum_{k=0}^{\infty} [\mathbf{R}\boldsymbol{x}(k) - \hat{\boldsymbol{x}}_r(k)]^{\mathrm{T}} [\mathbf{R}\boldsymbol{x}(k) - \hat{\boldsymbol{x}}_r(k)], \tag{4.8}$$

is chosen as the performance criterion. Achieving this minimization ensures that the model reduction is optimal in the time domain. The recursive least-squares estimation algorithm discussed in the previous chapter is particularly well-suited for this task, as it effectively minimizes the loss of precision in the model reduction process.

4.2.2 Determination of the Weighting Matrix

The weighting matrix $\mathbf{R}$ plays a pivotal role in model reduction. For slowly varying, large inertia systems with predominantly real dominant eigenvalues, the time-domain response is typically monotonic and non-oscillatory. In such cases, $\mathbf{R}$ can often be selected by retaining a few key state variables or their linear combinations, leading to satisfactory results [13, 14]. However, for high-order systems with complex modes, a different approach is necessary.

To ensure an effective reduced model, a method based on modal analysis is recommended for determining the matrix $\mathbf{R}$. The general form of $\mathbf{R}$ can be derived using the following linear transformation:

$$\mathbf{U}\tilde{\boldsymbol{x}} = \boldsymbol{x}. \tag{4.9}$$

This transformation converts the original system in Eq. (4.1) into its modal canonical form, where $\mathbf{U}$ is the modal matrix composed of the eigenvectors of the system matrix $\mathbf{A}$. The system in its modal canonical form is then expressed as:

$$\dot{\tilde{\boldsymbol{x}}} = \mathbf{U}^{-1}\mathbf{A}\mathbf{U}\tilde{\boldsymbol{x}} + \mathbf{U}^{-1}\mathbf{B}\boldsymbol{u} = \boldsymbol{\Lambda}\tilde{\boldsymbol{x}} + \boldsymbol{\Gamma}\boldsymbol{u}, \tag{4.10}$$

where

$$\mathbf{\Lambda} = \mathbf{U}^{-1}\mathbf{A}\mathbf{U} = \begin{bmatrix} \mathbf{\Lambda}_1 & 0 \\ 0 & \mathbf{\Lambda}_2 \end{bmatrix}. \tag{4.11}$$

Here, $\mathbf{\Lambda}$ is the eigenvalue matrix of $\mathbf{A}$, with $\mathbf{\Lambda}_1$ 和 $\mathbf{\Lambda}_2$ being block diagonal matrices of dimensions $r \times r$ and $(n-r) \times (n-r)$, respectively.

Assuming that the eigenvalues corresponding to the dominant modes of $\mathbf{A}$ are arranged within $\mathbf{\Lambda}_1$, the matrix $\mathbf{A}_r$ of the reduced model will correspond to these dominant modes. This means that the eigenvectors corresponding to the r dominant modes are contained in the first r columns of the modal matrix $\mathbf{U}$. From these relationships, the system can be partitioned as follows:

$$\dot{\tilde{\boldsymbol{x}}} = \begin{bmatrix} \dot{\tilde{\boldsymbol{x}}}_1 \\ \dot{\tilde{\boldsymbol{x}}}_2 \end{bmatrix} = \begin{bmatrix} \mathbf{\Lambda}_1 & 0 \\ 0 & \mathbf{\Lambda}_2 \end{bmatrix} \begin{bmatrix} \tilde{\boldsymbol{x}}_1 \\ \tilde{\boldsymbol{x}}_2 \end{bmatrix} + \begin{bmatrix} \mathbf{\Gamma}_1 \\ \mathbf{\Gamma}_2 \end{bmatrix} \boldsymbol{u}, \tag{4.12}$$

and

$$\tilde{\boldsymbol{x}} = \mathbf{U}^{-1}\boldsymbol{x} = \mathbf{V}\boldsymbol{x} = \begin{bmatrix} \mathbf{V}_1 \\ \mathbf{V}_2 \end{bmatrix} \boldsymbol{x} = \begin{bmatrix} \mathbf{V}_1\boldsymbol{x} \\ \mathbf{V}_2\boldsymbol{x} \end{bmatrix}. \tag{4.13}$$

Thus, the reduced state equation becomes:

$$\dot{\tilde{\boldsymbol{x}}}_1 = \mathbf{\Lambda}_1 \tilde{\boldsymbol{x}}_1 + \mathbf{\Gamma}_1 \boldsymbol{u}. \tag{4.14}$$

Assuming the reduced state vector $\boldsymbol{x}_r$ is approximately $\tilde{\boldsymbol{x}}_1$, we have:

$$\boldsymbol{x}_r \approx \tilde{\boldsymbol{x}}_1 = \mathbf{R}\boldsymbol{x}. \tag{4.15}$$

This leads to the conclusion that:

$$\mathbf{R} = \mathbf{V}_1. \tag{4.16}$$

The choice of $\mathbf{V}_1$ offers some flexibility. By selecting $\mathbf{V}_1$ appropriately, the reduced model can retain the desired dynamic characteristics and stability of the original system. Typically, the dominant modes of the original high-order system should be preserved to ensure that the time-domain response of the reduced model closely approximates that of the original system.

Determining which modes are dominant is itself a complex topic within modal analysis and often depends on practical considerations and the engineer's experience. While this chapter focuses on model reduction, the selection of dominant modes, as discussed by Hughes [15] in his work on space structure vibration modes, is a crucial step.

In practice, certain criteria are useful for selecting modes to retain. Let the complex eigenvalues of matrix $\mathbf{A}$ be $\lambda_1, \lambda_2, \ldots, \lambda_n$, where

$$\lambda_i = \sigma_i + j\omega_i, \tag{4.17a}$$

$$\rho_i = \sqrt{\sigma_i^2 + \omega_i^2}, \quad i = 1, 2, \ldots, n. \tag{4.17b}$$

Here:

- ρ_i is the magnitude (or modulus) of the complex eigenvalue λ_i. Geometrically, if we think of λ_i as a point in the complex plane at coordinates (σ_i, ω_i), then ρ_i is the distance of that point from the origin.
- A small σ_i corresponds to large time constants, indicating "slow" components in the time response.
- A large ω_i corresponds to high-frequency components in the transient response, which typically decay quickly.
- In a physical system, poles are caused by devices that store energy. The overall time response is determined by the poles of the system, while its shape is influenced by both the poles and zeros.

Combining modal analysis with parameter estimation for model reduction adds an additional layer of physical insight. The dynamic response of many systems primarily depends on the low-frequency components (dominant eigenvalues), while finer details or small fluctuations depend on characteristics within a specific frequency band. An oscillatory component usually involves four pieces of information: oscillation frequency, damping, amplitude, and phase. The first two are typically encoded in the poles (eigenvalues) of the system, while the latter two are influenced by both poles and zeros. Retaining only the dominant complex poles via **R** might not fully capture the amplitude and phase information of oscillatory components, but the time-domain approximation provided by least-squares estimation can compensate for this loss.

In real-world scenarios, engineers may encounter unstable state-space equations with eigenvalues having positive real parts. If the methods discussed do not yield satisfactory results, it may be necessary to decompose the model to isolate unstable modes and then reduce the stable parts [16, 17]. Additionally, when dealing with large systems that are not fully controllable or observable, attention must be given to the uncontrollable and unobservable parts during model reduction.

4.2.3 Steps for Model Reduction

Based on the algorithms presented, the model reduction process can be summarized into the following five steps:

1. *Data Acquisition:* Collect the sampled data of the system input $\boldsymbol{u}(k)$ and state $\boldsymbol{x}(k)$ from simulations or experiments on the target system.

2. *Weighting Matrix Determination:* Define the model reduction weighting matrix **R** based on the system parameter matrix **A**, ensuring it meets the desired quality requirements for the reduced model.
3. *Data Processing:* Apply the matrix **R** to the sampled state data $\boldsymbol{x}(k)$ to obtain the reduced state vector $\boldsymbol{x}_r(k)$.
4. *Model Estimation:* Use the processed data and the recursive least-squares estimation method to derive the reduced-order state-space model.
5. *Model Verification:* Validate the reduced-order model. If the reduced model does not meet the required performance criteria, repeat steps 2 through 4 as necessary.

This procedure is illustrated in Fig. 4.1.

In the field of large-scale dynamic systems, models are sometimes described by transfer functions or transfer function matrices in the frequency domain. The

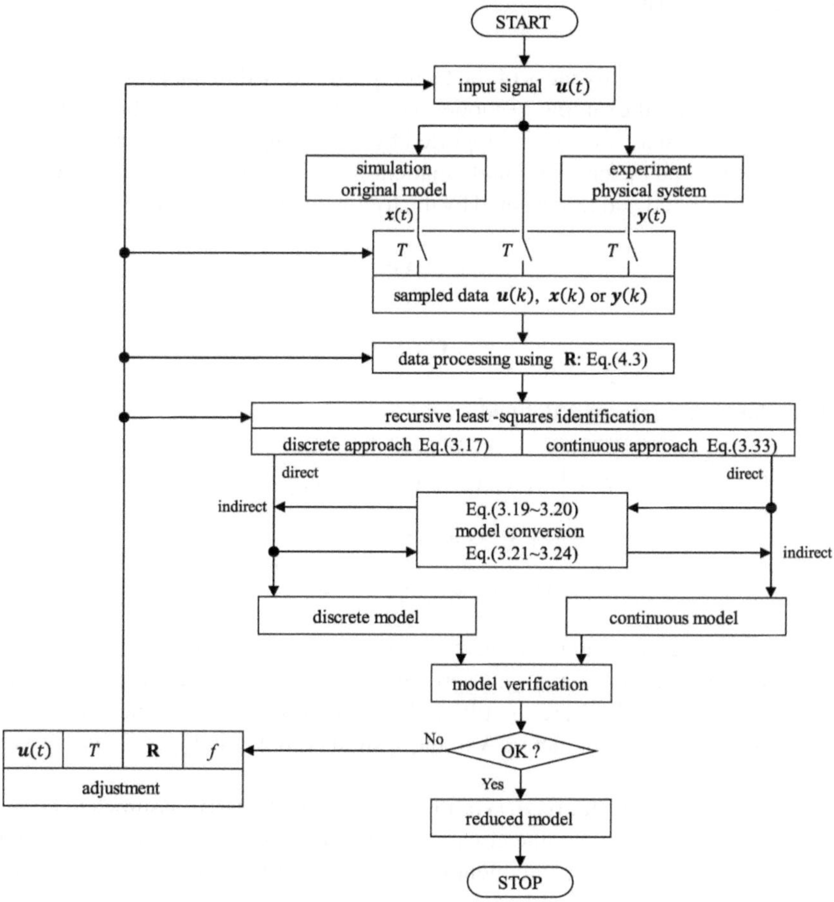

Fig. 4.1 Procedure for model reduction

method presented in this chapter can be effectively applied to reduce the order of any large-scale continuous- or discrete-time dynamic system, whether it is described by time-domain state-space equations or frequency-domain transfer function matrices, through appropriate model conversion.

It is important to note that obtaining an optimal reduced model often involves an iterative process of trial and error. Several factors, such as the choice of the matrix **R**, the selection of the input signal $\boldsymbol{u}(t)$, and the setting of the sampling interval T, play a critical role in achieving accurate results [18, 19]. Prior knowledge and insights about the original system, including its physical structure, stability, and dynamic characteristics, are invaluable in guiding the model reduction process and improving the quality of the reduced model.

4.3 An Example

To demonstrate the effectiveness of the technique discussed in the previous sections, we present a practical example of model reduction for a large rotor-bearing system [20]. This system features a rotor supported by two tilting-pad bearings at each end, with five tilting-pads each. The mathematical model, derived using finite-element analysis, is linearized and represented by the following 26th-order state-space model:

$$\dot{\boldsymbol{x}} = \mathbf{A}\boldsymbol{x}, \tag{4.18a}$$

$$y = \mathbf{C}\boldsymbol{x} = x_3. \tag{4.18b}$$

Here, the state vector $\boldsymbol{x}$ is defined as:

$$\begin{aligned} \boldsymbol{x} = \Big[& x_1\ x_2\ x_3\ x_4\ x_5\ \ y_1\ y_2\ y_3\ y_4\ y_5 \\ & \dot{x}_1\ \dot{x}_2\ \dot{x}_3\ \dot{x}_4\ \dot{x}_5\ \ \dot{y}_1\ \dot{y}_2\ \dot{y}_3\ \dot{y}_4\ \dot{y}_5 \\ & \phi_{11}\ \phi_{12}\ \phi_{13}\ \phi_{51}\ \phi_{52}\ \phi_{53} \Big]^{\mathrm{T}}. \end{aligned} \tag{4.19}$$

In this state vector, x_i and y_i (where $i = 1, \ldots, 5$) represent the displacement of the rotor axis in the horizontal and vertical directions at respective locations, and ϕ_{1j}, ϕ_{5j} (where $j = 1, 2, 3$) are the tilt angles of the tilting-pads in the bearing. The nondimensional coefficient matrix **A** of the system is shown in Fig. 4.2, with $\boldsymbol{y}$ as the output variable.

Since the external excitation $\boldsymbol{u}(t)$ of the model in Eq. (4.18) is assumed to be zero, a modified version of the parameter estimation algorithm from the previous chapter is applied, where the input component $\boldsymbol{u}(t)$ is excluded from the vector. Otherwise, the solution to Eq. (3.33) would not exist due to singularity issues.

Consider the free-decay response of the original model, with the initial values of the state-variables set as follows:

$$
\mathbf{A} = \left[\begin{array}{cccccccccccccccccccccccccc}
0 & 0 & 0 & 0 & 0 & 0 & 0 & 0 & 0 & 0 & 1 & 0 & 0 & 0 & 0 & 0 & 0 & 0 & 0 & 0 & 0 & 0 & 0 & 0 & 0 & 0 \\
0 & 0 & 0 & 0 & 0 & 0 & 0 & 0 & 0 & 0 & 0 & 1 & 0 & 0 & 0 & 0 & 0 & 0 & 0 & 0 & 0 & 0 & 0 & 0 & 0 & 0 \\
0 & 0 & 0 & 0 & 0 & 0 & 0 & 0 & 0 & 0 & 0 & 0 & 1 & 0 & 0 & 0 & 0 & 0 & 0 & 0 & 0 & 0 & 0 & 0 & 0 & 0 \\
0 & 0 & 0 & 0 & 0 & 0 & 0 & 0 & 0 & 0 & 0 & 0 & 0 & 1 & 0 & 0 & 0 & 0 & 0 & 0 & 0 & 0 & 0 & 0 & 0 & 0 \\
0 & 0 & 0 & 0 & 0 & 0 & 0 & 0 & 0 & 0 & 0 & 0 & 0 & 0 & 1 & 0 & 0 & 0 & 0 & 0 & 0 & 0 & 0 & 0 & 0 & 0 \\
0 & 0 & 0 & 0 & 0 & 0 & 0 & 0 & 0 & 0 & 0 & 0 & 0 & 0 & 0 & 1 & 0 & 0 & 0 & 0 & 0 & 0 & 0 & 0 & 0 & 0 \\
0 & 0 & 0 & 0 & 0 & 0 & 0 & 0 & 0 & 0 & 0 & 0 & 0 & 0 & 0 & 0 & 1 & 0 & 0 & 0 & 0 & 0 & 0 & 0 & 0 & 0 \\
0 & 0 & 0 & 0 & 0 & 0 & 0 & 0 & 0 & 0 & 0 & 0 & 0 & 0 & 0 & 0 & 0 & 1 & 0 & 0 & 0 & 0 & 0 & 0 & 0 & 0 \\
0 & 0 & 0 & 0 & 0 & 0 & 0 & 0 & 0 & 0 & 0 & 0 & 0 & 0 & 0 & 0 & 0 & 0 & 1 & 0 & 0 & 0 & 0 & 0 & 0 & 0 \\
0 & 0 & 0 & 0 & 0 & 0 & 0 & 0 & 0 & 0 & 0 & 0 & 0 & 0 & 0 & 0 & 0 & 0 & 0 & 1 & 0 & 0 & 0 & 0 & 0 & 0 \\
-1.4855 & 2.9368 & -2.3730 & 0.3735 & -0.0584 & 1.4067 & 0 & 0 & 0 & 0 & -2.8528 & 0 & 0 & 0 & 0 & -0.0012 & 0 & 0 & 0 & 0 & 0.0003 & -0.3696 & 0.3705 & 0 & 0 & 0 \\
1.5625 & -7.3517 & 8.1631 & -2.8131 & 0.4399 & 0 & 0 & 0 & 0 & 0 & 0 & 0 & 0 & 0 & 0 & 0 & 0 & 0 & 0 & 0 & 0 & 0 & 0 & 0 & 0 & 0 \\
-1.4152 & 9.1505 & -12.8571 & 6.7219 & -1.5981 & 0 & 0 & 0 & 0 & 0 & 0 & 0 & 0 & 0 & 0 & 0 & 0 & 0 & 0 & 0 & 0 & 0 & 0 & 0 & 0 & 0 \\
0.1795 & -2.5411 & 5.4167 & -4.6401 & 1.5848 & 0 & 0 & 0 & 0 & 0 & 0 & 0 & 0 & 0 & 0 & 0 & 0 & 0 & 0 & 0 & 0 & 0 & 0 & 0 & 0 & 0 \\
-0.0731 & 1.0342 & -3.3520 & 4.1251 & -2.4669 & 0 & 0 & 0 & 0 & 1.7716 & 0 & 0 & 0 & 0 & -3.5902 & 0 & 0 & 0 & 0 & -0.0010 & 0 & 0 & 0 & 0.0003 & -0.4656 & 0.4663 \\
-1.5345 & 0 & 0 & 0 & 0 & -2.2046 & 2.9368 & -2.3730 & 0.3735 & -0.0584 & -0.0012 & 0 & 0 & 0 & 0 & -3.2043 & 0 & 0 & 0 & 0 & 0.6928 & 0.1202 & 0.1203 & 0 & 0 & 0 \\
0 & 0 & 0 & 0 & 0 & 1.5625 & -7.3517 & 8.1631 & -2.8131 & 0.4399 & 0 & 0 & 0 & 0 & 0 & 0 & 0 & 0 & 0 & 0 & 0 & 0 & 0 & 0 & 0 & 0 \\
0 & 0 & 0 & 0 & 0 & -1.4152 & 9.1505 & -12.8571 & 6.7219 & -1.5981 & 0 & 0 & 0 & 0 & 0 & 0 & 0 & 0 & 0 & 0 & 0 & 0 & 0 & 0 & 0 & 0 \\
0 & 0 & 0 & 0 & 0 & 0.1795 & -2.5411 & 5.4167 & -4.6401 & 1.5848 & 0 & 0 & 0 & 0 & 0 & 0 & 0 & 0 & 0 & 0 & 0 & 0 & 0 & 0 & 0 & 0 \\
0 & 0 & 0 & 0 & -1.8555 & -0.0731 & 1.0342 & -3.3520 & 4.1251 & -3.2506 & 0 & 0 & 0 & 0 & -0.0010 & 0 & 0 & 0 & 0 & -3.8580 & 0 & 0 & 0 & 0.8334 & 0.1513 & 0.1514 \\
2.2205 & 0 & 0 & 0 & 0 & -0.6755 & 0 & 0 & 0 & 0 & 2.0000 & 0 & 0 & 0 & 0 & 2.3007 & 0 & 0 & 0 & 0 & -1.1099 & 0 & 0 & 0 & 0 & 0 \\
1.2201 & 0 & 0 & 0 & 0 & 0.7892 & 0 & 0 & 0 & 0 & -0.4653 & 0 & 0 & 0 & 0 & 2.2537 & 0 & 0 & 0 & 0 & 0 & -0.5639 & 0 & 0 & 0 & 0 \\
-0.5173 & 0 & 0 & 0 & 0 & -1.3629 & 0 & 0 & 0 & 0 & 1.7103 & 0 & 0 & 0 & 0 & -1.5468 & 0 & 0 & 0 & 0 & 0 & 0 & -0.5684 & 0 & 0 & 0 \\
0 & 0 & 0 & 0 & 2.1364 & 0 & 0 & 0 & 0 & -0.7018 & 0 & 0 & 0 & 0 & 2.0000 & 0 & 0 & 0 & 0 & 2.2089 & 0 & 0 & 0 & -1.0682 & 0 & 0 \\
0 & 0 & 0 & 0 & 1.2135 & 0 & 0 & 0 & 0 & 0.7551 & 0 & 0 & 0 & 0 & -0.4302 & 0 & 0 & 0 & 0 & 2.2430 & 0 & 0 & 0 & 0 & -0.5466 & 0 \\
0 & 0 & 0 & 0 & -0.5333 & 0 & 0 & 0 & 0 & -1.3296 & 0 & 0 & 0 & 0 & 1.6733 & 0 & 0 & 0 & 0 & -1.5588 & 0 & 0 & 0 & 0 & 0 & -0.5501
\end{array}\right]
$$

Fig. 4.2 Nondimensional system matrix **A** (stable)

$$x^{26}(0) = [0\ 0\ 1\ 0\ 0\quad 0\ 0\ 1\ 0\ 0\quad 0\ 0\ 0\ 0\ 0\quad 0\ 0\ 0\ 0\ 0\quad 0\ 0\ 0\quad 0\ 0\ 0]^{\mathrm{T}}. \tag{4.20}$$

The 26 eigenvalues of the original system are listed in Table 4.1. By selecting 14, 8, and 6 eigenvalues from the original list (as shown in Table 4.1), we can obtain three reduced models of dimensions 14, 8, and 6, respectively, using their corresponding modal matrix **U**, reduction weighting matrix **R**, and sampling data according to the procedure described earlier. The initial values of the reduced state vectors are:

$$\begin{aligned} x^{14}(0) = \big[& -2.09932\ \ 2.30143\ \ -0.70903\ \ -3.10788 \\ & 0.04950\ \ -0.06189\ \ -0.84207\ \ 0.03514 \\ & -0.02017\ \ 0.00421\ \ 0.46030\ \ 0.34064 \\ & 1.38056\ \ 0.14335 \big]^{\mathrm{T}}, \end{aligned} \tag{4.21a}$$

$$\begin{aligned} x^{8}(0) = \big[& -2.09932\ \ 2.30143\ \ -0.70903\ \ -3.10788 \\ & 0.46030\ \ 0.34064\ \ 1.38056\ \ 0.14335 \big]^{\mathrm{T}}, \end{aligned} \tag{4.21b}$$

$$x^{6}(0) = \big[-0.02017\ \ 0.00421\ \ 0.46030\ \ 0.34064\ \ 1.38056\ \ 0.14335 \big]^{\mathrm{T}}. \tag{4.21c}$$

As shown in Table 4.1, the eigenvalues of the reduced models closely match those of the original system. This highlights the effectiveness of the discussed approach, as the neglected modes are incorporated into the reduced models, resulting in synthetic, combined modes.

Figure 4.3 provides an eigenvalue plot of the original and reduced models, while Fig. 4.4 compares the free-decay responses y of the original and reduced models. Figure 4.5 shows the amplitude spectrum of y, computed using a fast Fourier transform (FFT) with $\Delta T = 0.3$ and $N = 2048$ data points. Here, NDF denotes the nondimensional frequency. The corresponding frequency values are given in Table 4.2. The calculation formulas are as follows:

$$\Delta F = \frac{1}{N \cdot \Delta T}, \quad F = (n-1) \cdot \Delta F, \quad \Omega = 2\pi F,$$

where ΔF is the frequency interval, ΔT is the time interval of data sampling, N is the total number of data points used for FFT, n is the sequential number of the n-th point in the amplitude spectrum, F is the frequency value of the n-th point, and Ω is the angular frequency corresponding to the eigenvalue.

From Tables 4.1 and 4.2, it is evident that the eigenvalues are accurately reflected in the frequency spectra. Figures 4.4 and 4.5 demonstrate that the model reduction results are satisfactory in both time-domain response and spectral analysis.

Table 4.1 Eigenvalues of the original model and its reduced counterparts

No	Real part	Imaginary part	Module
Original 26th order system			
1	−0.03837	4.70680	4.70696
2	−0.03837	−4.70680	4.70696
3	−0.03919	4.70399	4.70415
4	−0.03919	−4.70399	4.70415
5	−3.87615	0.00000	3.87615
6	−3.42579	0.00000	3.42579
7	−2.89310	0.00000	2.89310
8	−3.43572	0.00000	3.43572
9	−0.16828	1.84683	1.85448
10	−0.16828	−1.84683	1.85448
11	−0.17356	1.84284	1.85100
12	−0.17356	−1.84284	1.85100
13	−0.39472	0.75165	0.84898
14	−0.39472	−0.75165	0.84898
15	−0.30597	0.64133	0.71058
16	−0.30597	−0.64133	0.71058
17	−0.24585	0.64953	0.69450
18	−0.24585	−0.64953	0.69450
19	−0.15719	0.53444	0.55707
20	−0.15719	−0.53444	0.55707
21	−0.46721	0.34525	0.34528
22	−0.46721	−0.34525	0.34528
23	−0.01293	0.36076	0.36100
24	−0.01293	−0.36076	0.36100
25	−0.60891	0.00000	0.60891
26	−0.59138	0.00000	0.59138
Reduced 14th order system			
1	−0.06739	5.68076	5.68116
2	−0.06739	−5.68076	5.68116
3	−0.06878	5.67595	5.67636
4	−0.06878	−5.67595	5.67636
15	−0.44619	0.66844	0.80368
16	−0.44619	−0.66844	0.80368
17	−0.27338	0.66487	0.71888
18	−0.27338	−0.66487	0.71888
19	−0.15104	0.46920	0.49291
20	−0.15104	−0.46920	0.49291

(continued)

Table 4.1 (continued)

No	Real part	Imaginary part	Module
21	−0.00468	0.34555	0.34559
22	−0.00468	−0.34555	0.34559
23	−0.01297	0.36112	0.36135
24	−0.01297	−0.36112	0.36135
Reduced 8th order system			
1	−0.06737	5.68077	5.68117
2	−0.06737	−5.68077	5.68117
3	−0.06873	5.67592	5.67634
4	−0.06873	−5.67592	5.67634
21	−0.00468	0.34555	0.34559
22	−0.00468	−0.34555	0.34559
23	−0.01296	0.36112	0.36135
24	−0.01296	−0.36112	0.36135
Reduced 6th order system			
19	−0.16121	0.53815	0.56178
20	−0.16121	−0.53815	0.56178
21	−0.00468	0.34555	0.34559
22	−0.00468	−0.34555	0.34559
23	−0.01297	0.36112	0.36135
24	−0.01297	−0.36112	0.36135

Additionally, different choices of the matrix **R** (i.e., different eigenvalue selections) lead to reduced models with varying characteristics, confirming the versatility and effectiveness of the introduced reduction technique.

4.4 Model Reduction Versus Modal Reduction: A Comparison

In the field of dynamic system modeling and control, high-order mathematical models frequently arise from first-principles physical modeling. While these models provide valuable insight into system behavior, their complexity can hinder real-time implementation, analysis, and simulation. To address this, model reduction techniques are employed to generate simplified representations that retain essential dynamic characteristics. Among these, modal reduction is particularly prominent in mechanical and structural applications due to its physical interpretability and focus on low-frequency dominant modes. Although modal reduction is a subset of the broader model reduction framework, the two approaches differ in methodology and scope.

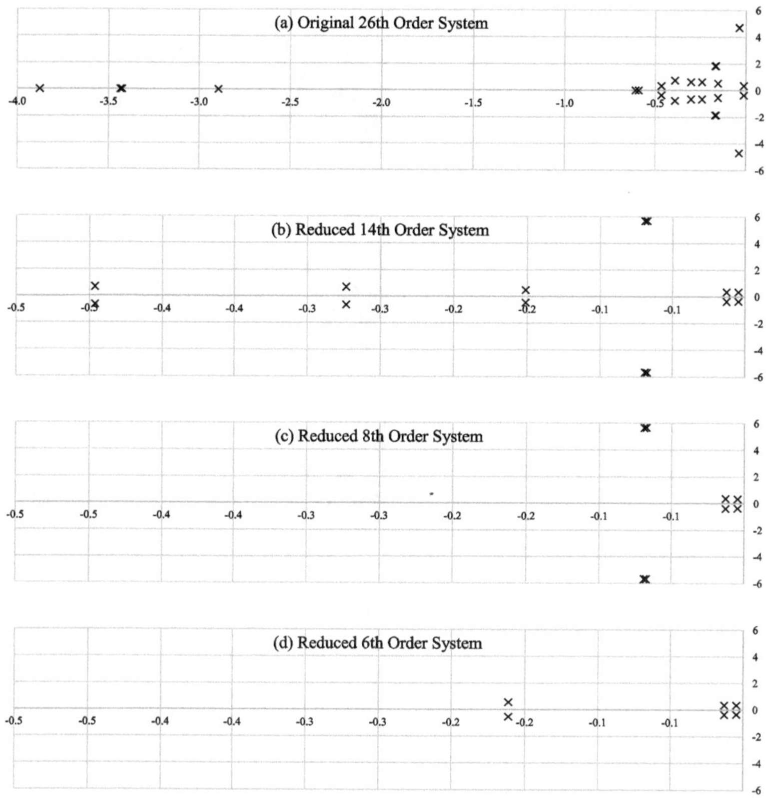

Fig. 4.3 Eigenvalue plot of the original model and the reduced models

This section presents a comparative overview of model reduction and modal reduction, highlighting their shared principles and distinct advantages in the context of control system design.

Model reduction refers to a broad class of mathematical methods that approximate a high-order dynamical system with a lower-order model, while preserving fidelity to its essential behavior within a specified domain (typically in time or frequency). The goal is to improve computational efficiency, control design tractability, and real-time implementability, without significantly sacrificing accuracy. Model reduction seeks a lower-order approximation of a system:

$$\text{Full-order model: } \dot{\boldsymbol{x}}(t) = \mathbf{A}\boldsymbol{x}(t) + \mathbf{B}\boldsymbol{u}(t), \quad \boldsymbol{y}(t) = \mathbf{C}\boldsymbol{x}(t),$$

$$\text{Reduced-order model: } \dot{\boldsymbol{x}}_r(t) = \mathbf{A}_r\boldsymbol{x}_r(t) + \mathbf{B}_r\boldsymbol{u}(t), \quad \boldsymbol{y}_r(t) = \mathbf{C}_r\boldsymbol{x}_r(t),$$

where the reduced system approximates the input–output behavior:

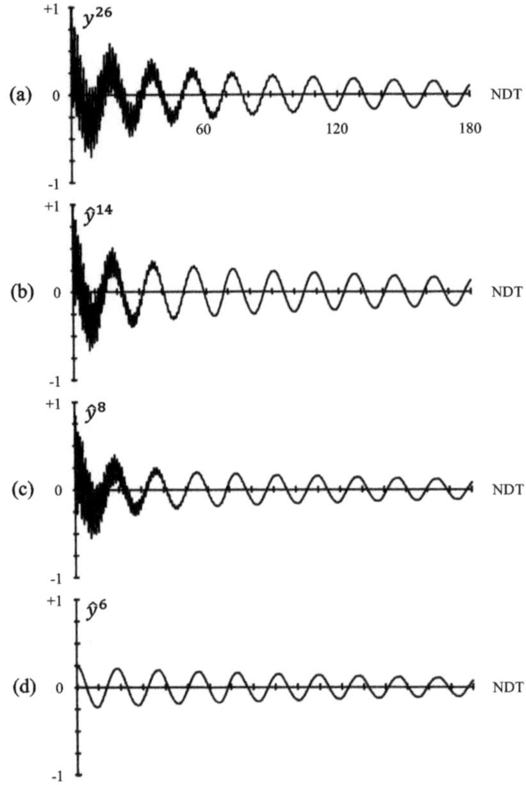

Fig. 4.4 Comparison of free-decay responses

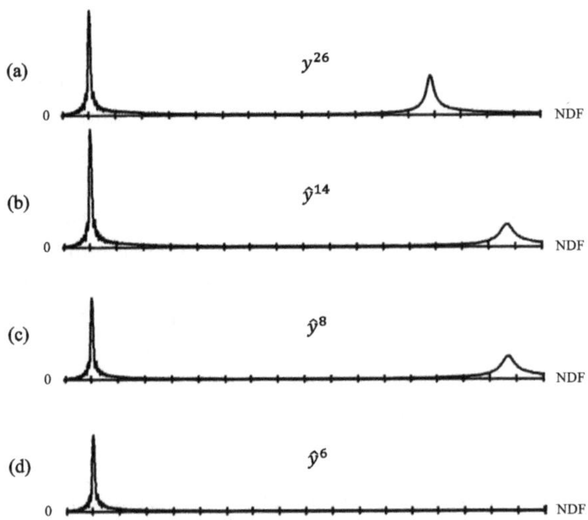

Fig. 4.5 Comparison of amplitude spectra

Table 4.2 Frequencies of the amplitude spectra

Figure 4.5	1st Peak			2nd Peak		
	n	Nondim. F	Nondim. Ω	n	Nondim. F	Nondim. Ω
(a)	35	0.05534	0.34770	461	0.74870	4.70421
(b)	35	0.05534	0.34770	556	0.90332	5.67573
(c)	35	0.05534	0.34770	556	0.90332	5.67573
(d)	35	0.05534	0.34770	/	/	/

$$\boldsymbol{y}_r(t) \approx \boldsymbol{y}(t), \quad \forall t.$$

Common techniques include: balanced truncation, Krylov subspace methods (e.g., moment matching), Proper Orthogonal Decomposition (POD), Hankel norm approximation, singular perturbation methods, and modal reduction. The main goal is to minimize the error norm (e.g., $\|\mathbf{G}(s) - \mathbf{G}_r(s)\|_\infty$) while preserving stability and key dynamic features.

In control system design, reduced-order models:

- Simplify controller synthesis (especially in high-order systems).
- Enable observer or estimator design.
- Support real-time embedded control through reduced computational load.
- Aid in robustness and performance analysis through simplified transfer functions or state-space representations.

Modal reduction is a specific type of model reduction method that uses the modal decomposition of a linear system. It exploits the fact that the dynamics of a LTI system can be described in terms of its eigenvalues (modes) and corresponding mode shapes (eigenvectors). Modal reduction leverages the modal properties of the system, particularly in a continuous-time second-order mechanical system:

$$\mathbf{M}\ddot{\boldsymbol{x}} + \mathbf{C}\dot{\boldsymbol{x}} + \mathbf{K}\boldsymbol{x} = \boldsymbol{f}(t).$$

By Eigen decomposing (transforming) the undamped system:

$$\mathbf{K}\boldsymbol{\phi}_i = \omega_i^2 \mathbf{M}\boldsymbol{\phi}_i.$$

The system can be expressed in modal coordinates:

$$\ddot{q}_i(t) + 2\zeta_i\omega_i\dot{q}_i(t) + \omega_i^2 q_i(t) = \boldsymbol{\phi}_i^{\mathrm{T}}\boldsymbol{f}(t).$$

This is a second-order differential equation describing the dynamics of the i-th modal coordinate $q_i(t)$, where.

- $q_i(t)$ is the modal amplitude.
- ζ_i is the damping ratio for mode i.
- ω_i is the natural frequency for mode i.
- $\boldsymbol{\phi}_i$ is the mode shape vector for mode i.
- $\boldsymbol{f}(t)$ is the external force vector.
- $\boldsymbol{\phi}_i^{\mathrm{T}}\boldsymbol{f}(t)$ projects the force onto the i-th mode.

This mass-normalized modal equation is the standard form of a damped second-order system in modal analysis. The system can be transformed into modal coordinates via eigenvalue decomposition by solving the equation

$$\left(\mathbf{K} - \omega_i^2\mathbf{M}\right)\boldsymbol{\phi}_i = 0.$$

Only the most influential (dominant) eigenmodes—typically those corresponding to low frequencies, high excitation levels, or low damping—are retained, while the rest are discarded to achieve a reduced-order model.

The modal reduction method involves:

- Diagonalizing or transforming the system to modal coordinates.
- Truncating high-frequency or weakly controllable/observable modes.
- Reconstructing a reduced-order system with dominant modes only.

In control system design, modal reduction is especially prevalent in:

- Mechanical and structural dynamics (e.g., vibration control, aerospace structures).
- Model-based control of flexible structures.
- Design of low-order observers and filters that focus on dominant modes.
- Analysis of resonance and damping characteristics.

A comparative summary of model reduction and modal reduction is provided in Table 4.3, highlighting both their conceptual overlap and methodological distinctions.

Despite their differences, model reduction and modal reduction share several important conceptual and practical elements:

1. *Purpose:* Both aim to simplify high-order dynamic models while preserving key system behavior, whether for simulation, analysis, or controller design.
2. *Approximation Criteria:* They both involve choosing which aspects of the system to preserve and which to neglect, often guided by:

    - Energy contribution.
    - Time/frequency domain impact.
    - Control-/observability Gramian analysis.

Table 4.3 Comparison of model reduction and modal reduction

Aspect	Model reduction (general)	Modal reduction
Scope	Broad class of techniques	Subset focused on modal decomposition
Approach	Mathematical, often input–output based	Physics-based, mode-shape driven
Basis of reduction	Mathematical properties (e.g., controllability, observability)	Physical modal characteristics (e.g., natural frequencies, mode shapes)
Mathematical tools	Balancing, projection, singular values, Krylov subspaces	Eigenvalue/eigenvector decomposition
Applicability (system type)	Linear and nonlinear systems	Primarily linear, second-order mechanical systems
Preserves	Input–output behavior	Dominant vibrational modes
Physical interpretation	May be abstract	Often has clear physical meaning (modes represent vibration patterns)
Use in control design	General purpose, applicable to various control structures	Typically in vibration control, structural damping, or observer design
Best for	• High-order LTI systems without a clear modal structure • Need to preserve input–output behavior across frequency (balanced truncation or Hankel approximation)	• Flexible structures, resonance, vibration suppression/control • Emphasis on physically meaningful behavior and mode shapes

3. *Projection-Based Framework:* Many reduction techniques (including modal) can be formulated using projection matrices. That is, the full system is projected onto a lower-dimensional subspace defined by selected basis vectors (e.g., dominant modes in modal reduction or balanced modes in balanced truncation).
4. *Use in Control Engineering:* Both are routinely applied for

   - Reduced-order controller design.
   - Observer or Kalman filter design.
   - Real-time model predictive control (MPC).
   - Robustness and sensitivity analysis.

Several key factors should be considered when applying model or modal reduction in control system design:

- *Model Accuracy Versus Complexity:* Modal reduction is often more intuitive and physically interpretable for mechanical systems. However, it may overlook global system dynamics that are better preserved in generalized model reduction techniques, such as balanced truncation or moment matching.

- *Robustness:* While modal reduction emphasizes dominant physical modes, it may be less effective in the presence of model uncertainties or unmodeled dynamics. In contrast, general model reduction methods frequently incorporate robustness considerations explicitly during the reduction process.
- *Computational Efficiency:* Modal reduction is typically more computationally efficient, particularly in structural and mechanical systems, due to the sparse or banded nature of mass and stiffness matrices. This makes it well-suited for real-time or embedded applications with limited computational resources.

Model reduction and modal reduction, while distinct in methodology and scope, share a common objective: simplifying complex system models to enable efficient analysis, simulation, and control. Modal reduction leverages the natural modes of vibration in mechanical and structural systems, making it particularly effective for capturing dominant low-frequency dynamics and offering valuable physical insight. General model reduction, by contrast, encompasses a wider array of techniques applicable to systems that may not possess a clear modal structure. These methods are often better suited for high-dimensional, computationally demanding applications requiring broad time-domain accuracy.

For control engineers, a clear understanding of both approaches is crucial. Modal reduction provides interpretability, while general model reduction often ensures robustness and numerical efficiency. As control systems grow in complexity and performance demands rise, hybrid strategies that combine modal intuition with advanced reduction algorithms are increasingly adopted—offering compact, accurate, and computationally tractable models tailored to specific control objectives.

4.5 Summary

Unlike many traditional modal reduction methods used in mechanics and modal analysis, this chapter presents a specialized approach for reducing models of large-scale, multi-modal systems. Developed from the aggregation method and augmented by techniques from system identification and modal analysis, this approach achieves high accuracy in both the time and frequency domains.

A key advantage of the method presented in this chapter is its ability to directly obtain a reduced model from a complex system, given adequate input and output data, without first needing to derive a high-order model and then reduce it. By applying the model reduction concepts discussed here and integrating them with the characteristics of mass–stiffness–damping systems, this method can be adapted for the online identification of actual systems, enabling the direct establishment of reduced models.

For example, by observing state variables at strategically chosen locations within the system and using the parameter estimation techniques outlined in Chap. 3, the reduced system

$$\mathbf{A}_r = \begin{bmatrix} 0 & \mathbf{I} \\ -(\mathbf{M}^{-1}\mathbf{K})_r & -(\mathbf{M}^{-1}\mathbf{C})_r \end{bmatrix},$$

can be directly identified. This approach is not only straightforward, overcoming the complexities of traditional modeling methods, but it also endows the reduced model with clear physical meaning, making it more suitable for practical applications.

References

1. Jamshidi M (1997) Large-scale systems: modeling, control and fuzzy logic. Prentice Hall
2. Antoulas AC (2005) Approximation of large-scale dynamical systems. SIAM
3. Mohammadpour J, Grigoriadis KM (2010) Efficient modeling and control of large-scale systems. Springer, Berlin
4. Benner P, Faßbender H (2021) Model order reduction: techniques and tools, encyclopedia of systems and control. Springer, Berlin, pp 1227–1234
5. Davison EJ (1966) A method for simplifying linear dynamic systems. IEEE Trans Autom Control 11(1):93–101
6. Davison EJ (1968) A new method for simplifying large linear dynamic systems. IEEE Trans Autom Control 13(2):214–215
7. Chidambara MR, Davison EJ (1967) On a method for simplifying linear dynamic systems. IEEE Trans Autom Control 12(1):119–121
8. Aoki M (1968) Control of large-scale dynamic systems by aggregation. IEEE Trans Autom Control 13(3):246–253
9. Aoki M (1978) Some approximation methods for estimation and control of large scale systems. IEEE Trans Autom Control 23(2):173–182
10. Soong TT (1977) On model reduction of large-scale systems. J Math Anal Appl 60(2):477–482
11. Kymn KO (1990) Aggregation in input-output models: a comprehensive review, 1946–1971. Econ Syst Res 2(1):65–93
12. Hildenbrand W (2018) Aggregation (theory), the new Palgrave dictionary of economics. Palgrave Macmillan, pp 152–164
13. Lastman GJ, Sinha NK, Rózsa P (1984) On the selection of states to be retained in a reduced-order model. IEE Proceed D 131(1):15–22
14. Bonvin D, Lastman GJ, Sinha NK (1985) Discussion on the selection of states to be retained in a reduced-order model. IEE Proceed D 132(2):86
15. Hughes PC (1987) Space structure vibration modes: how many exist? Which ones are important? IEEE Control Syst Mag 7(1):22–28
16. Hirzinger G, Kreisselmeier G (1975) On optimal approximation of high-order linear systems by low-order models. Int J Control 22(3):399–408
17. Siret JM, Michailesco G, Bertrand P (1977) Optimal approximation of high-order systems subject to polynomial inputs. Int J Control 26(6):963–971

18. Sinha NK, Lastman GJ (1982) Identification of continuous-time multivariable systems from sampled data. Int J Control 35(1):117–126
19. Zhu HA (1988) A novel method of model reduction based on system parameter identification. In: Preprints of the 8th IFAC/IFORS symposium on identification and system parameter estimation, PRC, vol 2, pp 910–915
20. Zhu HA, Xie YB, Yu L (1988) Model reduction for rotor-bearing systems. In: Proceedings of the 6th international modal analysis conference, USA, vol 1, pp 798–804

Chapter 5
Controllability and Observability of MKC Systems

System analysis typically encompasses two aspects: qualitative analysis and quantitative analysis. In qualitative analysis, we are interested in the general properties and universal characteristics of the system. In quantitative research, we focus on the precise response of the system to specific inputs and initial conditions, as well as the exact behavior of the system. In this chapter, we will discuss two fundamental qualitative properties of linear dynamic systems—controllability and observability—particularly in the context of large-scale mass–stiffness–damping systems governed by second-order matrix differential equations. These systems represent a class of complex, interconnected large dynamic systems with significant real-world implications, such as mechanical structures, aerospace vehicles, and automotive suspension systems. Understanding controllability and observability in these systems is crucial for the design of effective control strategies and for ensuring system performance and safety.

5.1 Introduction

Since Kalman [1] introduced the concepts of controllability and observability in his groundbreaking paper in 1960, extensive and in-depth research has been conducted on these properties for various types of systems. These include systems with lumped and distributed parameters, linear and nonlinear dynamics, time-invariant and time-varying characteristics, continuous- and discrete-time domains, deterministic and stochastic behaviors, among others. This body of work has yielded numerous valuable results [2–4].

Controllability and observability are fundamental concepts in control theory, especially in the study of complex systems. To ensure completeness of the discussion, the relevant definitions and theorems are provided here [3, 4].

H. Zhu, *Model-Based Control of Mass–Stiffness–Damping Systems*, Advances in Industrial Control, https://doi.org/10.1007/978-3-031-97592-9_5

5.1.1 Controllability

We focus specifically on the state controllability of linear dynamic systems represented in state-space form. Since controllability is a property inherent to the state equations, output equations are not directly involved in this context.

Consider the following n-dimensional linear time-varying (LTV) state equation:

$$\dot{\boldsymbol{x}}(t) = \mathbf{A}(t)\boldsymbol{x}(t) + \mathbf{B}(t)\boldsymbol{u}(t), \tag{5.1}$$

where $\boldsymbol{x}(t)$ is $n \times 1$ state vector; $\boldsymbol{u}(t)$ is $l \times 1$ input (control) vector. $\mathbf{A}(t)$ and $\mathbf{B}(t)$ are $n \times n$ and $n \times l$ matrices, respectively, with elements that are functions of time t, defined over the time domain $(-\infty, +\infty)$. The state space of this system is an n-dimensional complex vector space, denoted by $\boldsymbol{\Sigma}$.

Definition 5.1 The state Eq. (5.1) is said to be state controllable at time t_0 if, for any initial state $\boldsymbol{x}(t_0) = \boldsymbol{x}_0$ and any final state $\boldsymbol{x}_1$ in the state space $\boldsymbol{\Sigma}$, there exists a control input $\boldsymbol{u}(t)$ defined over a finite time interval $[t_0, t_1]$ (with $t_1 > t_0$) that transfers the state from $\boldsymbol{x}_0$ to $\boldsymbol{x}_1$. If this condition cannot be met, the system in Eq. (5.1) is said to be uncontrollable at t_0.

In the literature, if a non-zero state can be transferred to the zero state 0, then the state is said to be controllable. If it is possible to reach another non-zero state from 0, the state is said to be reachable. To simplify the following discussion, this book does not make such a distinction in its definitions. Additionally, in the literature, Eq. (5.1) is often referred to as being fully controllable. For simplicity, the term "fully" is omitted in this book.

It should be noted that the above definition only requires that the control input $\boldsymbol{u}(t)$ be able to transfer any state in the state space $\boldsymbol{\Sigma}$ to any other state within a finite period of time; it does not specify what kind of trajectory the state transition should follow. Moreover, there are no constraints imposed on the control input $\boldsymbol{u}(t)$. In other words, its magnitude can be arbitrarily large.

Now, consider the following n-dimensional linear time-invariant (LTI) system

$$\dot{\boldsymbol{x}} = \mathbf{A}\boldsymbol{x} + \mathbf{B}\boldsymbol{u}, \tag{5.2}$$

where $\boldsymbol{x}$ is $n \times 1$ state vector; $\boldsymbol{u}$ is $l \times 1$ control input vector. $\mathbf{A}$ and $\mathbf{B}$ are $n \times n$ and $n \times l$ real constant system matrix and constant input matrix, respectively. For time-invariant dynamic systems, the time domain of interest extends from the present moment to infinity, i.e., $[0, +\infty)$.

If a LTI system is controllable, then it is controllable for all $t_0 \geq 0$, and any state can be transferred to any other state within any finite, non-zero time interval. Therefore, in the study of the controllability of LTI systems, the notation of t_0 and t_1 is often omitted.

Theorem 5.1 *The n-dimensional LTI state Eq.* (5.2) *is controllable if and only if any one of the following equivalent conditions holds:*

i. *In the complex domain* $\mathbb{C}$, *within* $[0, +\infty)$, *all the rows of* $e^{-\mathbf{A}t}\mathbf{B}$ *(and necessarily,* $e^{\mathbf{A}t}\mathbf{B}$*) are linearly independent. (Although all the elements of* $\mathbf{A}$ *and* $\mathbf{B}$ *are real numbers, it is still correct to consider them as elements over the complex domain.)*

ii. *In* $\mathbb{C}$, *all the rows of* $(s\mathbf{I} - \mathbf{A})^{-1}\mathbf{B}$ *are linearly independent.*

iii. *The controllability Gramian*

$$\mathbf{W}_c(t) = \int_0^t e^{\mathbf{A}\tau}\mathbf{B}\mathbf{B}^* e^{\mathbf{A}^*\tau} \mathrm{d}\tau,$$

is non-singular for any $t > 0$.

iv. *The* $n \times (n \cdot l)$ *controllability matrix*

$$\mathcal{C} = \left[\, \mathbf{B} \mid \mathbf{A}\mathbf{B} \mid \mathbf{A}^2\mathbf{B} \mid \cdots \mid \mathbf{A}^{n-1}\mathbf{B} \,\right], \tag{5.3}$$

has full rank n.

v. *For every eigenvalue* λ *of* $\mathbf{A}$ *(and necessarily, each* λ *in* $\mathbb{C}$*), the* $n \times (n + l)$ *complex matrix*

$$[\, \lambda\mathbf{I} - \mathbf{A} \mid \mathbf{B} \,],$$

has rank n. *Or equivalently,* $(s\mathbf{I} - \mathbf{A})$ *and* $\mathbf{B}$ *are left coprime.*

When transferring a non-zero initial state $\boldsymbol{x}(0)$ to the zero state, the required control input $\boldsymbol{u}(t)$. may become impractically large if the time interval is very short. If no constraints are imposed on $\boldsymbol{u}(t)$, the transfer can be achieved in arbitrarily short times, but with potentially unbounded control effort. In practice, control inputs are subject to constraints, and the minimum time to achieve state transfer may be limited.

If a linear dynamic system is controllable, there will typically be many different control inputs $\boldsymbol{u}$ that can transfer $\boldsymbol{x}(t_0)$ to $\boldsymbol{x}_1$ at time t_1, since the trajectory of the state transition between $\boldsymbol{x}(t_0)$ and $\boldsymbol{x}_1$ inot specified. Among the infinite possible control inputs that can accomplish the task of transferring $\boldsymbol{x}(t_0)$ to $\boldsymbol{x}_1$, one may seek an optimal control according to certain constraints and performance criteria, such as minimizing the total control energy:

$$J = \int_{t_0}^{t_1} \|\boldsymbol{u}(t)\|^2 \mathrm{d}t.$$

This leads to the concept of minimum-energy control.

5.1.2 Observability

The concept of observability is dual to that of controllability. While controllability addresses the ability to steer the state of the system using control inputs, observability concerns the ability to infer the state of the system from output measurements. If a dynamic system is observable, then all of its modes can be observed in the output. These two concepts are defined under the premise that complete knowledge of the dynamic equation is available, meaning the matrices **A**,**B**, and **C** are already known.

Consider the following n-dimensional LTV dynamic system

$$\dot{\boldsymbol{x}}(t) = \mathbf{A}(t)\boldsymbol{x}(t) + \mathbf{B}(t)\boldsymbol{u}(t), \tag{5.4a}$$

$$\boldsymbol{y}(t) = \mathbf{C}(t)\boldsymbol{x}(t), \tag{5.4b}$$

where $\boldsymbol{y}(t)$ is the $p \times 1$ output vector; $\mathbf{C}(t)$ is the $p \times n$ output matrix, with elements that are functions of time t, defined over the time domain $(-\infty, +\infty)$.

Definition 5.2 The dynamic system (5.4) is said to be state observable at t_0 if, given the control input $\boldsymbol{u}(t)$ and the output $\boldsymbol{y}(t)$ over a finite time interval $[t_0, t_1]$ (with $t_1 > t_0$), it is possible to uniquely determine the initial state $\boldsymbol{x}(t_0)$. If this is not possible, the system is said to be unobservable at t_0.

Now, consider the following LTI dynamic system

$$\dot{\boldsymbol{x}} = \mathbf{A}\boldsymbol{x} + \mathbf{B}\boldsymbol{u}, \tag{5.5a}$$

$$\boldsymbol{y} = \mathbf{C}\boldsymbol{x}, \tag{5.5b}$$

where **A**, **B**, and **C** are, respectively, $n \times n$, $n \times l$, and $p \times n$ constant matrices. The time interval of interest is $[0, +\infty)$. Similar to the previous discussion on controllability, if a LTI dynamic system is observable, it is observable at each $t_0 \geq 0$, and the estimation of the initial state can be completed within any finite, non-zero time interval. Therefore, in the study of observability of LTI dynamic systems, the notation for t_0 and t_1 is often omitted.

Theorem 5.2 *The n-dimensional LTI system (5.5) is observable if and only if any one of the following equivalent conditions holds:*

i. *In the complex domain $\mathbb{C}$, within $[0, +\infty)$, all the columns of $\mathbf{C}e^{\mathbf{A}t}$ are linearly independent.*
ii. *In $\mathbb{C}$, all the columns of $\mathbf{C}(s\mathbf{I} - \mathbf{A})^{-1}$ are linearly independent.*
iii. *The observability Gramian*

$$\mathbf{W}_o(t) = \int_0^t e^{\mathbf{A}^*\tau}\mathbf{C}^*\mathbf{C}e^{\mathbf{A}\tau}\mathrm{d}\tau,$$

is non-singular for any $t > 0$.

iv. *The* $(n \cdot p) \times n$ *observability matrix*

$$\mathcal{O} = \begin{bmatrix} \mathbf{C} \\ \mathbf{CA} \\ \mathbf{CA}^2 \\ \vdots \\ \mathbf{CA}^{n-1} \end{bmatrix}, \tag{5.6}$$

has full rank n.

v. *For every eigenvalue* λ *of* $\mathbf{A}$ *(and necessarily, each* λ *in* $\mathbb{C}$*), the* $(n+p) \times n$ *complex matrix*

$$\begin{bmatrix} \lambda\mathbf{I} - \mathbf{A} \\ \mathbf{C} \end{bmatrix},$$

has rank n. *Or equivalently,* $(s\mathbf{I} - \mathbf{A})$ *and* $\mathbf{C}$ *are right coprime.*

Observability ensures that all the modes (states) of the system can be reconstructed from the output measurements over a finite time interval. In control system design, observability is crucial for the implementation of state observers or estimators, such as the Luenberger observer or Kalman filter, which are used when not all states are directly measurable.

Understanding the controllability and observability of dynamic systems is essential for the design and analysis of control systems. These properties determine whether it is possible to control the system to achieve desired performance and whether the internal states can be accurately estimated from available measurements. In complex, interconnected systems like large-scale mass–spring–damper networks, these concepts play a critical role in ensuring effective control and observation strategies.

5.1.3 Duality Principle of Controllability and Observability

From the above discussion, it is evident that controllability and observability are closely related, both conceptually and in the mathematical form of their criteria. This intrinsic connection is encapsulated in what Kalman proposed as the principle of duality [1]. The duality principle not only reveals the reciprocal relationship between these two fundamental characteristics of control systems but also highlights the intrinsic connection between the two primary problems in control theory: the control (regulation) problem and the observation (estimation) problem.

From the above discussion, it can be seen that controllability and observability are very similar both conceptually and in the mathematical form of their criteria.

This intrinsic connection is what Kalman proposed as the duality principle [1]. The principle of duality not only reveals the dual relationship between these two fundamental characteristics of control systems but also highlights the intrinsic connection between the two primary issues in control theory: the control problem and the estimation problem.

The duality between controllability and observability implies that mathematical techniques and criteria developed for one can often be adapted for the other. This means that results regarding controllability can be transformed into corresponding results about observability, and vice versa.

Theorem 5.3 (Duality Theorem) *Consider the dynamic system given by Eq.* (5.4) *and its dual system defined by:*

$$\dot{z}(t) = -\mathbf{A}^*(t)z(t) + \mathbf{C}^*(t)v(t), \tag{5.7a}$$

$$w(t) = \mathbf{B}^*(t)\mathbf{z}(t), \tag{5.7b}$$

where $\mathbf{A}^*$, $\mathbf{B}^*$, *and* $\mathbf{C}^*$ *are the complex conjugate transposes of* $\mathbf{A}$, $\mathbf{B}$, *and* $\mathbf{C}$, *respectively.*

The system in Eq. (5.4) is controllable (or observable) at time t_0 *if and only if its dual system in Eq. (5.7) is observable (or controllable) at time* t_0.

In other words, Theorem 5.3 states that a necessary and sufficient condition for the system (5.4) to be state controllable (observable) is that its dual system (5.7) is state observable (controllable). According to this principle, the controllability (observability) of a system can be studied by examining the observability (controllability) of its dual system.

This duality principle is powerful because it allows us to leverage the mathematical tools and criteria established for one property to analyze the other. For example, the criteria for controllability, such as the controllability matrix rank condition or the controllability Gramian, have corresponding dual criteria for observability involving the observability matrix and observability Gramian.

The proofs for Theorems 5.1–5.3 can be found in most textbooks on linear systems theory and control, such as in [3, 4], and will therefore not be repeated here.

5.2 Analysis of Control Properties in MKC Systems

In this section, we will analyze the controllability and observability of a typical class of complex interconnected dynamic systems known as mass–stiffness–damping systems. These systems are commonly described by second-order matrix differential equations, which effectively reveal and characterize the physical properties and motion mechanisms of a wide range of dynamic systems. Our primary focus here will be on the issue of controllability, and we will present the corresponding theoretical results related to these systems.

5.2.1 General Properties of MKC Systems

Consider a dynamic system described by the following LTI second-order matrix differential equations:

$$\mathbf{M}\ddot{x} + \mathbf{C}\dot{x} + \mathbf{K}x = \mathbf{F}f, \tag{5.8a}$$

$$y = \mathbf{D}x + \mathbf{V}\dot{x}, \tag{5.8b}$$

where **M**, **C**, and **K** are $n \times n$ real constant matrices representing the mass (inertia), damping, and stiffness of the system, respectively. **F** is an $n \times l$ real constant input matrix representing the control input distribution. **D** and **V** are $p \times n$ real constant matrices representing displacement and velocity measurements, respectively. x is the $n \times 1$ displacement vector. f is the $l \times 1$ control force vector. y is the $p \times 1$ measurement output vector. $\dot{x}$ and $\ddot{x}$ are the first and second derivatives of x with respect to time, representing velocity and acceleration, respectively.

The mass matrix **M** is typically positive definite. In active control contexts, the input matrix **F** is generally of full rank and often structured such that each of its l columns contains only one non-zero element.

To facilitate analysis, we rewrite Eq. (5.8a) by pre-multiplying both sides by $\mathbf{M}^{-1}$:

$$\ddot{x} + \mathbf{M}^{-1}\mathbf{C}\dot{x} + \mathbf{M}^{-1}\mathbf{K}x = \mathbf{M}^{-1}\mathbf{F}f, \tag{5.9a}$$

$$y = \mathbf{D}x + \mathbf{V}\dot{x}. \tag{5.9b}$$

Next, we transform the second-order differential equation into a state-space representation by defining the state vector z:

$$z = \begin{bmatrix} x \\ \dot{x} \end{bmatrix}.$$

The state-space representation becomes:

$$\dot{z} = \begin{bmatrix} 0 & \mathbf{I} \\ -\mathbf{M}^{-1}\mathbf{K} & -\mathbf{M}^{-1}\mathbf{C} \end{bmatrix} z + \begin{bmatrix} 0 \\ \mathbf{M}^{-1}\mathbf{F} \end{bmatrix} f = \mathbf{A}\,z + \mathbf{B}f, \tag{5.10a}$$

$$y = [\,\mathbf{D} \mid \mathbf{V}\,]\,z = \mathbf{C}\,z, \tag{5.10b}$$

where $\mathbf{A}$ is the $2n \times 2n$ system matrix; $\mathbf{B}$ is the $2n \times l$ input matrix; $\mathbf{C}$ is the $p \times 2n$ output matrix.

We have now obtained the state-space representation of the mass–stiffness–damping system, which aligns with the standard form of linear state-space equations. This allows us to proceed with the analysis of state controllability for such systems.

For systems described by Eqs. (5.10a) and (5.10b) with known coefficient matrices $\mathbf{M}, \mathbf{K}, \mathbf{C}, \mathbf{F}, \mathbf{D}$, and $\mathbf{V}$, a common method for determining controllability and observability is to use the criteria provided by Theorems 5.1 and 5.2. Specifically, we utilize the controllability matrix $\mathcal{C}$ and the observability matrix $\mathcal{O}$:

$$\mathcal{C} = \left[\, \mathbf{B} \mid \mathbf{AB} \mid \mathbf{A}^2\mathbf{B} \mid \cdots \mid \mathbf{A}^{2n-1}\mathbf{B} \,\right],$$

$$\mathcal{O} = \left[\, \mathbf{C}^{\mathrm{T}} \mid (\mathbf{A}^{\mathrm{T}})\mathbf{C}^{\mathrm{T}} \mid (\mathbf{A}^{\mathrm{T}})^2\mathbf{C}^{\mathrm{T}} \mid \cdots \mid (\mathbf{A}^{\mathrm{T}})^{2n-1}\mathbf{C}^{\mathrm{T}} \,\right]^{\mathrm{T}}.$$

These matrices have dimensions $2n \times (2n \cdot l)$ and $(2n \cdot p) \times 2n$, respectively. Recognizing that the rank of a matrix is invariant under linear non-singular transformations, we can simplify calculations using the following properties:

$$\operatorname{rank}(\mathcal{C}) = \operatorname{rank}\left(\mathcal{C}\mathcal{C}^{\mathrm{T}}\right), \tag{5.11a}$$

$$\operatorname{rank}(\mathcal{O}) = \operatorname{rank}\left(\mathcal{O}^{\mathrm{T}}\mathcal{O}\right). \tag{5.11b}$$

Equation (5.11) allows us to assess controllability and observability by computing the rank of smaller, symmetric matrices, which simplifies the computational process for lower-order systems.

However, for complex large-scale systems, directly computing the rank of these matrices becomes computationally intensive due to their size. To address this challenge, we introduce a new criterion specifically tailored for mass–stiffness–damping systems.

From Theorem 5.1, the equivalent condition (v) states that the mass–stiffness–damping system is controllable if and only if:

$$\operatorname{rank}[\, s\mathbf{I} - \mathbf{A} \mid \mathbf{B} \,] = 2n, \tag{5.12}$$

where s is any complex number, and $\mathbf{I}$ is the identity matrix of appropriate dimensions.

Considering the specific structural characteristics of the matrices $\mathbf{A}$ and $\mathbf{B}$ in Eq. (5.10a), we expand Eq. (5.12):

$$\operatorname{rank}\begin{bmatrix} s\mathbf{I} & -\mathbf{I} & 0 \\ \mathbf{M}^{-1}\mathbf{K} & s\mathbf{I} + \mathbf{M}^{-1}\mathbf{C} & \mathbf{M}^{-1}\mathbf{F} \end{bmatrix} = 2n. \tag{5.13}$$

The matrix in Eq. (5.13) is of size $2n \times (n + n + l)$, or $2n \times (2n + l)$. We observe the following:

- The first n columns corresponding to $s\mathbf{I}$ are linearly independent (therefore, rank of the first n columns is n) when $s \neq 0$.
- The next n columns corresponding to $-\mathbf{I}$ are also linearly independent.
- The last l columns correspond to $\mathbf{M}^{-1}\mathbf{F}$ and have rank l due to the full rank of $\mathbf{F}$.

To simplify the rank computation, we notice a correspondence between the first n columns and the second n columns of the matrix in Eq. (5.13). Specifically, each column in the first n columns is related to a column in the second n columns. Utilizing this observation, Eq. (5.13) can be rewritten as follows:

$$\text{rank}\begin{bmatrix} s\mathbf{I} & -\mathbf{I} & 0 \\ \mathbf{M}^{-1}\mathbf{K} & s\mathbf{I}+\mathbf{M}^{-1}\mathbf{C} & \mathbf{M}^{-1}\mathbf{F} \end{bmatrix}$$

$$= \text{rank}\begin{bmatrix} s\mathbf{I} & -s\mathbf{I} & 0 \\ \mathbf{M}^{-1}\mathbf{K} & s^2\mathbf{I}+s\mathbf{M}^{-1}\mathbf{C} & \mathbf{M}^{-1}\mathbf{F} \end{bmatrix}, \tag{5.14}$$

$$= \text{rank}\begin{bmatrix} s\mathbf{I} & 0 & 0 \\ \mathbf{K} & s^2\mathbf{M}+s\mathbf{C}+\mathbf{K} & \mathbf{F} \end{bmatrix} = 2n. \tag{5.15}$$

From Eq. (5.15), a simplified criterion can be derived.

Corollary 5.1 *The mass–stiffness–damping system described by Eq.* (5.8a) *is controllable if and only if, for some s in the complex plane, the following condition is satisfied:*

$$\text{rank}\left[\, s^2\mathbf{M}+s\mathbf{C}+\mathbf{K} \mid \mathbf{F} \,\right] = n.$$

That is, the augmented matrix formed by $\left(s^2\mathbf{M}+s\mathbf{C}+\mathbf{K}\right)$ *and* $\mathbf{F}$ *must have full rank n. Any linear dependencies in the columns of* $\left(s^2\mathbf{M}+s\mathbf{C}+\mathbf{K}\right)$ *must be compensated by the corresponding columns in* $\mathbf{F}$. *Here, l represents the number of control inputs.*

In other words, the number of linearly dependent columns in the matrix $\left(s^2\mathbf{M}+s\mathbf{C}+\mathbf{K}\right)$ should not exceed l, and the corresponding column vectors in $\mathbf{F}$ should be able to eliminate this dependency. The proof of Corollary 5.1 can be directly derived from the equivalent condition (v) of Theorem 5.1.

This result follows directly from the equivalent condition (v) of Theorem 5.1 applied to the specific structure of the mass–stiffness–damping system. By reducing the rank condition to the second block row in Eq. (5.15), we focus on an $n \times (n+l)$ matrix, simplifying the analysis.

In practice, using Corollary 5.1 as a criterion involves evaluating the rank condition at specific values of the complex variable s, typically the eigenvalues $s = \lambda_i$ $(i = 1, \ldots, 2n)$ of the system matrix $\mathbf{A}$, which correspond to the natural frequencies or poles of the system. This approach simplifies the problem of determining controllability from a $2n \times (2n+l)$ system to checking the rank of an $n \times (n+l)$ matrix for each eigenvalue.

This criterion offers computational advantages over traditional methods, such as [2], especially for large-scale systems:

- Reduced Computational Complexity: The problem is reduced from analyzing a $2n \times 2n$ matrix to an $n \times (n + l)$ matrix, significantly lowering the computational burden.
- Simpler Implementation: The method is easier to program and implement, facilitating practical applications in control system design.

However, similar to other controllability criteria, this method requires knowledge of the eigenvalues λ_i of the system. As the system dimension n increases, computing these eigenvalues becomes challenging due to the complexity of solving high-order characteristic equations. This is a common limitation shared by various methods for determining controllability and observability in large-scale systems.

By deriving a simplified controllability criterion specific to mass–stiffness–damping systems, we have provided a practical method for assessing controllability that reduces computational effort. While this method still requires eigenvalue computation, it facilitates analysis for systems where directly computing the controllability matrix is infeasible due to system size. Understanding the controllability of such systems is crucial for designing effective control strategies that ensure desired system performance and stability.

5.2.2 Special Properties of MKC Systems

As an example, we analyze the controllability of an important class of mass–stiffness–damping systems: rotor-bearing systems.

Consider a rotor system discretized into n point-masses, supported by oil films in journal bearings. The bearings contribute to the dynamic behavior through supporting forces from the oil film. Thus, the entire rotor-bearing system has $2n$ degrees-of-freedom: n in the x-direction and n in the y-direction. For illustrative purposes, let $n = 5$ and assume the rotor is supported by two journal bearings located at both ends. We neglect gyroscopic torques in this analysis. The displacement vector is arranged as follows:

$$\boldsymbol{x} = \begin{bmatrix} x_1 \ x_2 \ x_3 \ x_4 \ x_5 \ \ y_1 \ y_2 \ y_3 \ y_4 \ y_5 \end{bmatrix}^{\mathrm{T}}. \tag{5.16}$$

The mass ($\mathbf{M}$), stiffness ($\mathbf{K}$), and damping ($\mathbf{C}$) matrices of the rotor-bearing system have the following characteristics:

1. The $2n \times 2n$ mass matrix $\mathbf{M}$ is diagonal:

$$\mathbf{M} = \mathrm{diag}\begin{bmatrix} m_1 \ m_2 \ m_3 \ m_4 \ m_5 \ \ m_1 \ m_2 \ m_3 \ m_4 \ m_5 \end{bmatrix}. \tag{5.17}$$

2. The stiffness matrix **K** is a $2n \times 2n$ matrix structured as:

$$
K = \begin{bmatrix} \mathbf{K}_{xx} & \mathbf{K}_{xy} \\ \mathbf{K}_{yx} & \mathbf{K}_{yy} \end{bmatrix}
= \begin{bmatrix}
k_{xx1} & \cdots & \cdots & \cdots & k_{xy1} & 0 & 0 & 0 & 0 \\
\cdots & \cdots & \cdots & \cdots & 0 & 0 & 0 & 0 & 0 \\
\cdots & \cdots & \cdots & \cdots & 0 & 0 & 0 & 0 & 0 \\
\cdots & \cdots & \cdots & \cdots & 0 & 0 & 0 & 0 & 0 \\
\cdots & \cdots & \cdots & k_{xx5} & 0 & 0 & 0 & 0 & k_{xy5} \\
k_{yx1} & 0 & 0 & 0 & 0 & k_{yy1} & \cdots & \cdots & \cdots \\
0 & 0 & 0 & 0 & 0 & \cdots & \cdots & \cdots & \cdots \\
0 & 0 & 0 & 0 & 0 & \cdots & \cdots & \cdots & \cdots \\
0 & 0 & 0 & 0 & 0 & \cdots & \cdots & \cdots & \cdots \\
0 & 0 & 0 & 0 & k_{yx5} & \cdots & \cdots & \cdots & k_{yy5}
\end{bmatrix}. \tag{5.18}
$$

- $\mathbf{K}_{xx}$ and $\mathbf{K}_{yy}$ are $n \times n$ submatrices representing stiffness in the x- and y-directions, respectively. These submatrices on the main diagonal of matrix **K** are generally non-zero due to shaft and bearing stiffness where "$\cdots$" denotes non-zero stiffness elements.
- $\mathbf{K}_{xy}$ and $\mathbf{K}_{yx}$ are $n \times n$ submatrices containing cross-stiffness terms arising from the oil-film forces in the bearings. These off-diagonal submatrices (upper right and lower left) are typically sparse but crucial for coupling the x and y dynamics.

3. The damping matrix **C** is also a $2n \times 2n$ matrix:

$$
\mathbf{C} = \begin{bmatrix} \mathbf{C}_{xx} & \mathbf{C}_{xy} \\ \mathbf{C}_{yx} & \mathbf{C}_{yy} \end{bmatrix}
= \begin{bmatrix}
c_{xx1} & 0 & 0 & 0 & 0 & c_{xy1} & 0 & 0 & 0 & 0 \\
0 & 0 & 0 & 0 & 0 & 0 & 0 & 0 & 0 & 0 \\
0 & 0 & 0 & 0 & 0 & 0 & 0 & 0 & 0 & 0 \\
0 & 0 & 0 & 0 & 0 & 0 & 0 & 0 & 0 & 0 \\
0 & 0 & 0 & 0 & c_{xx5} & 0 & 0 & 0 & 0 & c_{xy5} \\
c_{yx1} & 0 & 0 & 0 & 0 & c_{yy1} & 0 & 0 & 0 & 0 \\
0 & 0 & 0 & 0 & 0 & 0 & 0 & 0 & 0 & 0 \\
0 & 0 & 0 & 0 & 0 & 0 & 0 & 0 & 0 & 0 \\
0 & 0 & 0 & 0 & 0 & 0 & 0 & 0 & 0 & 0 \\
0 & 0 & 0 & 0 & c_{yx5} & 0 & 0 & 0 & 0 & c_{yy5}
\end{bmatrix}. \tag{5.19}
$$

- The rotor itself usually has negligible internal damping; therefore, **C** is extremely sparse, containing damping elements primarily at the bearing locations.
- $\mathbf{C}_{xx}$ and $\mathbf{C}_{yy}$ represent damping in the x- and y-directions.
- $\mathbf{C}_{xy}$ and $\mathbf{C}_{yx}$ represent cross-damping terms due to the bearing oil film.

Given the characteristics of **M**, **K**, and **C** in rotor-bearing systems:

- Mass Matrix **M**: Diagonal and positive definite.
- Stiffness Matrix **K**: Includes cross-stiffness terms $\mathbf{K}_{xy}$ and $\mathbf{K}_{yx}$ due to bearing oil-film forces.
- Damping Matrix **C**: Sparse, with damping primarily at the bearings, including cross-damping terms.

Because of the coupling introduced by the cross-stiffness and cross-damping terms, the matrix $\left(s^2\mathbf{M} + s\mathbf{C} + \mathbf{K}\right)$ retains these coupling characteristics.

Corollary 5.2 *A LTI rotor-bearing system with concentrated parameters, supported by journal bearings and characterized by matrices* **M**, **K**, *and* **C** *as described in* (5.17)*–*(5.19)*, is generally controllable.*

Since the cross-coupling terms in **K** and **C** are non-zero (due to the bearing oil-film forces), the matrix $\left(s^2\mathbf{M} + s\mathbf{C} + \mathbf{K}\right)$ will have full rank except for possible dependencies that can be compensated by the control input matrix **F**. By applying control forces in either the x- or y-directions, we can eliminate any linear dependencies in the rank condition. Therefore, according to the controllability criterion, the rotor-bearing system is controllable. From Corollary 5.1 and the above analysis, the proof of Corollary 5.2 is evident.

For rotor-bearing systems with n point-masses (i.e., $2n$ degrees-of-freedom), the rank of the matrix in Eq. (5.15) is always $2 \times 2n$, provided that control forces are applied. This holds regardless of the specific locations and number of control inputs, as long as they satisfy the rank condition.

While Corollary 5.2 is stated for a single-span rotor supported at both ends, the same reasoning applies to multi-span rotors with multiple supports. The presence of coupling terms and appropriately placed control inputs ensures controllability.

Similar conclusions about the controllability of rotor-bearing systems have been mentioned by Stanway and Burrows [5], Burrows and Sahinkaya [6], though without detailed elaboration.

Consider a rotor-bearing mass–stiffness–damping system with five point-masses supported by journal bearings at both ends [7]. The 20×20 nondimensional system matrix **A** is shown in Fig. 5.1.

The submatrices $-\mathbf{M}^{-1}\mathbf{K}$ and $-\mathbf{M}^{-1}\mathbf{C}$ within **A** exhibit the characteristics described earlier for **K** and **C** in Eqs. (5.18) and (5.19). While the original stiffness matrix **K** is approximately symmetric (except for bearing stiffness), this symmetry is altered by the multiplication with $-\mathbf{M}^{-1}$. Using the controllability criteria from Theorem 5.1, we can verify the correctness of Corollary 5.2 for this system.

$$
\mathbf{A}=\begin{bmatrix}
0&0&0&0&0&0&0&0&0&0&1&0&0&0&0&0&0&0&0&0\\
0&0&0&0&0&0&0&0&0&0&0&1&0&0&0&0&0&0&0&0\\
0&0&0&0&0&0&0&0&0&0&0&0&1&0&0&0&0&0&0&0\\
0&0&0&0&0&0&0&0&0&0&0&0&0&1&0&0&0&0&0&0\\
0&0&0&0&0&0&0&0&0&0&0&0&0&0&1&0&0&0&0&0\\
0&0&0&0&0&0&0&0&0&0&0&0&0&0&0&1&0&0&0&0\\
0&0&0&0&0&0&0&0&0&0&0&0&0&0&0&0&1&0&0&0\\
0&0&0&0&0&0&0&0&0&0&0&0&0&0&0&0&0&1&0&0\\
0&0&0&0&0&0&0&0&0&0&0&0&0&0&0&0&0&0&1&0\\
0&0&0&0&0&0&0&0&0&0&0&0&0&0&0&0&0&0&0&1\\
-1.1240&2.0784&-1.6795&0.2644&-0.0413&1.1127&0&0&0&0&-2.2548&0&0&0&0&-0.0440&0&0&0&0\\
1.1058&-5.2033&5.7775&-1.9910&0.3113&0&0&0&0&0&0&0&0&0&0&0&0&0&0&0\\
-1.0014&6.4744&-9.0970&4.7561&-1.1307&0&0&0&0&0&0&0&0&0&0&0&0&0&0&0\\
0.1271&-1.7995&3.8359&-3.2859&1.1223&0&0&0&0&0&0&0&0&0&0&0&0&0&0&0\\
-0.0517&0.7319&-2.3724&2.9196&-1.8648&0&0&0&0&1.7786&0&0&0&0&-3.5826&0&0&0&0&-0.0352\\
-1.2285&0&0&0&0&-1.5472&2.0784&-1.6795&0.2644&-0.0413&-0.0414&0&0&0&0&-2.4740&0&0&0&0\\
0&0&0&0&0&1.1058&-5.2033&5.7775&-1.9910&0.3113&0&0&0&0&0&0&0&0&0&0\\
0&0&0&0&0&-1.0014&6.4744&-9.0970&4.7561&-1.1307&0&0&0&0&0&0&0&0&0&0\\
0&0&0&0&0&0.1271&-1.7995&3.8359&-3.2859&1.1223&0&0&0&0&0&0&0&0&0&0\\
0&0&0&0&-1.6471&-0.0517&0.7319&-2.3724&2.9196&-2.2112&0&0&0&0&-0.0326&0&0&0&0&-3.3154
\end{bmatrix}
$$

Fig. 5.1 Nondimensional matrix **A** of a 5-point-mass rotor-bearing system

The structural characteristics of rotor-bearing systems inherently determine their controllability. The presence of cross-coupling stiffness and damping terms, along with appropriately applied control inputs, ensures that the controllability conditions are satisfied. This understanding is crucial for designing effective control strategies for such systems.

5.3 Structural Properties of State Controllability

With the advancement of systems science and the deepening of scientific research, both the subjects under study and the mathematical models describing them have become increasingly complex. As the order of dynamic systems increases, system analysis is increasingly plagued by issues such as ill-conditioning and indeterminate parameter values. This phenomenon is particularly pronounced in the study of controllability and observability of large-scale complex dynamic systems, where numerical instability and computational errors become significant concerns. Consequently, substantial efforts have been invested in addressing these challenges.

As previously discussed, the concept of controllability is crucial in the study of dynamic systems. However, among the numerous existing criteria for controllability, the numerical computation of the rank condition of matrices—especially in high-order cases—often presents ill-conditioned numerical problems. This frequently leads to numerical instability and computational errors. To overcome this issue, Lin [8] introduced the concepts of structure and structural controllability for single-input LTI systems using graph theory in 1974. He proved that even for systems with indeterminate parameter values, their structural controllability could be accurately determined. His work quickly prompted a series of subsequent studies and further developments [9–13].

An important special case in the study of composite systems is structural controllability. In this context, the subsystems of a composite system are often controllable subsystems of the same order [14]. Complex mass–stiffness–damping systems exemplify such composite systems. In this section, we build upon existing results of structural controllability to demonstrate the structural controllability of mass–stiffness–damping systems and present the corresponding findings.

5.3.1 General Properties of Structural Controllability

Before discussing the structural properties of state controllability for mass–stiffness–damping systems, it is necessary to introduce some relevant concepts, definitions, and theorems [8–14].

Consider the following LTI system

$$\dot{\boldsymbol{x}}(t) = \mathbf{A}\boldsymbol{x}(t) + \mathbf{B}\boldsymbol{u}(t), \tag{5.20}$$

where $\boldsymbol{x}(t)$ is an $n \times 1$ state vector, and $\boldsymbol{u}(t)$ is an $l \times 1$ input vector. The elements of the matrices $\mathbf{A}$ and $\mathbf{B}$ can be either fixed zeros (0), determined by the physical structure of the system, or arbitrary free parameters (denoted by "$\cdots$").

If, for every fixed zero element (0) in the matrix pair $\{\mathbf{A}, \mathbf{B}\}$, the corresponding element in another matrix pair $\left\{\tilde{\mathbf{A}}, \tilde{\mathbf{B}}\right\}$ is also fixed (0), and vice versa, then these two matrix pairs are said to have the same structure.

A matrix whose elements are specified as either fixed zeros or arbitrary parameters ("$\cdots$") is called a structural matrix.

A linear system composed of the matrix pair $\{\mathbf{A}, \mathbf{B}\}$ is referred to as a structural system.

If there is a one-to-one correspondence between the positions of the fixed zero (0) elements and the arbitrary non-zero elements ("$\cdots$") in the structural matrices of two systems $\{\mathbf{A}, \mathbf{B}\}$ and $\left\{\tilde{\mathbf{A}}, \tilde{\mathbf{B}}\right\}$, then these two systems are said to be structurally equivalent.

Definition 5.3 If the system $\{\mathbf{A}, \mathbf{B}\}$ has a structurally equivalent system that is controllable in the usual sense, then the system $\{\mathbf{A}, \mathbf{B}\}$ is said to be structurally controllable.

This definition implies that even if the specific numerical values of the parameters in the system $\{\mathbf{A}, \mathbf{B}\}$ make it uncontrollable, the system is still structurally controllable if, by adjusting the arbitrary parameters ("$\cdots$"), it can become controllable.

In other words, the system $\{\mathbf{A}, \mathbf{B}\}$ has the potential for controllability based on its physical structure.

A property of matrices is particularly useful for the following discussion: the generic rank of an $n \times q$ matrix $\mathbf{A}$, where $n \leq q$.

Definition 5.4 The generic rank of a structural matrix $\mathbf{A}$, denoted as grank($\mathbf{A}$), is the maximum rank that $\mathbf{A}$ can achieve as a function of its free parameters ("$\cdot\cdot\cdot$").

Clearly, the maximum possible generic rank of an $n \times q$ matrix $\mathbf{A}$ is n, which is the smaller of n and q. This means that by appropriately choosing the values of the free parameters, $\mathbf{A}$ can attain a rank up to n.

It is well known that there is a one-to-one correspondence between quadratic form matrices and directed graphs. Therefore, a general composite system $\{\mathbf{C}, \mathbf{A}, \mathbf{B}\}$ can typically be represented by a directed graph composed of $p + n + l$ vertices, where p is the dimension of the output vector.

Definition 5.5 Let $\mathrm{M}(\mathbf{A}, \mathbf{B})$ denote the directed graph of the composite system $\{\mathbf{A}, \mathbf{B}\}$, and $\mathrm{M}(\mathbf{C}, \mathbf{A})$ represent the directed graph of the composite system $\{\mathbf{C}, \mathbf{A}\}$, where the vertices corresponding to the input $\boldsymbol{u}$ and the output $\boldsymbol{y}$ are omitted from $\mathrm{M}(\mathbf{A}, \mathbf{B})$ and $\mathrm{M}(\mathbf{C}, \mathbf{A})$, respectively.

A composite system $\{\mathbf{A}, \mathbf{B}\}$ is called input-connectable if there exists an arborescence (a directed tree), not necessarily unique, rooted at $\boldsymbol{u}$ in the directed graph $\mathrm{M}(\mathbf{A}, \mathbf{B})$.

A composite system $\{\mathbf{C}, \mathbf{A}\}$ is called output connectable if there exists an arborescence rooted at $\boldsymbol{y}$ in the inverse graph $\mathrm{M}^*(\mathbf{C}, \mathbf{A})$.

If a composite system $\{\mathbf{C}, \mathbf{A}, \mathbf{B}\}$ is both input and output connectable, it is called connectable.

The importance of these concepts and definitions lies in their ability to determine whether a system is structurally controllable. We present the following theorem on the structural controllability of large-scale systems.

Theorem 5.4 *The system $\{\mathbf{A}, \mathbf{B}\}$ is structurally controllable if and only if the following condition is satisfied: The system $\{\mathbf{A}, \mathbf{B}\}$ is input-connectable, and*

$$\mathrm{grank}([\ \mathbf{A} \mid \mathbf{B}\ ]) = n. \tag{5.21}$$

This condition is simple and intuitive, making it a commonly used and effective criterion. In the following discussion, we will develop theories and methods regarding the controllability of mass–stiffness–damping systems based on Theorem 5.4.

5.3.2 *Structural Controllability of MKC Systems*

We now turn our attention to the structural controllability of mass–stiffness–damping systems. For a system with n degrees-of-freedom and l inputs, the state-space representation can be expressed as:

$$[\,\mathbf{A} \mid \mathbf{B}\,] = \left[\begin{array}{cc|c} 0 & \mathbf{I} & 0 \\ -\mathbf{M}^{-1}\mathbf{K} & -\mathbf{M}^{-1}\mathbf{C} & \mathbf{M}^{-1}\mathbf{F} \end{array}\right]. \tag{5.22}$$

Here, $\mathbf{A}$ is a $2n \times 2n$ matrix, and $\mathbf{B}$ is a $2n \times l$ matrix. The augmented matrix $[\,\mathbf{A} \mid \mathbf{B}\,]$ has dimensions $2n \times (2n + l)$. The three block columns in $[\,\mathbf{A} \mid \mathbf{B}\,]$ correspond to submatrices of sizes $2n \times n$, $2n \times n$, and $2n \times l$, respectively.

Systems classified as not input-connectable exhibit the property that certain states cannot be influenced by the control inputs due to the inherent structure of the system. Physically, this means the system lacks direct or indirect pathways (information channels) for the control inputs to affect those states. Conversely, a system is said to be input-connectable if there exist information channels—though not necessarily unique—through which the control inputs can reach and influence every state of the system.

Mass–stiffness–damping systems with n degrees-of-freedom can typically be decomposed into n second-order subsystems. The off-diagonal elements in the stiffness matrix $\mathbf{K}$ and damping matrix $\mathbf{C}$ represent the cross-interactions (coupling) between these subsystems. For example, consider a system with two degrees-of-freedom and two inputs:

$$\begin{bmatrix} m_1 & 0 \\ 0 & m_2 \end{bmatrix}\begin{bmatrix} \ddot{x}_1 \\ \ddot{x}_2 \end{bmatrix} + \begin{bmatrix} c_{11} & c_{12} \\ c_{21} & c_{22} \end{bmatrix}\begin{bmatrix} \dot{x}_1 \\ \dot{x}_2 \end{bmatrix} + \begin{bmatrix} k_{11} & k_{12} \\ k_{21} & k_{22} \end{bmatrix}\begin{bmatrix} x_1 \\ x_2 \end{bmatrix} = \begin{bmatrix} f_1 \\ f_2 \end{bmatrix}. \tag{5.23}$$

The dynamic structure of this system is illustrated in Fig. 5.2. Each second-order subsystem is individually controllable, and the cross-couplings between the two subsystems are represented by the off-diagonal elements in $\mathbf{K}$ and $\mathbf{C}$, which are symmetric. Therefore, as long as the cross-stiffness (k_{12}, k_{21}) and cross-damping (c_{12}, c_{21}) coefficients are not all zero, the control inputs can influence all states, enabling reachability from one subsystem to another. Conversely, if the subsystems are isolated (i.e., the off-diagonal elements are zero), the inputs cannot influence all states from one subsystem to another. Generalizing this observation leads to the following lemma.

Lemma 5.1 *A necessary and sufficient condition for a mass–stiffness–damping system with a single input and n degrees-of-freedom to be input-connectable (i.e., for all states to be reachable) is that no subsystem is isolated; that is, for each subsystem, the cross-stiffness and cross-damping coefficients representing cross-couplings with other subsystems are not all zero.*

One advantage of using graph theory [8] is that it allows us to represent and analyze the structural information of the system, facilitating the development of new algebraic assessment methods. Based on Lemma 5.1 and employing Boolean algebra and Boolean matrices [15], we have developed an algebraic algorithm to assess the controllability of single-input mass–stiffness–damping systems.

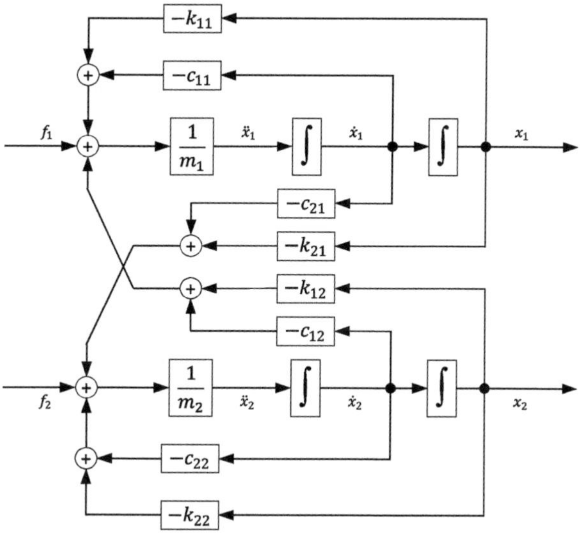

Fig. 5.2 Block diagram of a 2-DoF mass–stiffness–damping system

Definition 5.6 Given a structural matrix $\mathbf{A}_S$, its corresponding Boolean matrix $\mathbf{A}_B$ is defined such that for each element a_{ij}:

- If a_{ij}^S is a free parameter (denoted by "$\cdot\cdot\cdot$"), then $a_{ij}^B = 1$ (logical "true").
- If $a_{ij}^S = 0$ (a fixed zero element), then $a_{ij}^B = 0$ (logical "false").

Accordingly, we have the following algorithm for assessing input connectivity of single-input mass–stiffness–damping systems:

Step 1: Construct the Combined Boolean Matrix

1.1. Create Boolean Matrices:

- Generate the Boolean matrices $\mathbf{K}_B$ and $\mathbf{C}_B$ corresponding to the stiffness matrix $\mathbf{K}$ and the damping matrix $\mathbf{C}$, respectively.
- In these Boolean matrices, assign "1" to elements that are free parameters (non-zero elements in $\mathbf{K}$ and $\mathbf{C}$) and "0" to fixed zero elements.

1.2. Combine the Boolean Matrices:

- Perform a logical "OR" ("+", Addition) operation on $\mathbf{K}_B$ and $\mathbf{C}_B$ to obtain the combined Boolean matrix $\mathbf{K}_B \cup \mathbf{C}_B$.
- Since $\mathbf{K}$ and $\mathbf{C}$ are structurally symmetric, $\mathbf{K}_B \cup \mathbf{C}_B$ will also be symmetric. This means that $(\mathbf{K}_B \cup \mathbf{C}_B)_{ij} = (\mathbf{K}_B \cup \mathbf{C}_B)_{ji}$.
- In subsequent steps, rows and columns of $\mathbf{K}_B \cup \mathbf{C}_B$ are treated equivalently due to this symmetry.

Step 2: Initial Assessment of Direct Connectivity

2.1. Identify the Input Subsystem:

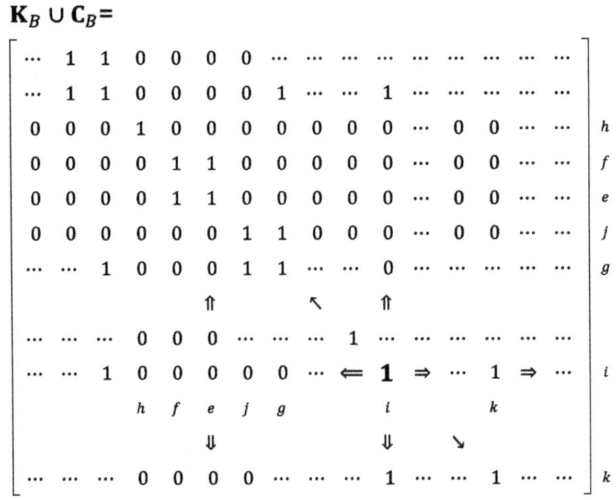

Fig. 5.3 Assessment of input connectivity

- Suppose the input is applied to the i-th subsystem (see Fig. 5.3).

2.2. Mark the Diagonal Element:

- Set the diagonal element $(\mathbf{K}_B \cup \mathbf{C}_B)_{ii}$ to "1" to indicate that the i-th subsystem is connected to the input.

2.3. Evaluate Direct Connections:

- Examine the i-th row of $\mathbf{K}_B \cup \mathbf{C}_B$.
- For each element $(\mathbf{K}_B \cup \mathbf{C}_B)_{ik}$:
  - If $(\mathbf{K}_B \cup \mathbf{C}_B)_{ik} = 1$, the k-th subsystem is directly input-connectable from the i-th subsystem. Record this information.
  - If $(\mathbf{K}_B \cup \mathbf{C}_B)_{ik} = 0$, the k-th subsystem is not directly input-connectable at this stage.

Step 3: Assessment of Indirect Connectivity

3.1. Assess Indirect Connections:

- For each subsystem j not directly input-connectable:
  - Examine the j-th row of $\mathbf{K}_B \cup \mathbf{C}_B$.
  - If any element $(\mathbf{K}_B \cup \mathbf{C}_B)_{jk} = 1$ and the k-th subsystem is already known to be directly or indirectly input-connectable, then the j-th subsystem is indirectly input-connectable through the k-th subsystem.
  - Update the records to reflect the indirect connectivity of the j-th subsystem.

3.2. Identify Isolated Subsystems:

- If all off-diagonal elements in the j-th row are "0" (i.e., $(\mathbf{K}_B \cup \mathbf{C}_B)_{jk} = 0$ for all $k \neq j$), then the j-th subsystem is isolated and not input-connectable.

- Mark the j-th subsystem as isolated in the records.

3.3. Iterate the Assessment:

- Repeat the indirect connectivity assessment for any newly identified subsystems until all subsystems have been evaluated.
- Continue this process until no additional subsystems can be classified as indirectly input-connectable.

Step 4: Finalize the Connectivity Assessment

- Compile the results to categorize each subsystem as:
  - Directly Input-Connectable.
  - Indirectly Input-Connectable.
  - Isolated (Not Input-Connectable).

The algorithm has the following features:

- *Symmetry Utilization:* Due to the symmetry of $\mathbf{K}_B \cup \mathbf{C}_B$, rows and columns are interchangeable in the assessment, simplifying the process.
- *Graph Representation:* This algorithm effectively traverses the connectivity graph of the system, where subsystems are nodes, and non-zero elements in $\mathbf{K}_B \cup \mathbf{C}_B$ represent edges.
- *Efficiency:* The algorithm can be efficiently implemented using standard graph traversal methods, such as depth-first search (DFS) or breadth-first search (BFS).
- *Multiple Inputs Extension:* While the algorithm is presented for a single-input system, it can be extended to multiple inputs by assessing each input separately and combining the results.

By using this algorithm, one can systematically determine the input connectivity of all subsystems in a single-input mass–stiffness–damping system. This assessment is crucial for evaluating the structural controllability of the system and for designing effective control strategies.

This algorithm is well-suited for computer implementation and is typically efficient. Multiple assessments may be required in the following situations:

- When there are multiple input-unconnectable subsystems.
- When the input is applied to an isolated subsystem.

An example of the assessment process is depicted in Fig. 5.4. For illustration, suppose the input is applied to the i-th subsystem:

- *Direct Connectivity:* The i-th subsystem and any subsystems directly connected to it (as identified in the i-th row of $\mathbf{K}_B \cup \mathbf{C}_B$) are directly input-connectable.
- *Indirect Connectivity:* Subsystems connected to directly input-connectable subsystems (but not directly connected to the i-th subsystem) are indirectly input-connectable.
- *Isolated Subsystems:* Subsystems that cannot be reached through any path from the i-th subsystem are isolated.

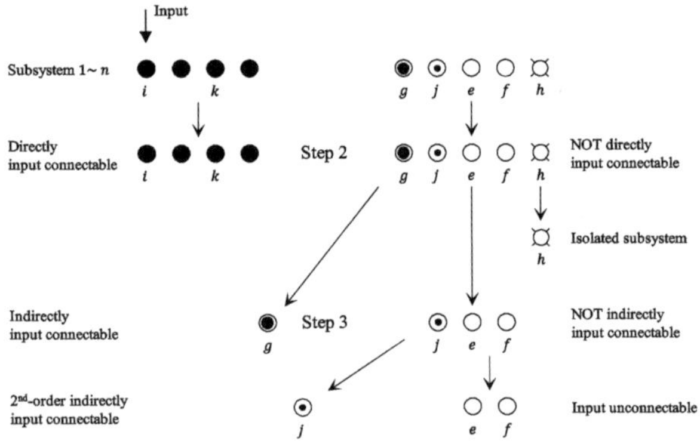

Fig. 5.4 Description of assessment algorithm for input connectivity

Executing this algorithm results in three possible outcomes for each subsystem:

- *Input-Connectable Subsystem:* The subsystem can be influenced by the input, either directly or indirectly.
- *Isolated Subsystem:* The subsystem cannot be influenced by the input and has no connections to other subsystems.
- *Interconnected but Input-Unconnectable Subsystem:* The subsystem is connected to others but is not reachable from the input.

The algorithm not only determines the input connectivity of the system but also provides details on the interconnections between multiple input-unconnectable subsystems, which is useful when considering the addition of control inputs.

The algorithm described for the single-input case can be extended to systems with multiple inputs using the superposition principle:

- Assess each input separately using the algorithm.
- Combine the results using a logical "OR" operation (logical sum in Boolean algebra), marking subsystems as "1" (true) if they are input-connectable through any input.
- Alternatively, subsystems not input-connectable via the i-th input can be reassessed using the $(i+1)$-th input.

The assessment process is easy to follow and implement, whether manually or through computational means.

If a mass–stiffness–damping system exhibits input disconnectivity, and since the structure of matrices **K** and **C** is generally fixed by the physical properties of the system, it may be necessary to modify the locations of the non-zero elements in the input matrix **F** or introduce additional control inputs to ensure the reachability of all states.

The simplified equivalent structure of the matrix in Eq. (5.22) can be represented as:

$$[\,\mathbf{A} \mid \mathbf{B}\,] = \begin{bmatrix} 0 & \mathbf{I} & 0 \\ \cdots & \cdots & \cdots \end{bmatrix}. \tag{5.24}$$

According to Theorem 5.4, a mass–stiffness–damping system with n degrees-of-freedom is structurally controllable if the generic rank of the matrix in Eq. (5.24) is $2n$. Based on the structure of the system and the properties of its structural matrices, the simplified equivalent condition derived from Eqs. (5.22) and (5.24) is:

$$\text{grank}([\,\mathbf{K} \mid \mathbf{F}\,]) = n. \tag{5.25}$$

This condition is independent of the damping matrix $\mathbf{C}$.

Using existing algorithms for calculating the generic rank, the generic rank of the matrix in Eq. (5.25) can be obtained efficiently, significantly reducing computational effort compared to conventional methods. Additionally, the properties of Boolean matrices can be used to derive corresponding computational methods.

Introducing additional independent inputs (columns in the input matrix $\mathbf{F}$) can eliminate linear dependencies between subsystems. Each independent input can eliminate the dependency of one subsystem, and additional inputs should be applied to the rows that increase independence.

Based on the preceding discussion, we can derive the following corollary.

Corollary 5.3 *A multi-input mass–stiffness–damping system is structurally controllable if and only if:*

- *There are no input-unconnectable (isolated) subsystems.*
- *The generic rank* $\text{grank}([\,\mathbf{K} \mid \mathbf{F}\,]) = n$.

The aforementioned principles can be directly applied to the analysis and design of mass–stiffness–damping systems. By judiciously assigning the non-zero entries in the input matrix $\mathbf{F}$ and introducing additional independent inputs if necessary, one can eliminate inaccessible or linearly dependent subsystems, thereby ensuring structural controllability.

For rotor-bearing systems with journal bearings, where physical cross-coupling in the x- and y-directions occurs at the bearing supports (as shown in the $\mathbf{K}$ and $\mathbf{C}$ matrices), the following corollary can be derived using Corollary 5.3:

Corollary 5.4 *Rotor-bearing systems with journal bearings that have cross-coupling at the bearing supports are generally structurally controllable.*

As an example, Fig. 5.5 shows the structural matrices $\mathbf{A}_S$ and $\mathbf{B}_S$ of a 26th-order rotor-bearing system with 5 point-masses and 4 inputs, supported at both ends by journal bearings with tilting-pads. The structural matrix $\mathbf{A}_S$ has a special structure due to the omission of the pad inertia in the tilting-pad bearings at both ends.

To construct the Boolean matrix $\mathbf{K}_B \cup \mathbf{C}_B$ for the 26th-order rotor-bearing system, follow these steps:

1. Combine Boolean Matrices of Stiffness and Damping:

$$
\mathbf{A}_S=\left[\begin{array}{cccccccccccccccccccc|cccccc}
0&0&0&0&0&0&0&0&0&0&1&0&0&0&0&0&0&0&0&0&0&0&0&0&0&0\\
0&0&0&0&0&0&0&0&0&0&0&1&0&0&0&0&0&0&0&0&0&0&0&0&0&0\\
0&0&0&0&0&0&0&0&0&0&0&0&1&0&0&0&0&0&0&0&0&0&0&0&0&0\\
0&0&0&0&0&0&0&0&0&0&0&0&0&1&0&0&0&0&0&0&0&0&0&0&0&0\\
0&0&0&0&0&0&0&0&0&0&0&0&0&0&1&0&0&0&0&0&0&0&0&0&0&0\\
0&0&0&0&0&0&0&0&0&0&0&0&0&0&0&1&0&0&0&0&0&0&0&0&0&0\\
0&0&0&0&0&0&0&0&0&0&0&0&0&0&0&0&1&0&0&0&0&0&0&0&0&0\\
0&0&0&0&0&0&0&0&0&0&0&0&0&0&0&0&0&1&0&0&0&0&0&0&0&0\\
0&0&0&0&0&0&0&0&0&0&0&0&0&0&0&0&0&0&1&0&0&0&0&0&0&0\\
0&0&0&0&0&0&0&0&0&0&0&0&0&0&0&0&0&0&0&1&0&0&0&0&0&0\\
\cdots&\cdots&\cdots&\cdots&\cdots&0&0&0&0&0&\cdots&0&0&0&0&\cdots&0&0&0&0&\cdots&\cdots&\cdots&0&0&0\\
\cdots&\cdots&\cdots&\cdots&\cdots&0&0&0&0&0&0&0&0&0&0&0&0&0&0&0&0&0&0&0&0&0\\
\cdots&\cdots&\cdots&\cdots&\cdots&0&0&0&0&0&0&0&0&0&0&0&0&0&0&0&0&0&0&0&0&0\\
\cdots&\cdots&\cdots&\cdots&\cdots&0&0&0&0&0&0&0&0&0&0&0&0&0&0&0&0&0&0&0&0&0\\
\cdots&\cdots&\cdots&\cdots&\cdots&0&0&0&0&0&0&0&0&0&\cdots&0&0&0&0&\cdots&0&0&0&\cdots&\cdots&\cdots\\
0&0&0&0&0&\cdots&\cdots&\cdots&\cdots&\cdots&\cdots&0&0&0&0&\cdots&0&0&0&0&\cdots&\cdots&\cdots&0&0&0\\
0&0&0&0&0&\cdots&\cdots&\cdots&\cdots&\cdots&0&0&0&0&0&0&0&0&0&0&0&0&0&0&0&0\\
0&0&0&0&0&\cdots&\cdots&\cdots&\cdots&\cdots&0&0&0&0&0&0&0&0&0&0&0&0&0&0&0&0\\
0&0&0&0&0&\cdots&\cdots&\cdots&\cdots&\cdots&0&0&0&0&0&0&0&0&0&0&0&0&0&0&0&0\\
0&0&0&0&0&\cdots&\cdots&\cdots&\cdots&\cdots&0&0&0&0&\cdots&0&0&0&0&\cdots&0&0&0&\cdots&\cdots&\cdots\\
\hline
\cdots&0&0&0&0&\cdots&0&0&0&0&\cdots&0&0&0&0&\cdots&0&0&0&0&\cdots&0&0&0&0&0\\
\cdots&0&0&0&0&\cdots&0&0&0&0&\cdots&0&0&0&0&\cdots&0&0&0&0&0&\cdots&0&0&0&0\\
\cdots&0&0&0&0&\cdots&0&0&0&0&\cdots&0&0&0&0&\cdots&0&0&0&0&0&0&\cdots&0&0&0\\
0&0&0&0&\cdots&0&0&0&0&\cdots&0&0&0&0&\cdots&0&0&0&0&\cdots&0&0&0&\cdots&0&0\\
0&0&0&0&\cdots&0&0&0&0&\cdots&0&0&0&0&\cdots&0&0&0&0&\cdots&0&0&0&0&\cdots&0\\
0&0&0&0&\cdots&0&0&0&0&\cdots&0&0&0&0&\cdots&0&0&0&0&\cdots&0&0&0&0&0&\cdots
\end{array}\right]
\qquad
\mathbf{B}_S=\left[\begin{array}{cccc}
0&0&0&0\\
0&0&0&0\\
0&0&0&0\\
0&0&0&0\\
0&0&0&0\\
0&0&0&0\\
0&0&0&0\\
0&0&0&0\\
0&0&0&0\\
0&0&0&0\\
\cdots&0&0&0\\
0&0&0&0\\
0&0&0&0\\
0&0&0&0\\
0&\cdots&0&0\\
0&0&\cdots&0\\
0&0&0&0\\
0&0&0&0\\
0&0&0&0\\
0&0&0&\cdots\\
\hline
0&0&0&0\\
0&0&0&0\\
0&0&0&0\\
0&0&0&0\\
0&0&0&0\\
0&0&0&0
\end{array}\right]
$$

Fig. 5.5 Structural matrices $\mathbf{A}_S$ and $\mathbf{B}_S$ of a 26th-order rotor-bearing system

- *Extract Relevant Boolean Matrices:* From the structural matrix $\mathbf{A}_S$, extract the Boolean matrices corresponding to the lower 16 rows (representing certain degrees-of-freedom) of the first 10 columns and the middle 10 columns. The first 10 columns correspond to displacement variables, and the middle 10 columns correspond to velocity variables.
- *Perform Logical Addition:* Perform a logical OR operation (element-wise logical addition) on these two Boolean matrices. This operation combines the structural information of both stiffness (**K**) and damping (**C**) cross-couplings.
- *Generate the Resulting Matrix:* The result is a combined 16×10 Boolean matrix that represents the connectivity due to stiffness and damping between the subsystems in the rotor-bearing system.

2. Append Control Input Boolean Matrix:

   - *Extract Control Input Matrix:* From $\mathbf{A}_S$, extract the Boolean matrix corresponding to the lower 16 rows of the last 6 columns. These columns represent the control input locations in the system.
   - *Append Matrices:* Append this 16×6 Boolean matrix to the right of the previously obtained 16×10 matrix.

$$
\mathbf{K}_B \cup \mathbf{C}_B = \left[\begin{array}{cccccccccc|cccccc}
1 & 1 & 1 & 1 & 1 & 0 & 0 & 0 & 0 & 0 & 1 & 1 & 1 & 0 & 0 & 0 \\
1 & 1 & 1 & 1 & 1 & 0 & 0 & 0 & 0 & 0 & 0 & 0 & 0 & 0 & 0 & 0 \\
1 & 1 & 1 & 1 & 1 & 0 & 0 & 0 & 0 & 0 & 0 & 0 & 0 & 0 & 0 & 0 \\
1 & 1 & 1 & 1 & 1 & 0 & 0 & 0 & 0 & 0 & 0 & 0 & 0 & 0 & 0 & 0 \\
1 & 1 & 1 & 1 & 1 & 0 & 0 & 0 & 0 & 0 & 0 & 0 & 0 & 1 & 1 & 1 \\
0 & 0 & 0 & 0 & 0 & 1 & 1 & 1 & 1 & 1 & 1 & 1 & 1 & 0 & 0 & 0 \\
0 & 0 & 0 & 0 & 0 & 1 & 1 & 1 & 1 & 1 & 0 & 0 & 0 & 0 & 0 & 0 \\
0 & 0 & 0 & 0 & 0 & 1 & 1 & 1 & 1 & 1 & 0 & 0 & 0 & 0 & 0 & 0 \\
0 & 0 & 0 & 0 & 0 & 1 & 1 & 1 & 1 & 1 & 0 & 0 & 0 & 0 & 0 & 0 \\
0 & 0 & 0 & 0 & 0 & 1 & 1 & 1 & 1 & 1 & 0 & 0 & 0 & 1 & 1 & 1 \\
\hline
1 & 0 & 0 & 0 & 0 & 1 & 0 & 0 & 0 & 0 & 1 & 0 & 0 & 0 & 0 & 0 \\
1 & 0 & 0 & 0 & 0 & 1 & 0 & 0 & 0 & 0 & 0 & 1 & 0 & 0 & 0 & 0 \\
1 & 0 & 0 & 0 & 0 & 1 & 0 & 0 & 0 & 0 & 0 & 0 & 1 & 0 & 0 & 0 \\
0 & 0 & 0 & 0 & 1 & 0 & 0 & 0 & 0 & 1 & 0 & 0 & 0 & 1 & 0 & 0 \\
0 & 0 & 0 & 0 & 1 & 0 & 0 & 0 & 0 & 1 & 0 & 0 & 0 & 0 & 1 & 0 \\
0 & 0 & 0 & 0 & 1 & 0 & 0 & 0 & 0 & 1 & 0 & 0 & 0 & 0 & 0 & 1
\end{array}\right]
$$

(a)

$$
\mathbf{K}_S = \left[\begin{array}{cccccccccc|cccccc}
\cdots & \cdots & \cdots & \cdots & \cdots & 0 & 0 & 0 & 0 & 0 & \cdots & \cdots & \cdots & 0 & 0 & 0 \\
\cdots & \cdots & \cdots & \cdots & \cdots & 0 & 0 & 0 & 0 & 0 & 0 & 0 & 0 & 0 & 0 & 0 \\
\cdots & \cdots & \cdots & \cdots & \cdots & 0 & 0 & 0 & 0 & 0 & 0 & 0 & 0 & 0 & 0 & 0 \\
\cdots & \cdots & \cdots & \cdots & \cdots & 0 & 0 & 0 & 0 & 0 & 0 & 0 & 0 & 0 & 0 & 0 \\
\cdots & \cdots & \cdots & \cdots & \cdots & 0 & 0 & 0 & 0 & 0 & 0 & 0 & 0 & \cdots & \cdots & \cdots \\
0 & 0 & 0 & 0 & 0 & \cdots & \cdots & \cdots & \cdots & \cdots & \cdots & \cdots & \cdots & 0 & 0 & 0 \\
0 & 0 & 0 & 0 & 0 & \cdots & \cdots & \cdots & \cdots & \cdots & 0 & 0 & 0 & 0 & 0 & 0 \\
0 & 0 & 0 & 0 & 0 & \cdots & \cdots & \cdots & \cdots & \cdots & 0 & 0 & 0 & 0 & 0 & 0 \\
0 & 0 & 0 & 0 & 0 & \cdots & \cdots & \cdots & \cdots & \cdots & 0 & 0 & 0 & 0 & 0 & 0 \\
0 & 0 & 0 & 0 & 0 & \cdots & \cdots & \cdots & \cdots & \cdots & 0 & 0 & 0 & \cdots & \cdots & \cdots \\
\hline
\cdots & 0 & 0 & 0 & 0 & \cdots & 0 & 0 & 0 & 0 & \cdots & 0 & 0 & 0 & 0 & 0 \\
\cdots & 0 & 0 & 0 & 0 & \cdots & 0 & 0 & 0 & 0 & 0 & \cdots & 0 & 0 & 0 & 0 \\
\cdots & 0 & 0 & 0 & 0 & \cdots & 0 & 0 & 0 & 0 & 0 & 0 & \cdots & 0 & 0 & 0 \\
0 & 0 & 0 & 0 & \cdots & 0 & 0 & 0 & 0 & \cdots & 0 & 0 & 0 & \cdots & 0 & 0 \\
0 & 0 & 0 & 0 & \cdots & 0 & 0 & 0 & 0 & \cdots & 0 & 0 & 0 & 0 & \cdots & 0 \\
0 & 0 & 0 & 0 & \cdots & 0 & 0 & 0 & 0 & \cdots & 0 & 0 & 0 & 0 & 0 & \cdots
\end{array}\right]
$$

(b)

Fig. 5.6 Matrices $\mathbf{K}_B \cup \mathbf{C}_B$ and $\mathbf{K}_S$ of a 26th-order rotor-bearing system

- *Generate the Final Boolean Matrix:* The resulting matrix is a 16×16 Boolean matrix $\mathbf{K}_B \cup \mathbf{C}_B$, which incorporates the structural information of stiffness, damping, and control inputs.
- *Visualize the Result:* This complete Boolean matrix is depicted in Fig. 5.6a.

By following these steps, we obtain the Boolean matrix $\mathbf{K}_B \cup \mathbf{C}_B$ that effectively represents the structural connectivity and potential controllability of the rotor-bearing system. This matrix is crucial for analyzing the structural controllability of the system and for identifying the pathways through which control inputs can influence various subsystems.

The process is explained as follows:

- Combining the Lower 16 Rows: The lower 16 rows of $\mathbf{A}_S$ correspond to specific degrees-of-freedom (e.g., rotational modes) that are of interest in the controllability analysis.
- Logical OR Operation: The logical OR operation effectively combines the connectivity information from both the stiffness and damping matrices, identifying where cross-couplings exist between subsystems.
- Appending the Control Input Matrix: Including the control input columns ensures that the resulting Boolean matrix accounts for how control inputs can influence the system, which is essential for assessing controllability.

This process provides a clear and logical method for constructing the Boolean matrix necessary for analyzing the structural controllability of complex mass–stiffness–damping systems like rotor-bearing assemblies.

From Fig. 5.6a, it is evident that regardless of whether the system is single-input or multi-input, and regardless of which subsystem an input is applied to, the 26th-order mass–stiffness–damping system is input-connectable. For instance:

- When a single input is applied to any of the first five subsystems, these subsystems are directly input-connectable.
- The 7th to 9th subsystems are indirectly input-connectable, and so on.

Comparing Fig. 5.6a with Fig. 5.5, we observe that the damping matrix **C** does not affect the form of $\mathbf{K}_B \cup \mathbf{C}_B$. This indicates that if the cross-stiffness is non-zero, the damping **C** is irrelevant to input connectivity (or reachability).

Since a structural matrix with all non-zero diagonal elements has full generic rank, Fig. 5.6b indicates that the generic rank of $\mathbf{K}_S$ is 16, which is full rank. Therefore, whether single-input or multi-input, this 26th-order rotor-bearing system satisfies the rank condition of Corollary 5.3.

From this example, we conclude that rotor-bearing systems using journal bearings with cross-coupling of stiffness or damping at the bearing supports are generally structurally controllable.

A 12th-order state-space model of a rotor-bearing system with 6 degrees-of-freedom was used in [11, 12] to illustrate issues related to structural controllability. However, since a different mathematical model was adopted there, no general conclusions or straightforward criteria were derived regarding the controllability of mass–stiffness–damping systems or rotor-bearing systems.

5.4 Summary

Building upon the classical theories of state controllability and structural controllability of linear systems, this chapter has provided a comprehensive analysis of the controllability of mass–stiffness–damping systems. We have introduced simplified criteria and developed efficient algorithms to assess controllability, making it more accessible for high-order and complex systems. As a practical application, our examination of rotor-bearing systems supported by journal bearings demonstrates that linear lumped-parameter rotor-bearing systems with cross-coupling at the bearing supports are generally structurally controllable.

The methods presented here are of significant importance for the modeling and real-time control of complex mass–stiffness–damping systems, such as large-scale structural frameworks, advanced robotic mechanisms, and intricate power transmission and distribution networks. These systems are prevalent in various engineering fields, and understanding their controllability is crucial for effective control system design and implementation.

Although this chapter primarily focuses on controllability, it is noteworthy that, according to the duality theorem, the results pertaining to controllability can be directly applied to the study of observability in mass–stiffness–damping systems. This suggests that similar simplified criteria and algorithms can be developed for observability analysis, thereby providing a more complete framework for the analysis and control of such systems.

In summary, the methodologies and insights presented in this chapter not only enhance our theoretical understanding of the controllability of complex dynamic systems but also offer practical tools for engineers and researchers. These contributions pave the way for more efficient design and real-time control of mass–stiffness–damping systems across various applications.

References

1. Kalman RE (1960) On the general theory of control systems. In: Proceedings of the 1st international IFAC congress on automatic and remote control, Moscow, USSR, vol 1, pp 491–502
2. Hughes PC, Skelton RE (1980) Controllability and observability of linear matrix-second-order systems. Trans ASME J Appl Mech 47(2):415–420
3. Kailath T (1980) Linear systems. Prentice Hall
4. Chen CT (1999) Linear system theory and design. Oxford University Press, Oxford
5. Stanway R, Burrows CR (1981) Active vibration control of a flexible rotor on flexibly-mounted journal bearings. Trans ASME J Dyn Syst Measur Control 103(4):383–388
6. Burrows CR, Sahinkaya MN (1983) Vibration control of multi-mode rotor-bearing systems. Proceed Royal Soc Lond Ser A Math Phys Sci 386(1790):77–94
7. Zhu HA, Xie YB, Yu L (1988) Model reduction for rotor-bearing systems. In: Proceedings of the 6th international modal analysis conference, USA, vol 1, pp 798–804
8. Lin CT (1974) Structural controllability. IEEE Trans Autom Control 19(3):201–208
9. Shields RW, Pearson JB (1976) Structural controllability of multiinput linear systems. IEEE Trans Autom Control 21(2):203–212
10. Lin CT (1977) System structure and minimal structure controllability. IEEE Trans Autom Control 5(22):855–862
11. Burrows CR, Sahinkaya MN (1981) A new algorithm for determining structural controllability. Int J Control 33(2):379–392
12. Burrows CR, Sahinkaya MN (1983) A modified algorithm for determining structural controllability. Int J Control 37(6):1417–1431
13. Datta BN (1984) A new criterion of controllability. IEEE Trans Autom Control 29(5):444–446
14. Jamshidi M (1997) Large-scale systems: modeling, control and fuzzy logic. Prentice Hall
15. Kim KH (1982) Boolean matrix theory and applications. Marcel Dekker

Part II
Basic Control

The Road Map: Basic Control

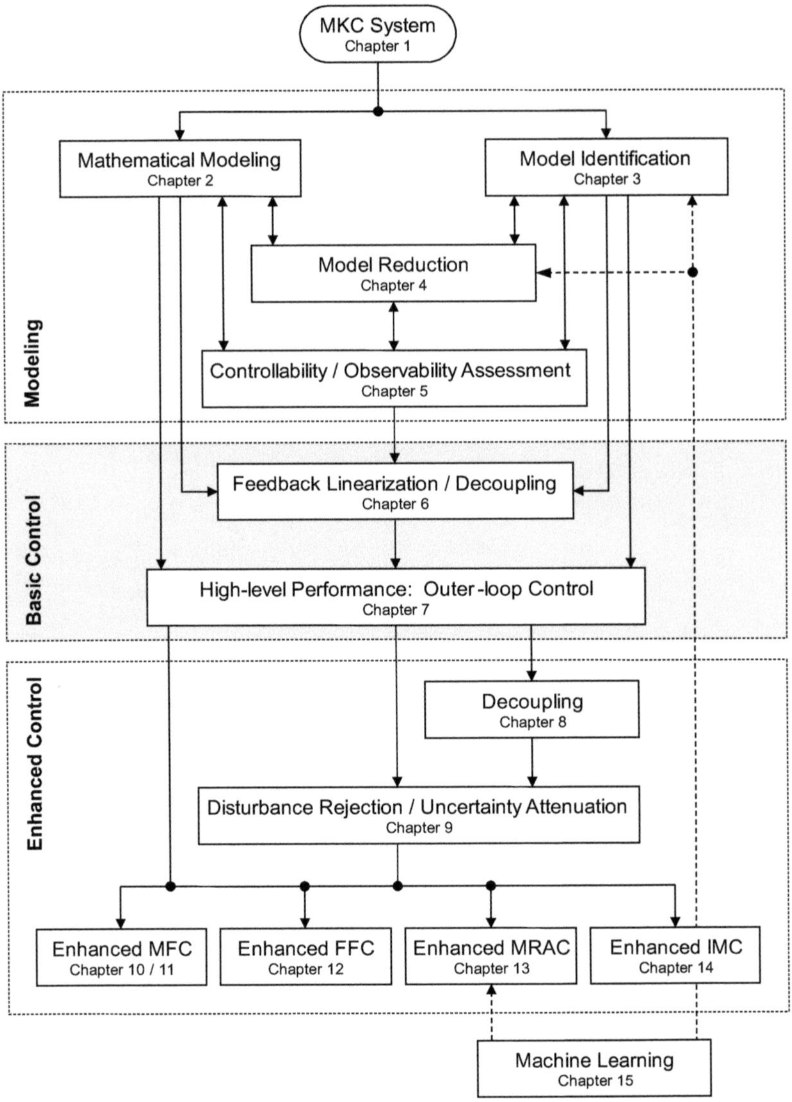

Chapter 6
Model-Based Feedback Linearization

Most practical mass–stiffness–damping systems exhibit significant nonlinear behavior. Over the past three decades, the field of nonlinear systems has advanced considerably [1, 2], largely due to the adoption of mathematical theories derived from differential geometry [3]. Although these geometric methods are still evolving, the emergence of differential algebra has enabled further insights into problems like input–output inversion and dynamic feedback decoupling, representing a significant breakthrough in the control community. Stabilization, which is one of the most critical challenges in nonlinear control, has also been approached with new and promising methodologies [4].

Despite these advances, finding techniques to control systems described by nonlinear models remains challenging. A major breakthrough occurred in the last 30 years with the development of feedback linearization [5]. This approach uses state and feedback transformations to convert a nonlinear system into an equivalent linear system, allowing established linear control strategies to be applied. Feedback linearization has been successfully implemented to address difficult tasks such as controlling robotic manipulators with complex nonlinear dynamics [6]. Its dual technique, linearization by output injection, is actively being used to tackle the problem of nonlinear state estimation [7].

This chapter focuses on model-based feedback linearization of mass–stiffness–damping systems, with an illustrative example involving robotic manipulators that include non-negligible motor dynamics. The dynamics considered here encompass both the mechanical behavior of the rigid links and the electrical behavior of the motor actuators, aiming for a comprehensive control design. As will be shown, the model-based algorithms presented do not rely on advanced mathematical tools, yet they achieve both simplicity and effectiveness in controlling these complex nonlinear systems.

H. Zhu, *Model-Based Control of Mass–Stiffness–Damping Systems*, Advances in Industrial Control, https://doi.org/10.1007/978-3-031-97592-9_6

6.1 Introduction

Parametric nonlinearity and structural nonlinearity are two different types of nonlinear behavior that can occur in physical systems, particularly in the fields of mass–stiffness–damping systems:

- Parametric nonlinearity occurs when the parameters of a system, such as stiffness, damping, or inertia, vary with the state or input of the system. In other words, the characteristics of the system depend on the state variables or input parameters of the system, making the behavior of the system nonlinear. The nonlinear behavior arises from the variation in the system parameters. The governing equations remain structurally linear, but the parameters within those equations are functions of the states or inputs of the system. It is common in systems where material properties, geometrical configurations, or other system parameters change with operating conditions (e.g., temperature, displacement). For mass–stiffness–damping systems, examples are:

  - *Nonlinear Spring Stiffness:* In mechanical systems, a spring may have a stiffness $k(x)$ that changes with displacement x, leading to a nonlinear force–displacement relationship:

    $$F(x) = k(x) \cdot x.$$

    Here, the system is linear if $k(x)$ is constant, but becomes nonlinear if $k(x)$ varies with x.
  - *Variable Damping:* In fluid systems, damping might depend on velocity or temperature, such as in a damper where the damping coefficient $c(v)$ varies with velocity v.

  Parametric nonlinearity can lead to phenomena such as parametric resonance, where certain variations in system parameters cause the system to become unstable or exhibit complex dynamics.
- Structural nonlinearity refers to a type of nonlinearity that arises due to the inherent structure or configuration of the system. The governing equations themselves are nonlinear, meaning the response of the system is not directly proportional to the input, regardless of the system parameters. The nonlinearity is embedded in the structure of the governing equations of the system. This type of nonlinearity is often related to the physical or geometrical properties of the system, such as large deformations, nonlinear constitutive laws, or boundary conditions. The response of the system is inherently nonlinear and cannot be described by linear equations even with constant parameters. For mass–stiffness–damping systems, examples are:

  - *Nonlinear Pendulum:* The dynamics of a simple pendulum, similar to a single-link robotic manipulator, are governed by a nonlinear equation:

$$\ddot{q}(t) + \frac{g}{l} \sin\left[q(t)\right] = 0,$$

where $q(t)$ is the angular displacement. The presence of the $\sin\left[q(t)\right]$ term introduces structural nonlinearity.

– *Large Deflections in Beams:* When a beam undergoes large deflections, the relationship between force and displacement becomes nonlinear due to geometric nonlinearities. The governing equation for such a beam includes nonlinear terms. The nonlinear equation is

$$\mathrm{M}(x) = \mathrm{E} \cdot \mathrm{I}\frac{\mathrm{d}^2 w}{\mathrm{d}^2 x} + \text{higher-order terms},$$

where $\mathrm{M}(x)$ is the bending moment, and $w(x)$ is the deflection, E is the Young's modulus (modulus of elasticity) of the material, describing its stiffness or flexibility, while I is the second moment of area (also known as the area moment of inertia), representing the geometric resistance of the cross-sectional shape to bending. The composite parameter E·I, often termed flexural rigidity or bending stiffness, combines the material property E and the cross-sectional geometry I into a single convenient term. This parameter is widely used in structural mechanics to characterize how effectively a beam or structural element resists bending deformation.

Structural nonlinearity can lead to behaviors such as bifurcations, chaotic motion, and complex dynamic responses that are fundamentally different from those in linear systems.

Key differences between parametric nonlinearity and structural nonlinearity are listed in Table 6.1. Understanding these differences is crucial in analyzing and designing control strategies for systems exhibiting nonlinear behavior.

Linearization is a process commonly used in control engineering to approximate a nonlinear system by a linear model. This is typically done around an operating condition to simplify the analysis and design of control systems. Validity of the linear approximation is often limited to a specific operating condition as well. The most common methods of linearization are outlined here:

- *Taylor Series Expansion:* This method involves expanding the nonlinear function into a Taylor series around a specific operating point and keeping only the linear terms (first-order approximation). The typical steps of this method are: (i) identify the operating point x_0; (ii) compute the first derivative of the nonlinear function $f(x)$ at x_0. Then, the linearized model around x_0 is given by:

$$f(x) \approx f(x_0) + \left.\frac{\mathrm{d}f(x)}{\mathrm{d}x}\right|_{x=x_0} (x - x_0).$$

This method is widely used in control systems to linearize a nonlinear dynamic system around a steady-state operating point.

Table 6.1 Key differences between parametric and structural nonlinearities

	Parametric nonlinearity	Structural nonlinearity
Origin of nonlinearity	Arises from the variation of system parameters with states or inputs The structure of the system remains linear	Arises from the inherent structure of the governing equations of the system, which are nonlinear
System equations	The system equations are linear in form but include parameters that are nonlinear functions of state variables or inputs	The system equations themselves are nonlinear, independent of parameter variations
Typical examples	Nonlinear spring stiffness, variable damping	Robotic manipulator, nonlinear pendulum, large deflections in beams
Effects on dynamics	Can cause complex behaviors like parametric resonance, but only when parameters vary with the state of the system	Can inherently cause phenomena such as chaos, bifurcations, and other complex dynamics without the need for parameter variation
Analysis and control	Often analyzed using methods like perturbation techniques, or compensated for using adaptive or robust control strategies	Requires nonlinear analysis techniques, such as phase-plane analysis, Lyapunov functions, or numerical simulations. Control often involves nonlinear control strategies

- *Jacobian Linearization:* This is an extension of the Taylor series method for systems with multiple inputs and outputs. The Jacobian matrix, which contains first-order partial derivatives, is used to linearize the system. For a system described by

$$\dot{\boldsymbol{x}} = \boldsymbol{f}(\boldsymbol{x}, \boldsymbol{u}),$$

  where $\boldsymbol{x}$ is the state vector and $\boldsymbol{u}$ is the input vector, compute the Jacobian matrices

$$\mathbf{A} = \left.\frac{\partial \boldsymbol{f}}{\partial \boldsymbol{x}}\right|_{\boldsymbol{x}_0, \boldsymbol{u}_0} \quad \text{and} \quad \mathbf{B} = \left.\frac{\partial \boldsymbol{f}}{\partial \boldsymbol{u}}\right|_{\boldsymbol{x}_0, \boldsymbol{u}_0},$$

  at the operating point $(\boldsymbol{x}_0, \boldsymbol{u}_0)$. The linearized system is then:

$$\Delta\dot{\boldsymbol{x}} = \mathbf{A}\Delta\boldsymbol{x} + \mathbf{B}\Delta\boldsymbol{u},$$

  where $\Delta\boldsymbol{x} = \boldsymbol{x} - \boldsymbol{x}_0$ and $\Delta\boldsymbol{u} = \boldsymbol{u} - \boldsymbol{u}_0$.

  This method is commonly used for linearizing nonlinear state-space models in control system design.
- *Small-Signal Linearization:* This method is used to linearize a nonlinear system by considering small deviations around an equilibrium point. It is essentially an application of the Taylor series method for small perturbations. This method

typically has two steps: (i) assume small perturbations around the operating point and (ii) linearize the system by considering only the first-order terms. This method is often used in electrical engineering, particularly in the analysis of circuits (e.g., small-signal analysis of transistors).
- *Piecewise Linearization:* This method approximates a nonlinear function by a series of linear segments. Each segment represents the behavior of the system over a specific range of inputs or states. Typically, this method has three steps: (i) divide the input or state space into regions; (ii) for each region, derive a linear model that approximates the nonlinear function; (iii) switch between these linear models based on the region in which the system operates. This method is useful in systems with known nonlinearities that can be segmented into regions, like power electronics or systems with threshold effects.
- *Gain Scheduling:* In this method, the parameters of a linear controller are adjusted based on the operating point or conditions of the nonlinear system. Typical steps of this method are: (i) linearize the system around multiple operating points; (ii) design a linear controller for each operating point; (iii) use a scheduling variable to interpolate or switch between controllers based on the current operating conditions. This method is common in aerospace and automotive control systems, where operating conditions (e.g., speed, altitude) vary widely and adaptive control is required.
- *Feedback Linearization:* This method involves transforming a nonlinear system into a linear system using a change of variables and feedback control. It typically involves: (i) identify the mathematical model including the nonlinear system dynamics; (ii) apply a coordinate transformation and state feedback to cancel out the nonlinearities. Then, the result is a linear system in the new coordinates. This method is useful in advanced control schemes, such as in robotic control and aerospace applications, where exact linearization is achievable.

Each of these methods has its own domain of applicability, depending on the nature of the system and the desired accuracy of the linear model.

A mass–stiffness–damping system typically comprises multiple subsystems—such as the plant, controllers, power amplifiers, actuators, and sensors—each introducing its own dynamics, potential nonlinearities, and cross-couplings. Figure 6.1 illustrates a general schematic in which exogenous inputs usually include command signals, disturbances, and sensor noise. These disturbances may result from fluctuations in power supply, nonlinearities in the driving mechanism, or variations in payload.

As a practical example, this chapter discusses model-based feedback linearization of robotic manipulators, which are a representative class of mass–stiffness–damping systems with significant structural nonlinearity. In robotic control, more comprehensive models that incorporate link flexibility [8], joint flexibility [9–11], actuator dynamics [12–15], and joint nonlinearities [16] are increasingly used in design to achieve better performance and to address pressing technological developments [17, 18]. Historically, designers often relied on local linearization followed by classical

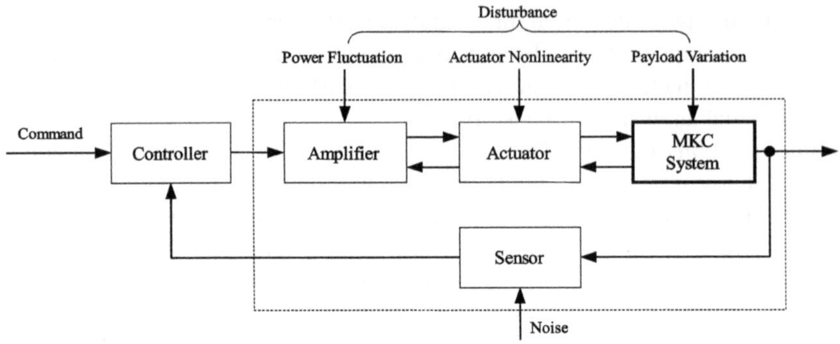

Fig. 6.1 Generic configuration of mass–stiffness–damping systems

linear control methods [19]. However, when the plant is highly nonlinear or undergoes broad dynamic excursions [7], such simplified approaches may be inadequate.

Fine motion control of robotic manipulators has become an important goal, motivated by advancements in robot morphology and newly defined tasks requiring high-speed maneuvers and precise end-effector trajectory tracking. Pursuing these objectives has fostered cross-disciplinary collaboration between the control and robotics communities, highlighting the complexities of manipulator dynamics. Simplified models remain common due to cost and real-time feasibility considerations. Nevertheless, the emergence of low-cost, high-speed computational hardware has sparked renewed interest in more "complete" models that account for additional dynamic and nonlinear phenomena once considered negligible [20]. Since many robotic systems are inherently nonlinear, nonlinear control theory appears especially relevant.

In practice, rigid-joint, rigid-link manipulators can be characterized by strongly coupled, nonlinear second-order differential equations [6, 19, 21]. Joint torques or forces act as inputs, while actuator dynamics is often omitted by treating actuators as pure torque sources. However, unlike other serial unmodeled dynamics (e.g., the computational delay of digital controllers [22] or the dynamic delay and nonlinearity in power amplifiers), actuator dynamics plays a major role in overall robotic behavior, since it interacts with the joint–link dynamics in both forward and feedback paths. Particularly at high speeds [23], an unmodeled fast actuator mode may be fully excited, influencing control outcomes. Empirical studies [24, 25] indicate that unmodeled drive system interactions often dominate the actual dynamics of a manipulator, rendering simple PD loops inadequate. Achieving high-speed accuracy thus depends on a thorough understanding of both the mechanical and drive system dynamics [16, 26, 27]. While some approaches introduce higher-order models of actuators and motors, many controller designs still rely on simplified linear representations.

In the past few years, there has also been notable progress in actuator technologies for both macro- and micro-scale robotics. In addition to the conventional electromagnetic, hydraulic, and pneumatic actuators (and natural muscle), new technologies such as shape memory alloy, piezoelectric, magnetostrictive, and polymeric actuators have emerged. Nevertheless, most commercial electric-drive manipulators still use DC servomotors. Consequently, for illustrative purposes, this chapter focuses on manipulators driven by armature-controlled DC motors, where the armature voltage serves as the control input.

The issue of actuator dynamics has been explored in several studies. Experimental evaluations [24, 25] show that unmodeled drive interactions can dominate manipulator behavior in high torque amplification systems, making feedback loops like PD insufficient to mitigate these effects. Successful high-speed manipulation of industrial robots hinges on accurately modeling both the manipulator links and the drive system [16]. Investigations also demonstrate that the more comprehensive the model used in dynamic control, the better the observed performance [26, 27]. For instance, general third-order robot dynamic models have been proposed, incorporating drive motor dynamics [26, 27], though many designs remain based on linearized single-axis concepts.

Some researchers have tackled extended robot dynamic models that account for manipulator–motor interactions in a fully coupled manner. In one approach, Freund's systematic nonlinear control theory [28] is employed in [12] to handle a more complete robot model that includes motor actuator dynamics. Since directly applying Freund's method to the overall motor-manipulator system can be challenging, the decomposition principle from large-scale system theory is used to separate the system. Alternatively, advanced differential-geometric methods [3] have been leveraged to design nonlinear feedback controllers that simultaneously incorporate manipulator and actuator dynamics [13]. Another technique [14] provides a design approach for developing robot tracking controllers, compensating for actuator dynamics under rigid-link, electrically driven conditions, ensuring robust exponential stability of link tracking errors even with parametric uncertainties. Additional research [15] proposes a nonlinear feedback law that achieves exponential convergence of both contact force and position tracking errors in constrained robots with significant motor actuator dynamics.

For rigid-joint, rigid-link manipulators driven by motor actuators, the system dynamics typically include:

- Mechanical dynamics of the joints and links.
- Electrical dynamics of the motor drives.
- Other nonlinear effects and uncertainties (e.g., Coulomb friction, viscous friction), which are challenging to model precisely.
- Potential external disturbances.

In low-speed scenarios, the electrical dynamics of the motor drive may not be dominant, though frictional effects in the actuators remain unavoidable. The algorithms introduced in this chapter target manipulator systems with non-negligible motor dynamics; additional uncertainties are treated as disturbances and will be addressed in subsequent chapters. This more comprehensive modeling strategy is motivated by the challenges it poses and its promise for advancing contemporary and future technology.

6.2 Dynamics of MKC Systems with Actuators

Because motor actuator dynamics can dominate other unmodeled dynamics in motor-driven manipulator systems [23, 24], and due to the complex interaction between actuators and link dynamics, this chapter focuses on the electromechanical dynamics of the coupled motor-manipulator system. All remaining residual dynamic effects are left unmodeled and treated separately.

6.2.1 *Mechanical Dynamics*

Let us first consider the dynamic equations of motion for mechanical manipulators. A typical n-link rigid-joint, rigid-link manipulator is modeled as a series of n rigid bodies, with one end fixed to the base and the other end free. These bodies are connected by revolute or prismatic joints, each driven by a torque or force actuator that includes friction and other nonlinear effects. According to the Lagrange–Euler formulation for mechanical manipulators [19, 21], the dynamics of such a rigid-joint manipulator can be expressed as the following highly nonlinear and strongly coupled second-order matrix differential equation:

$$\mathbf{D}(\boldsymbol{q})\ddot{\boldsymbol{q}} + \boldsymbol{h}(\dot{\boldsymbol{q}}, \boldsymbol{q}) + \boldsymbol{\tau}_D = \boldsymbol{\tau}_C, \tag{6.1}$$

and

$$\boldsymbol{h}(\dot{\boldsymbol{q}}, \boldsymbol{q}) = \mathbf{C}(\dot{\boldsymbol{q}}, \boldsymbol{q})\dot{\boldsymbol{q}} + \boldsymbol{g}(\boldsymbol{q}) + \boldsymbol{f}(\dot{\boldsymbol{q}}, \boldsymbol{q}), \tag{6.2}$$

where

- $\ddot{\boldsymbol{q}}, \dot{\boldsymbol{q}}, \boldsymbol{q} \in \mathbb{R}^n$ represent the acceleration, velocity, and position vectors of the n joints, respectively.
- $\mathbf{D}(\boldsymbol{q}) \in \mathbb{R}^{n\times n}$ is the symmetric, positive-definite inertia matrix.
- $\boldsymbol{h}(\dot{\boldsymbol{q}}, \boldsymbol{q}) \in \mathbb{R}^n$ is a vector of coupling torques, comprising:
  - $\mathbf{C}(\dot{\boldsymbol{q}}, \boldsymbol{q})\dot{\boldsymbol{q}} \in \mathbb{R}^n$, the vector of centrifugal and Coriolis torques,
  - $\boldsymbol{g}(\boldsymbol{q}) \in \mathbb{R}^n$, the vector of gravitational torques,
  - $\boldsymbol{f}(\dot{\boldsymbol{q}}, \boldsymbol{q}) \in \mathbb{R}^n$, the vector of modeled friction or damping torques.
- $\boldsymbol{\tau}_D \in \mathbb{R}^n$ represents external disturbance torques, along with any unmodeled nonlinear effects.
- $\boldsymbol{\tau}_C \in \mathbb{R}^n$ is the control torque vector.

Because frictional terms (e.g., Coulomb or viscous friction) at the joints, encapsulated by $\boldsymbol{f}(\dot{\boldsymbol{q}}, \boldsymbol{q})$, are often too uncertain to model precisely, this book treats such friction as part of the unmodeled dynamics. Consequently, these terms are merged into $\boldsymbol{\tau}_D$.

Despite the complexity of the mechanical dynamics in (6.1), it is not especially difficult to devise a nonlinear control law—such as the model-based linearization and decoupling method provided in Appendix A (also known as the computed-torque control method in robotics)—to "cancel out" the nonlinearities and decouple the system. Because the said algorithm (Algorithm A.1) is frequently referenced in this book, a concise introduction is provided in Appendix A. This method has become a widely accepted "recipe" within the robotics community, inspiring numerous variations across different applications.

However, when the electrical dynamics of the motor drives is taken into account, the interaction between the joint-link mechanics and the actuator dynamics introduces significantly greater complexity. Directly applying computed-torque control to the full motor-manipulator system may lead to unsatisfactory performance, especially under rapid operation, where unmodeled coupling effects cannot be neglected.

6.2.2 Electrical Dynamics

Motor actuators generate the torques or forces required to move the joints of a manipulator system. For an n-joint manipulator, the dynamic model of the n motor actuators can be derived using Kirchhoff's voltage law around the armature winding circuits of the n motors. This leads to the following voltage equation:

$$\left(\mathbf{L}_a \frac{\mathrm{d}\boldsymbol{i}_a}{\mathrm{d}t} + \mathbf{R}_a \boldsymbol{i}_a\right) + \mathbf{K}_E \frac{\mathrm{d}\boldsymbol{p}}{\mathrm{d}t} = \boldsymbol{u}_a, \tag{6.3}$$

where

- $\boldsymbol{i}_a, \boldsymbol{u}_a, \boldsymbol{p} \in \mathbb{R}^n$ denote the armature winding current, the applied armature voltage, and the rotor position of the n control motors, respectively.
- $\mathbf{L}_a, \mathbf{R}_a \in \mathbb{R}^{n \times n}$ are diagonal matrices containing the armature winding inductances and resistances.
- $\mathbf{K}_E \in \mathbb{R}^{n \times n}$ is a diagonal matrix of back-EMF constants $k_{Ei} > 0$, $(i = 1, \ldots, n)$.

The electrical dynamics of each motor couples to the mechanical dynamics via an algebraic torque equation. In general, the torque developed at the shaft of the i-th motor is proportional to the product of the armature winding current i_{ai} and the field winding current i_{fi}. Under armature-controlled DC motors, the field winding current remains constant, so the developed torque vector is:

$$\boldsymbol{\tau}_a = \mathbf{K}_T \boldsymbol{i}_a, \tag{6.4}$$

where $\boldsymbol{\tau}_a \in \mathbb{R}^n$ is the vector of motor shaft torques, and $\mathbf{K}_T \in \mathbb{R}^{n \times n}$ is a diagonal matrix of torque constants $k_{Ti} > 0$, $(i = 1, \ldots, n)$.

In Eqs. (6.1) and (6.3), the two position vectors $\boldsymbol{q}$ and $\boldsymbol{p}$ (representing the joint shafts and motor shafts, respectively) satisfy:

$$\boldsymbol{p} = \mathbf{N}\boldsymbol{q}, \tag{6.5}$$

where $\mathbf{N} \in \mathbb{R}^{n\times n}$ is a diagonal matrix representing the transmission ratios between the high-speed motor shafts (with angular velocity $\dot{\boldsymbol{p}}$) and the joint shafts (with angular velocity $\dot{\boldsymbol{q}}$). In a direct-drive manipulator, $\mathbf{N}$ reduces to the identity matrix $\mathbf{I}$.

Assuming there is no backlash or elastic deformation in the transmission, the work done by the motor shaft equals the work done by the joint shaft:

$$\tau_{Ci} \cdot q_i = \tau_{ai} \cdot p_i, \quad i = 1, \ldots, n. \tag{6.6}$$

From Eqs. (6.5) and (6.6), the control torque vector at the joint shafts is:

$$\boldsymbol{\tau}_C = \mathbf{N}\boldsymbol{\tau}_a. \tag{6.7}$$

Combining Eqs. (6.4) and (6.7) yields:

$$\boldsymbol{\tau}_C = (\mathbf{N}\mathbf{K}_T)\boldsymbol{i}_a. \tag{6.8}$$

Here, it is assumed that any effective inertias of the actuators and associated friction effects have already been translated (or “reflected”) to the joint shafts. Equation (6.8) confirms that the control torques for driving the manipulator are governed by a simple algebraic relationship with the motor armature currents.

By combining these equations, one obtains the block diagram depicting the electromechanical dynamics of a motor-manipulator system, shown in Fig. 6.2. If the conventional computed-torque method (i.e., the Algorithm A.1) is applied directly to this plant, the resulting performance may be unsatisfactory, as illustrated in the following example.

Example 6.1 This example demonstrates how unmodeled motor dynamics can degrade the performance of the computed-torque method (Algorithm A.1). For simplicity, we focus on the single-link manipulator described in Sect. 2.4.1. Although

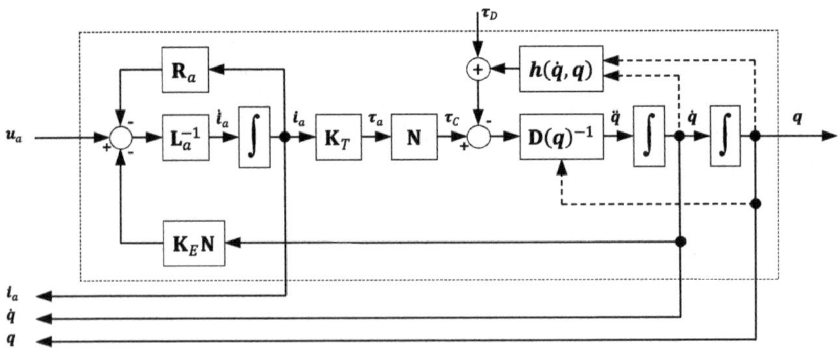

Fig. 6.2 Block diagram of the electromechanical dynamics

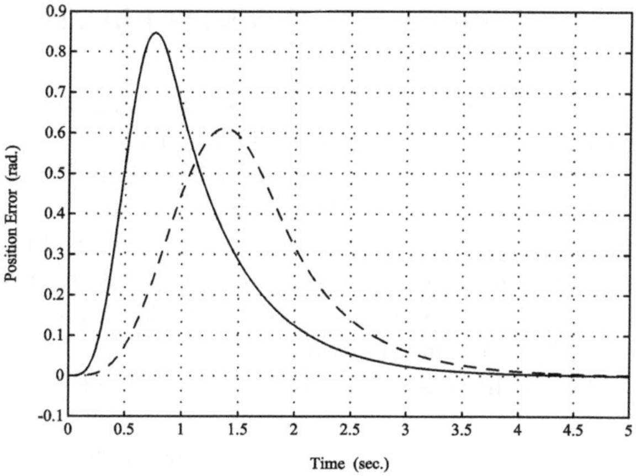

Fig. 6.3 Effect of the unmodeled motor dynamics

the steady-state gain of the motor drive has been included in the model, the dynamic characteristics of the motor remain unmodeled. The velocity and position feedback gains are $k_v = 20$ and $k_p = 100$, respectively. The joint of the manipulator moves from $q_0 = \pi/2$ to $(q_0 + q_f)$, where $q_f = 2.0$ rad. The desired trajectory is specified in Appendix C.1, and the resulting tracking errors are plotted in Fig. 6.3 {Dashed line: Low speed. Solid line: High speed}.

Speed Variations.

- Low speed: The link travels from q_0 to $(q_0 + q_f)$ over $t \in [0, 2]$, i.e., $t_f = 2$.
- High speed: The same movement occurs over $t \in [0, 1]$, i.e., $t_f = 1$.

From Fig. 6.3, it is clear that even though the steady-state gain of the motor is accounted for, the transient tracking error under conventional computed-torque control remains large. This is because the dynamic behavior of the motor is not modeled. For multi-link manipulators, cross-coupled interactions can further degrade performance.

In addition, the example illustrates that unmodeled motor dynamics can have an even more pronounced effect during high-speed operation.

Figure 6.2 highlights the electromechanical complexity of the motor-manipulator system, where mechanical and electrical subsystems closely interact in a feedback loop. Since the more comprehensive model presented above captures these interactions, it provides a basis for designing more effective controllers than conventional computed-torque methods alone can achieve.

6.3 Preliminaries of Feedback Linearization

The complexity of controlling many nonlinear systems has motivated the development of feedback linearization. One of its key features is incorporating a complete dynamic model of the nonlinear plant, making it possible to integrate more sophisticated control features. The core idea is that, in a suitable coordinate system, a specific nonlinear state feedback can be designed to produce an inner closed-loop system with linear dynamics [29]. Once this inner-loop feedback has been established, an outer-loop control law can be designed in the new coordinates to meet the desired performance objectives.

As a preliminary [3], consider a single-input nonlinear system:

$$\dot{\boldsymbol{x}} = \boldsymbol{f}(\boldsymbol{x}) + \boldsymbol{g}(\boldsymbol{x})u, \tag{6.9}$$

where $\boldsymbol{x} \in \mathbb{R}^l$ is the state vector, and $\boldsymbol{f}(\boldsymbol{x})$ and $\boldsymbol{g}(\boldsymbol{x})$ are smooth vector fields on $\mathbb{R}^l$ with $\boldsymbol{f}(0) = 0$. The scalar $u \in \mathbb{R}$ is the control input. Feedback linearizability of the system means that there exist:

1. A diffeomorphism $\mathbf{T} : U \to \mathbb{R}^l$, defined on some neighborhood $U \subset \mathbb{R}^l$ containing the origin.
2. A nonlinear feedback control law of the form

$$u = \alpha(\boldsymbol{x}) + \beta(\boldsymbol{x})v, \quad \beta(\boldsymbol{x}) \neq 0 \text{ on } U, \tag{6.10}$$

such that the new state variables

$$\boldsymbol{y} = \mathbf{T}(\boldsymbol{x}), \tag{6.11}$$

obey a linear and controllable system of equations:

$$\dot{\boldsymbol{y}} = \mathbf{A}\boldsymbol{y} + \boldsymbol{b}v, \tag{6.12}$$

where

$$\mathbf{A} = \begin{bmatrix} 0 & 1 & 0 & \cdots & 0 & 0 \\ 0 & 0 & 1 & \cdots & 0 & 0 \\ \vdots & \vdots & \vdots & \ddots & \vdots & \vdots \\ 0 & 0 & 0 & \cdots & 0 & 1 \\ 0 & 0 & 0 & \cdots & 0 & 0 \end{bmatrix}, \quad \boldsymbol{b} = \begin{bmatrix} 0 \\ 0 \\ \vdots \\ 0 \\ 1 \end{bmatrix}. \tag{6.13}$$

In other words, applying the nonlinear transformation in Eq. (6.11) and the nonlinear control law in Eq. (6.10) to the nonlinear system in Eq. (6.9) produces the linear, controllable system described by Eq. (6.12).

Here, the diffeomorphism $\mathbf{T}(\boldsymbol{x})$ serves as a nonlinear change of coordinates. The principle of feedback linearization is that, under this transformation, there exists a nonlinear control law that cancels the nonlinearities of the plant. When $U = \mathbb{R}^l$, the feedback linearization is said to be global. The necessary and sufficient conditions on the vector fields $\boldsymbol{f}$ and $\boldsymbol{g}$ in Eq. (6.9) for the existence of such a transformation—ensuring the system is feedback linearizable—are provided in [5, 6].

In the past three decades, differential-geometric methods have significantly advanced nonlinear control design [30, 31]. In robotics, these methods have also gained traction for managing complex robotic dynamics [32]. Their chief merit lies in offering a rigorous, mathematically unified framework for analyzing and synthesizing nonlinear control systems.

Overall, many existing methods of feedback linearization follow a common theme: they transform the nonlinear system into an equivalent linear one—often through a sequence of state coordinate transformations and/or an input transformation via nonlinear state feedback. Instead of these indirect approaches based on the general form (6.9), we will next introduce a direct, model-based method for feedback linearizing the motor-manipulator system, leveraging the more complete model described in the preceding sections.

6.4 Model-Based Methods

It is well-established that conventional feedback linearization typically requires an integrated dynamic model of the plant—such as the form in Eq. (6.9)—to be derived on $\mathbb{R}^l$. Concretely, for an n-joint manipulator with n motors, one would need to develop an $\mathbb{R}^{3n}$ model (i.e., $l = 2n + n$) before carrying out feedback linearization. Both the model derivation and the subsequent linearization steps can be quite cumbersome. However, in the specific motor-manipulator setup discussed above, there exist direct approaches that yield simpler and more compact solutions, thus streamlining the control design process.

6.4.1 *Model-Based Linearization and Decoupling*

As shown earlier, motor-manipulator systems exhibit highly nonlinear, strongly cross-coupled dynamics. This complexity invalidates most classical linear control methods. To address these challenges, we present a direct transformation strategy that decouples and linearizes the electromechanical interactions in motor-manipulator systems. Since [13] established that such systems are feedback linearizable using differential-geometric methods, this aspect will not be revisited here.

From Fig. 6.2, the primary obstacle to decoupling and linearizing the system is the negative feedback path ($\mathbf{K}_E\mathbf{N}$) linking electrical and mechanical dynamics. To eliminate this path, we adopt the control law:

$$\boldsymbol{u}_a := \boldsymbol{u}_b + \left(\hat{\mathbf{K}}_E\mathbf{N}\right)\dot{\boldsymbol{q}}, \tag{6.14}$$

where $\boldsymbol{u}_b \in \mathbb{R}^n$ is an intermediate control vector. Throughout this derivation, parameters denoted with a hat symbol "$\widehat{(\bullet)}$" represent estimated plant values, assumed accurate enough that $\widehat{(\bullet)} \rightarrow (\bullet)$.

Given this assumption, substituting Eq. (6.14) into Eq. (6.3) yields:

$$\mathbf{L}_a\frac{\mathrm{d}\boldsymbol{i}_a}{\mathrm{d}t} + \mathbf{R}_a\boldsymbol{i}_a = \boldsymbol{u}_b. \tag{6.15}$$

Thus, the undesirable feedback interaction ($\mathbf{K}_E\mathbf{N}$) between the electrical and mechanical subsystems is canceled, reducing the input–output relationship from $\boldsymbol{u}_b$ to $\boldsymbol{i}_a$ to a set of first-order lag terms (i.e., delayed integrators).

We next refine $\boldsymbol{u}_b$ to further simplify the dynamics:

$$\boldsymbol{u}_b := \hat{\mathbf{L}}_a\boldsymbol{u}_{bc} + \hat{\mathbf{R}}_a\boldsymbol{i}_a. \tag{6.16}$$

Let $\boldsymbol{u}_{bc}$ be defined as:

$$\boldsymbol{u}_{bc} := \left(\mathbf{N}\hat{\mathbf{K}}_T\right)^{-1}\boldsymbol{u}_c, \tag{6.17}$$

where $\boldsymbol{u}_c \in \mathbb{R}^n$ is another intermediate control vector. Substituting Eqs. (6.16) and (6.17) into Eq. (6.15) gives:

$$\frac{\mathrm{d}\boldsymbol{i}_a}{\mathrm{d}t} = \left(\mathbf{N}\hat{\mathbf{K}}_T\right)^{-1}\boldsymbol{u}_c. \tag{6.18}$$

This result indicates that the mapping from $\boldsymbol{u}_c$ to $\boldsymbol{i}_a$ has been reduced to a set of first-order pure integrators with a constant coefficient matrix $\left(\mathbf{N}\hat{\mathbf{K}}_T\right)^{-1}$. In practical implementations, the armature current $\boldsymbol{i}_a$ required by the control law can either be measured directly using a sensor installed in the motor armature circuit or retrieved from a dedicated servo-drive output, if such functionality is available in the hardware.

To account for the intrinsic nonlinear coupling in the mechanical dynamics of the manipulator, we temporarily ignore the disturbance term $\boldsymbol{\tau}_D$ in Eq. (6.1), leaving:

$$\mathbf{D}(\boldsymbol{q})\ddot{\boldsymbol{q}} + \boldsymbol{h}(\dot{\boldsymbol{q}}, \boldsymbol{q}) = \boldsymbol{\tau}_C. \tag{6.19}$$

We must now substitute the control torque $\boldsymbol{\tau}_C$ for the intermediate control $\boldsymbol{u}_c$. Combining Eqs. (6.8) and (6.18), and differentiating Eq. (6.19) with respect to time, yields:

$$\mathbf{D}(\boldsymbol{q})\frac{\mathrm{d}^3\boldsymbol{q}}{\mathrm{d}t^3} + \frac{\mathrm{d}\mathbf{D}(\boldsymbol{q})}{\mathrm{d}t}\ddot{\boldsymbol{q}} + \frac{\mathrm{d}\boldsymbol{h}(\dot{\boldsymbol{q}}, \boldsymbol{q})}{\mathrm{d}t} = \frac{\mathrm{d}\boldsymbol{\tau}_C}{\mathrm{d}t} = (\mathbf{N}\mathbf{K}_T)\frac{\mathrm{d}\boldsymbol{i}_a}{\mathrm{d}t} = \boldsymbol{u}_c. \tag{6.20}$$

To decouple and linearize the system, define:

$$\boldsymbol{u}_c := \hat{\mathbf{D}}(\boldsymbol{q})\boldsymbol{u} + \left[\hat{\dot{\mathbf{D}}}(\boldsymbol{q})\hat{\ddot{\boldsymbol{q}}} + \hat{\dot{\boldsymbol{h}}}(\dot{\boldsymbol{q}}, \boldsymbol{q})\right], \tag{6.21}$$

where $\boldsymbol{u} \in \mathbb{R}^n$ is the outer-loop feedback control. Substituting Eq. (6.21) into Eq. (6.20) gives:

$$\frac{\mathrm{d}^3\boldsymbol{q}}{\mathrm{d}t^3} = u. \tag{6.22}$$

Hence, the originally coupled nonlinear system is now transformed into a set of n decoupled linear systems, each functioning like a pure third-order integrator. Note that the estimated acceleration $\hat{\ddot{\boldsymbol{q}}}$ is used instead of the actual $\ddot{\boldsymbol{q}}$ because acceleration signals are generally not measured in manipulator systems.

At this stage, each subsystem behaves like an open-loop triple integrator, allowing the use of conventional or advanced linear control strategies on these decoupled single-input, single-output subsystems. The detailed design of this outer-loop feedback control will be explored in the following chapters. The direct transformation algorithm described above can be summarized as follows:

Algorithm 6.1 (*Perfect Linearization and Decoupling*) Assume that the electromechanical dynamics of a mechanical manipulator system (see Fig. 6.2) is accurately modeled and that no external disturbances are present. By using the serial-compensating controls in Eqs. (6.14), (6.16), (6.17), and (6.21):

$$\boldsymbol{u}_a := \boldsymbol{u}_b + \left(\hat{\mathbf{K}}_E\mathbf{N}\right)\dot{\boldsymbol{q}},$$

$$\boldsymbol{u}_b := \hat{\mathbf{L}}_a\boldsymbol{u}_{bc} + \hat{\mathbf{R}}_a\boldsymbol{i}_a,$$

$$\boldsymbol{u}_{bc} := \left(\mathbf{N}\hat{\mathbf{K}}_T\right)^{-1}\boldsymbol{u}_c,$$

$$\boldsymbol{u}_c := \hat{\mathbf{D}}(\boldsymbol{q})\boldsymbol{u} + \left[\hat{\dot{\mathbf{D}}}(\boldsymbol{q})\hat{\ddot{\boldsymbol{q}}} + \hat{\dot{\boldsymbol{h}}}(\dot{\boldsymbol{q}}, \boldsymbol{q})\right],$$

the mapping from the control input $\boldsymbol{u}$ to the position output $\boldsymbol{q}$ becomes a set of independent triple-integrators.

Conceptually, the algorithm can be described in four main steps:

1. Cancel Undesirable Motor-Mechanical Feedback
   Negate the unwanted feedback interactions via Eq. (6.14).
2. Convert Motor Current Dynamics to First-Order Integrators
   Use Eq. (6.18) to reduce the input–output relationship from $\boldsymbol{u}_c$ to $\boldsymbol{i}_a$ into a set of pure integrators.
3. Compensate for Manipulator Nonlinearities
   Differentiate Eq. (6.19) and define $\boldsymbol{u}_c$ using Eqs. (6.20) and (6.21), thereby handling intrinsic coupling in the mechanical dynamics of the manipulator.

4. Achieve Decoupled Triple Integrators
   Finalize the transformation so that each axis behaves like a simple triple integrator (Eq. 6.22), facilitating the design of conventional linear controllers.

This direct transformation approach eliminates the need to construct a large-scale $\mathbb{R}^{3n}$ integrated model for subsequent conventional feedback linearization, offering a more concise and practical solution for motor-manipulator control. Although the resulting algorithm is compact, the real-time computation required can be substantial—an unavoidable aspect of perfect feedback linearization, whether implemented explicitly or implicitly. The performance of the algorithm is demonstrated in the following example.

Example 6.2 We apply Algorithm 6.1 to the trajectory-tracking control of a two-link manipulator, as described in Sect. 2.4.2. The specific motion task appears in Appendix C.2. For the outer-loop control, we define:

$$\boldsymbol{u} := \boldsymbol{r} - \mathbf{K}_a\hat{\ddot{\boldsymbol{q}}} - \mathbf{K}_v\dot{\boldsymbol{q}} - \mathbf{K}_p\boldsymbol{q},$$

where

$$\boldsymbol{r}(t) := \dddot{\boldsymbol{q}}_d(t) + \mathbf{K}_a\ddot{\boldsymbol{q}}_d(t) + \mathbf{K}_v\dot{\boldsymbol{q}}_d(t) + \mathbf{K}_p\boldsymbol{q}_d(t),$$

and $\boldsymbol{q}_d(t)$ is the desired joint-space trajectory. In this example, the feedback gains are set to $\mathbf{K}_a = 30\ \mathbf{I}$, $\mathbf{K}_v = 300\ \mathbf{I}$, and $\mathbf{K}_p = 1000\ \mathbf{I}$. Here, $\mathbf{I}$ represents the identity matrix.

Because direct measurement of joint acceleration is typically unavailable in manipulator systems, we employ the following approximate estimator [6]:

$$\mathbf{T}_{pf}\frac{\mathrm{d}\hat{\ddot{\boldsymbol{q}}}}{\mathrm{d}t} + \hat{\ddot{\boldsymbol{q}}} = \frac{\mathrm{d}\dot{\boldsymbol{q}}}{\mathrm{d}t},$$

where $\hat{\ddot{\boldsymbol{q}}}$ is the estimated acceleration and $\mathbf{T}_{pf}$ is a small time constant (or diagonal matrix) in the pre-filter.

Figure 6.4 {Dotted line: Upper link. Dashed line: Lower link} illustrates the tracking errors when the manipulator is controlled by Algorithm 6.1, using the estimated acceleration $\hat{\ddot{\boldsymbol{q}}}$. Theoretically, if exact accelerations were known, the tracking error would be zero. Nonetheless, the observed tracking error is quite small, indicating that the nonlinear electromechanical dynamics are effectively compensated.

From the error responses in Fig. 6.4, it is evident that the residual tracking error in the upper link primarily arises from the inaccuracy in estimating $\ddot{\boldsymbol{q}}$. In this example, $\mathbf{T}_{pf} = 0.001\ \mathbf{I}$. Reducing $\mathbf{T}_{pf}$ further (i.e., lowering the filter time constants) can diminish the unwanted error even more.

The lower-link positioning error is partly excited by the error dynamics of the upper link, reflecting the inherent cross-coupling in the two-link manipulator system.

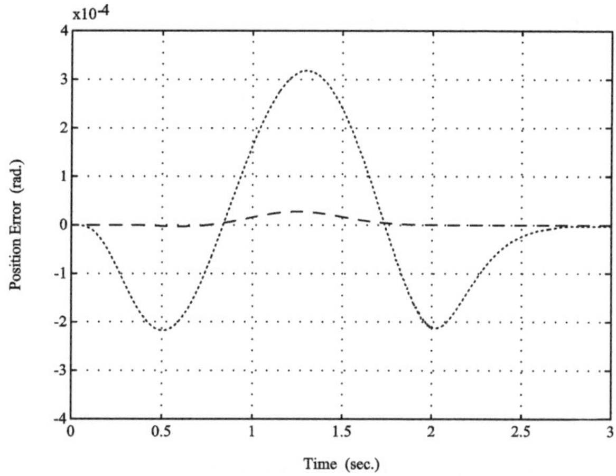

Fig. 6.4 Performance of Algorithm 6.1

6.4.2 *A Simplified Scheme*

As discussed, motor actuator dynamics can be reduced to a set of first-order lag terms with poles at $\left(-\mathbf{L}_a^{-1}\mathbf{R}_a\right)$ by applying the control law in Eq. (6.14). In other words, after this step, the transfer function from $\boldsymbol{u}_b$ to $\boldsymbol{i}_a$ behaves like delayed integrators.

If one is willing to trade off some control performance for simpler implementation, a pole-shifting method can be introduced, leading to the approximate linearization approach described below. Suppose we use the lead compensator:

$$\boldsymbol{u}_b : \mathbf{T}_{pf}\frac{\mathrm{d}\boldsymbol{u}_b}{\mathrm{d}t} + \boldsymbol{u}_b = \hat{\mathbf{L}}_a\frac{\mathrm{d}\boldsymbol{u}_{bc}}{\mathrm{d}t} + \hat{\mathbf{R}}_a\boldsymbol{u}_{bc}, \tag{6.23}$$

where $\boldsymbol{u}_{bc} \in \mathbb{R}^n$ is an intermediate control vector, and $\mathbf{T}_{pf} \in \mathbb{R}^{n\times n}$ is a diagonal pre-filter time-constant matrix. By making $\mathbf{T}_{pf}$ much smaller than $\left(\hat{\mathbf{R}}_a^{-1}\hat{\mathbf{L}}_a\right)$, this lead compensator effectively shifts the undesirable poles at $\left(-\hat{\mathbf{L}}_a^{-1}\hat{\mathbf{R}}_a\right)$ of the original system to $\left(-\mathbf{T}_{pf}^{-1}\right)$, while introducing low-frequency zeros $\left(-\hat{\mathbf{L}}_a^{-1}\hat{\mathbf{R}}_a\right)$and high-frequency poles $\left(-\mathbf{T}_{pf}^{-1}\right)$.

The control law in Eq. (6.23) is designed to cancel poles. Substituting Eq. (6.23) into Eq. (6.15), and assuming perfect pole–zero cancelation along with negligible residual dynamics from $\left(-\mathbf{T}_{pf}^{-1}\right)$, we obtain:

$$\boldsymbol{i}_a \approx \boldsymbol{u}_{bc}, \tag{6.24}$$

in an approximate (non-physical) sense. This simplification stems from balancing control accuracy against computational complexity.

Because this method also compensates for the intrinsically coupled nonlinear dynamics, we again ignore the disturbing and uncertain term $\boldsymbol{\tau}_D$ in Eq. (6.1), reducing it to Eq. (6.19). Define:

$$\boldsymbol{u}_{bc} := \left(\mathbf{N}\hat{\mathbf{K}}_T\right)^{-1}\boldsymbol{u}_c, \tag{6.25}$$

and let $\boldsymbol{u}_c \in \mathbb{R}^n$ be:

$$\boldsymbol{u}_c := \hat{\mathbf{D}}(\boldsymbol{q})\boldsymbol{u} + \hat{\boldsymbol{h}}(\dot{\boldsymbol{q}}, \boldsymbol{q}), \tag{6.26}$$

where $\boldsymbol{u} \in \mathbb{R}^n$ is the outer-loop control to be determined. Combining these definitions yields:

$$\dot{\boldsymbol{q}} \approx u. \tag{6.27}$$

Hence, the originally coupled nonlinear plant is effectively transformed into a decoupled linear system of second-order integrators, as the transfer function from $\boldsymbol{u}$ to $\boldsymbol{q}$ approximates a set of double integrators.

At this stage, each of the n subsystems behaves like an open-loop double integrator, enabling the use of standard linear feedback techniques for the outer-loop controller. We will analyze the resulting closed-loop behavior of Eq. (6.27) in subsequent chapters. The pre-compensation algorithm described above can be summarized as follows:

Algorithm 6.2 Assume the electromechanical system of a mechanical manipulator (Fig. 6.2) is perfectly modeled and free from external disturbances. Then, applying the serial-compensating controls in Eqs. (6.14), (6.23), (6.25), and (6.26):

$$\boldsymbol{u}_a := \boldsymbol{u}_b + \left(\hat{\mathbf{K}}_E\mathbf{N}\right)\dot{\boldsymbol{q}},$$

$$\boldsymbol{u}_b : \mathbf{T}_{pf}\frac{\mathrm{d}\boldsymbol{u}_b}{\mathrm{d}t} + \boldsymbol{u}_b = \hat{\mathbf{L}}_a\frac{\mathrm{d}\boldsymbol{u}_{bc}}{\mathrm{d}t} + \hat{\mathbf{R}}_a\boldsymbol{u}_{bc},$$

$$\boldsymbol{u}_{bc} := \left(\mathbf{N}\hat{\mathbf{K}}_T\right)^{-1}\boldsymbol{u}_c,$$

$$\boldsymbol{u}_c := \hat{\mathbf{D}}(\boldsymbol{q})\boldsymbol{u} + \hat{\boldsymbol{h}}(\dot{\boldsymbol{q}}, \boldsymbol{q}),$$

transforms the input–output behavior from $\boldsymbol{u}$ to $\boldsymbol{q}$ into a set of decoupled double integrators. Note that $\boldsymbol{u}_c$ here closely resembles the computed-torque control from Algorithm A.1, except that motor dynamics is now properly accounted for. This facilitates reusing many existing techniques initially designed for computed-torque systems.

The performance of Algorithm 6.2 is demonstrated in the following example.

Example 6.3 We apply Algorithm 6.2 to the same two-link manipulator from Example 6.2, with identical motion tasks. For closed-loop control, define:

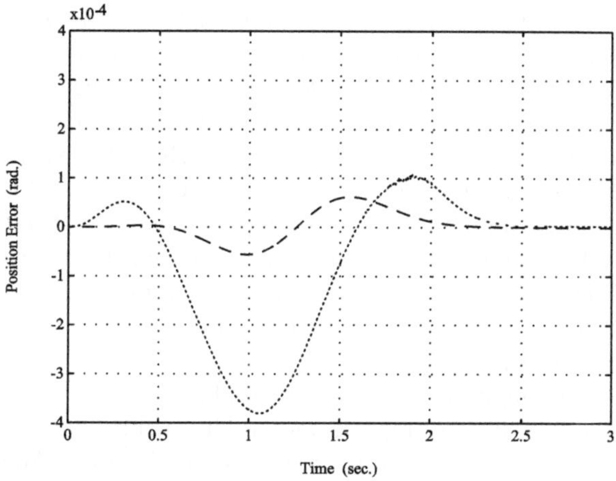

Fig. 6.5 Performance of Algorithm 6.2

$$\boldsymbol{u} := \boldsymbol{r} - \mathbf{K}_v\dot{\boldsymbol{q}} - \mathbf{K}_p\boldsymbol{q},$$

where

$$\boldsymbol{r}(t) := \ddot{\boldsymbol{q}}_d(t) + \mathbf{K}_v\dot{\boldsymbol{q}}_d(t) + \mathbf{K}_p\boldsymbol{q}_d(t),$$

is the reference input, $\mathbf{K}_v = 20\ \mathbf{I}$ and $\mathbf{K}_p = 100\ \mathbf{I}$.

Figure 6.5 {Dotted line: Upper link. Dashed line: Lower link} presents the tracking errors. The small residual errors indicate that motor dynamics is effectively compensated. These small errors stem from residual dynamics in $\boldsymbol{u}_b$, governed by $\mathbf{T}_{pf} = 0.001\ \mathbf{I}$. Reducing $\mathbf{T}_{pf}$ further decreases this undesired effect.

6.5 Summary

This chapter addresses the model-based feedback linearization and decoupling of nonlinear, cross-coupled mass–stiffness–damping systems that include actuator dynamics. As an illustrative example, it presents model-based feedback linearization and decoupling for motor-actuated robotic manipulators, highlighting algorithms that jointly account for the mechanical link dynamics and the electrical motor dynamics. By applying appropriate nonlinear transformations, the strongly coupled, highly nonlinear electromechanical behavior of a motor-manipulator system can be linearized and decoupled without relying on advanced mathematical tools.

Algorithm 6.1 achieves near-perfect decoupling and linearization under ideal assumptions that there is no parameter mismatch, no unmodeled dynamics, and no external disturbances. Its derivation is compact, drawing directly on the physical characteristics of the plant system by examining the signal flow from the motor voltage $\boldsymbol{u}_a$ to the joint position $\boldsymbol{q}$, including the three integrators, four gains [$\mathbf{L}_a^{-1}$, $\mathbf{K}_T$, $\mathbf{N}$, $\mathbf{D}(\boldsymbol{q})^{-1}$], and three main feedback paths [$\mathbf{R}_a$, $(\mathbf{K}_E\mathbf{N})$, $\boldsymbol{h}(\dot{\boldsymbol{q}}, \boldsymbol{q})$]. This step-by-step transformation provides significant physical insight. Although it is primarily designed for motor-manipulator systems, the underlying concept can also be applied to other manipulator architectures, such as flexible-joint systems [33], as well as to broader classes of nonlinear, cross-coupled mass–stiffness–damping systems.

Algorithm 6.2 offers a simplified version that trades some performance for reduced complexity through pole–zero cancelation. Although this compromises strict linearizability and can make the system more sensitive to parameter variations and disturbances, it remains practical in many real-world scenarios where engineering tradeoffs are acceptable.

Throughout, the chapter highlights that model-based control using more comprehensive system representations inevitably demands substantial computational resources, especially for real-time applications. Additionally, the presented algorithms assume a perfectly modeled plant without external disturbances; in reality, mismatches in parameters and unmodeled dynamics, such as payload changes and flexible modes, may significantly degrade performance. Future chapters will examine methods for compensating these uncertainties, while continuing to use the serial pre-compensators derived here to yield systems that are "almost decoupled" and "almost linearized", thereby facilitating robust control design in practical motor-manipulator implementations.

References

1. Levine J (1989) Remarks on some worked out applications of nonlinear control theory. In: Descusse J, Fliess M, Isidori A, Leborgne D (eds) New trends in nonlinear control theory. Springer, Berlin, pp 445–458
2. Khalil HK (2014) Nonlinear systems. Pearson Education, London
3. Isidori A (1989) Nonlinear control systems: an introduction. Springer
4. Descusse J, Fliess M, Isidori A, Leborgne D (1989) New trends in nonlinear control theory. In: Proceedings of an international conference on nonlinear systems. Springer
5. Vidyasagar M (1993) Nonlinear systems analysis. Prentice-Hall
6. Spong MW, Vidyasagar M (1989) Robot dynamics and control. Wiley
7. Levis AH (1987) Challenges to control: a collective view. IEEE Trans Autom Control 32(4):275–285
8. Book WJ (1990) Modeling, design, and control of flexible manipulator arms: a tutorial review. In: Proceedings of the 29th IEEE conference on decision and control, pp 500–506
9. Spong MW (1987) Modeling and control of elastic joint robots. Trans ASME J Dyn Syst Measur Control 109(4):310–319
10. Tomei P (1991) Positioning control of flexible joint robots. In: de Wit CC (ed) Advanced robot control. Springer, pp 216–228
11. Mrad F, Ahmad S (1992) Control of flexible joint robots. Robot Comput Integr Manufact 9(2):137–144
12. Beekmann RW, Lee KY (1988) Nonlinear robotic control including drive motor interactions. In: Proceedings of the 1988 American control conference, pp 1333–1338
13. Tarn TJ, Bejczy AK, Yun X, Li Z (1991) Effect of motor dynamics on nonlinear feedback robot arm control. IEEE Trans Robot Autom 7(1):114–122

14. Dawson DM, Qu Z, Carroll JJ (1992) Tracking control of rigid-link electrically-driven robot manipulators. Int J Control 56(5):991–1006
15. Krishnan H, McClamroch NH (1994) Tracking in nonlinear differential-algebraic control systems with applications to constrained robot systems. Automatica 30(12):1885–1897
16. Marilier T, Richard JA (1989) Non-linear mechanic and electric behavior of a robot axis with a "harmonic-drive" gear. Robot Comput Integr Manufact 5(2/3):129–136
17. Gu YL, Loh RNK, Coleman N, Lin CF (1992) Control of weapon pointing systems based on robotic formulation. In: Proceedings of the 1992 American control conference, vol 1, pp 413–419
18. Sepehri N, Lawrence PD (1992) Model-based sensor-based velocity control of teleoperated heavy-duty hydraulic machines. In: Proceedings of the 1992 IEEE/RSJ international conference on intelligent robots and systems, vol 2, pp 859–864
19. Schilling RJ (1990) Fundamentals of robotics: analysis and control. Prentice-Hall
20. Arimoto S (1990) Design of robot control systems. Adv Robot 4(1):79–97
21. Craig JJ (2022) Introduction to robotics: mechanics and control. Pearson Education
22. Franklin GF, Powell JD, Workman ML (1998) Digital control of dynamic systems. Addison-Wesley
23. Leahy MB (1990) Model-based control of industrial manipulators: an experimental analysis. J Robot Syst 7(5):741–758
24. Leahy MB, Saridis GN (1989) Compensation of industrial manipulator dynamics. Int J Robot Res 8(4):73–84
25. Leahy MB, Valavanis KP, Saridis GN (1989) Evaluation of dynamic models for PUMA robot control. IEEE Trans Robot Autom 5(2):242–245
26. Good MC, Sweet LM, Strobel KL (1985) Dynamic models for control system design of integrated robot and drive systems. Trans ASME J Dyn Syst Measur Control 107(1):53–59
27. Chen Y (1989) Replacing a PID controller by a lag-lead compensator for a robot: a frequency response approach. IEEE Trans Robot Autom 5(2):174–182
28. Freund E (1982) Fast nonlinear control with arbitrary pole-placement for industrial robots and manipulators. Int J Robot Res 1(1):65–78
29. Spong MW (1989) On the force control problem for flexible joint manipulators. IEEE Trans Autom Control 34(1):107–111
30. Kravaris C, Kantor JC (1990) Geometric methods for nonlinear process control. 1. Background. Ind Eng Chem Res 29(12):2295–2310
31. Kravaris C, Kantor JC (1990) Geometric methods for nonlinear process control. 2. Controller synthesis. Ind Eng Chem Res 29(12):2310–2323
32. Berger RM, ElMaraghy HA (1992) Feedback linearization control of flexible joint robots. Robot Comput Integr Manufact 9(3):239–246
33. Zhu HA, Teo CL, Hong GS, Poo AN (1993) A simple scheme of nonlinear decoupling control for robotic manipulators with flexible joints. In: Proceedings of TENCON'93 IEEE region 10 international conference on computers, communications and automation, pp 165–168

Chapter 7
Synthesis of Outer-Loop Controllers

Outer-loop controllers, which determine the desired closed-loop characteristics of a system, are essential components of any practical control scheme. They govern high-level performance objectives such as reference tracking, robustness, and stability margins. In this chapter, we explore straightforward, model-based designs for both outer-loop and feedforward controllers aimed at mass–stiffness–damping systems. By leveraging the underlying physics and structural properties of these systems, the introduced controllers integrate naturally with inner-loop compensators and thereby streamline the overall design, ensuring that critical performance requirements are met in real-world applications.

7.1 Introduction

In addition to the inner-loop serial pre-compensators responsible for feedback linearization, outer-loop controllers are crucial for the closed-loop control of pre-compensated (i.e., decoupled and linearized) mass–stiffness–damping systems. Exploiting the structural properties of these pre-compensated systems allows for simple state-feedback controllers in the transformed coordinates, effectively specifying high-level performance objectives. For simplicity, this book also employs state-feedback controllers in the outer loop, with pole-placement guiding the selection of feedback gains.

Many mass–stiffness–damping systems, such as robotic manipulators, demand precise positioning accuracy. Unfortunately, external disturbances and internal uncertainties—including friction, payload variations, and other non-ideal factors—often introduce significant steady-state errors. A common practice to address such errors involves integral compensation in the control law [1]. Well-known schemes include the PID structure [2, 3], whose effectiveness and simplicity have led to widespread adoption [4, 5]. Nonetheless, the additional integrator in standard PID sometimes

H. Zhu, *Model-Based Control of Mass–Stiffness–Damping Systems*, Advances in Industrial Control, https://doi.org/10.1007/978-3-031-97592-9_7

degrades transient performance by introducing problematic zeros [6], potentially causing overshoot or oscillatory behavior.

Motivated by the benefits and limitations of PD and PID controllers, this chapter introduces a model-based controller with integral compensation for mass–stiffness–damping system control. This two-degrees-of-freedom scheme maintains the flexibility to assign both the poles and zeros of the resulting transfer function, enabling designers to achieve desired closed-loop performance while mitigating the steady-state errors that often arise from system uncertainties and disturbances.

7.2 State-Feedback Control

In this book, state-feedback controllers serve as outer-loop controllers for the serially pre-compensated mass–stiffness–damping systems that have been decoupled and linearized via the algorithms in Chap. 6. Since feedback linearization is known to permit robust control laws that accommodate parametric uncertainty [7, 8], and because many authors have already contributed extensive work on robust state-feedback design for various mass–stiffness–damping systems, including robotic manipulators [9], the robust design of feedback gains is not discussed further here. Instead, this section outlines the relevant control structures.

With the decoupling and linearization algorithms in Chap. 6, the "ideal" dynamics of a pre-compensated mass–stiffness–damping system is reduced, under perfect conditions, to either

$$\dddot{\boldsymbol{q}} = \boldsymbol{u}, \tag{7.1}$$

or

$$\ddot{\boldsymbol{q}} = \boldsymbol{u}, \tag{7.2}$$

assuming no external disturbances or model inaccuracies. Suppose the states $\boldsymbol{q}$, $\dot{\boldsymbol{q}}$, and $\ddot{\boldsymbol{q}}$ of these systems are all available for feedback. In the triple-integrator system of Eq. (7.1), a PDD^2 (position–velocity–acceleration) state-feedback control law,

$$\boldsymbol{u} := \boldsymbol{r}_1 - \mathbf{K}_a\ddot{\boldsymbol{q}} - \mathbf{K}_v\dot{\boldsymbol{q}} - \mathbf{K}_p\boldsymbol{q}, \tag{7.3}$$

may be adopted, while a PD (position–velocity) state-feedback control law,

$$\boldsymbol{u} := \boldsymbol{r}_2 - \mathbf{K}_v\dot{\boldsymbol{q}} - \mathbf{K}_p\boldsymbol{q}, \tag{7.4}$$

applies to the double-integrator system of Eq. (7.2). Here, $\mathbf{K}_a$, $\mathbf{K}_v$, and $\mathbf{K}_p \in \mathbb{R}^{n\times n}$ are diagonal feedback-gain matrices with positive elements, and $\boldsymbol{r}_1, \boldsymbol{r}_2 \in \mathbb{R}^n$ denote reference commands. These control laws produce the respective command-to-position transfer functions

$$\boldsymbol{Q} = \left(\mathbf{I}s^3 + \mathbf{K}_a s^2 + \mathbf{K}_v s + \mathbf{K}_p\right)^{-1}\mathbf{R}_1, \tag{7.5}$$

for Eq. (7.1), and

$$\boldsymbol{Q} = \left(\mathbf{I}s^2 + \mathbf{K}_v s + \mathbf{K}_p\right)^{-1}\mathbf{R}_2, \tag{7.6}$$

for Eq. (7.2), where s is the Laplace operator. Methods for robust feedback-gain design are well documented [10–12] and can achieve strong closed-loop performance with high-gain strategies [8, 13]. One of the most straightforward approaches is pole-placement using fixed or adaptive gains [14–16]. This book employs a fixed-gain pole-placement scheme (Appendix B) for simplicity.

When performing set-point motion, one can define the reference command $\boldsymbol{r}$ (be it $\boldsymbol{r}_1$ or $\boldsymbol{r}_2$) as

$$\boldsymbol{r} := \mathbf{K}_p\left(\boldsymbol{q}_f - \boldsymbol{q}_i\right), \tag{7.7}$$

where $\boldsymbol{q}_i, \boldsymbol{q}_f \in \mathbb{R}^n$ are the initial and final positions of the mass–stiffness–damping system. For trajectory tracking, feedforward controllers (pre-filters) supply the references

$$\boldsymbol{r}_1(t) := \dddot{\boldsymbol{q}}_d(t) + \mathbf{K}_a\ddot{\boldsymbol{q}}_d(t) + \mathbf{K}_v\dot{\boldsymbol{q}}_d(t) + \mathbf{K}_p\boldsymbol{q}_d(t), \tag{7.8}$$

for the triple-integrator system in Eq. (7.1), or

$$\boldsymbol{r}_2(t) := \ddot{\boldsymbol{q}}_d(t) + \mathbf{K}_v\dot{\boldsymbol{q}}_d(t) + \mathbf{K}_p\boldsymbol{q}_d(t), \tag{7.9}$$

for the double-integrator system in Eq. (7.2). Here, $\boldsymbol{q}_d(t) \in \mathbb{R}^n$ denotes the desired trajectory, assumed to be l-times continuously differentiable, with $l \geq 3$ for $\boldsymbol{r}_1$ and $l \geq 2$ for $\boldsymbol{r}_2$. Under perfect modeling and state measurement, these references lead to

$$\dddot{\boldsymbol{e}}(t) + \mathbf{K}_a\ddot{\boldsymbol{e}}(t) + \mathbf{K}_v\dot{\boldsymbol{e}}(t) + \mathbf{K}_p\boldsymbol{e}(t) = 0, \tag{7.10}$$

for Eq. (7.1), or

$$\ddot{\boldsymbol{e}}(t) + \mathbf{K}_v\dot{\boldsymbol{e}}(t) + \mathbf{K}_p\boldsymbol{e}(t) = 0, \tag{7.11}$$

for Eq. (7.2), where $\boldsymbol{e}(t) = \boldsymbol{q}_d(t) - \boldsymbol{q}(t)$ is the tracking error. These equations imply

$$\boldsymbol{q}(t) = \boldsymbol{q}_d(t) \quad \text{or} \quad \boldsymbol{Q}(s) = \boldsymbol{Q}_d(s), \tag{7.12}$$

i.e., perfect tracking is achieved. Realistically, though, errors arise from imperfect modeling and state observations, as seen in Example 6.2 (where $\hat{\boldsymbol{q}}$ is estimated),

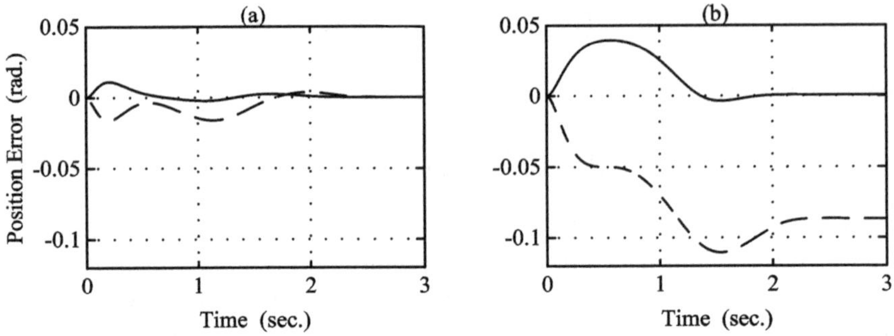

Fig. 7.1 Tracking errors due to payload variation

Example 6.3 (residual dynamics caused by $\mathbf{T}_{pf}$), and the following Example 7.1 (payload variation).

Example 7.1 In this example, the setup remains the same as in Examples 6.2 and 6. 3, except for a heavy payload—unaccounted for in the pre-compensators—added to the upper end of the upper link of the two-link manipulator. As shown in Fig. 7.1 {a Algorithm 6.1, b Algorithm 6.2. Dashed line: Upper link. Solid line: Lower link}, this additional mass significantly increases the tracking errors, underscoring the influence of parametric uncertainty on overall system performance.

From the above example, it is evident that the control scheme derived from Algorithm 6.2 suffers from considerable steady-state error under payload variations, whereas the Algorithm 6.1—based system does not. The key distinction lies in the additional (innate) integrator present in the forward path of Algorithm 6.1—specifically, the integrator located between the control input $\boldsymbol{u}$ and the joint where the disturbance $\boldsymbol{\tau}_D$ (i.e., the payload effect) acts. This extra integration stage naturally attenuates constant disturbances and thereby minimizes steady-state error.

To improve the steady-state performance of Algorithm 6.2—based systems—or similarly, those employing the computed-torque control (Algorithm A.1) for robotic control—one can introduce integral compensation. This adjustment provides the missing integration mechanism needed to counteract constant disturbances, such as payload changes, and helps restore the accuracy lost to unmodeled dynamic effects.

7.3 State-Feedback with Integral Compensation

As demonstrated, the pre-compensated double-integrator system produced by Algorithm 6.2 can exhibit steady-state error when exposed to internal parameter variations or external disturbances. The same shortcoming also arises in the conventional computed-torque control (Algorithm A.1). Systems of this type can generally be described by

$$\ddot{q} = u - d, \tag{7.13}$$

where $d \in \mathbb{R}^n$ represents equivalent disturbances stemming from internal and external uncertainties. To eliminate steady-state error in Eq. (7.13), one viable strategy is to incorporate an integral component into the outer-loop control u. Because steady-state accuracy is often a primary concern in mass–stiffness–damping systems such as robotic manipulators, the following subsections focus on using integral compensation for set-point motion scenarios.

7.3.1 Additional Integral Component

In conventional practice, the PID control law [9, 17, 18] can be used to eliminate steady-state errors in the system of Eq. (7.13). This approach is represented by

$$u := \mathbf{K}_v \frac{\mathrm{d}e}{\mathrm{d}t} + \mathbf{K}_p e + \mathbf{K}_i \int e \mathrm{d}t, \tag{7.14}$$

where $\mathbf{K}_i \in \mathbb{R}^{n\times n}$ is a diagonal gain matrix with positive elements, and $e \in \mathbb{R}^n$ is the error vector defined by

$$e := r - q, \tag{7.15}$$

with $r \in \mathbb{R}^n$ denoting the reference input. Under the PID control of Eq. (7.14), the command response of the system is:

$$Q = \left[\left(\mathbf{I}s^3 + \mathbf{K}_v s^2 + \mathbf{K}_p s + \mathbf{K}_i\right)^{-1}\left(\mathbf{K}_v s^2 + \mathbf{K}_p s + \mathbf{K}_i\right)\right] R, \tag{7.16}$$

where R is also chosen as the desired position for set-point motion, as in Eq. (7.7). For trajectory tracking, one can generate R via a feedforward controller (pre-filter) that is the inverse of the transfer function $[\bullet]$ in Eq. (7.16).

Meanwhile, the disturbance response of the PID-controlled system is:

$$Q^D_{R=0} = \left[\left(\mathbf{I}s^3 + \mathbf{K}_v s^2 + \mathbf{K}_p s + \mathbf{K}_i\right)^{-1}(\mathbf{I}s)\right] D, \tag{7.17}$$

indicating that a constant disturbance D is ultimately mitigated by the integral ("I") term. This is the primary advantage of PID, and similar to previously discussed controllers, one can place the closed-loop poles of the system arbitrarily by adjusting $\mathbf{K}_v$, $\mathbf{K}_p$, and $\mathbf{K}_i$.

Despite this, the overshoot (or oscillation) in the closed-loop response can still be significant if the PID control law is used, even with carefully chosen poles. This overshoot is particularly problematic in motion tasks such as insertion operation of robotic manipulators, where large deviations can cause severe collisions [19]. In high-speed maneuvers, these undesirable transients may become even more pronounced.

By examining the command transfer function [•] in Eq. (7.16), it becomes evident that the system contains $3n$ poles and $2n$ zeros. The crux of the matter is that a controller designer cannot freely assign all $5n$ elements (the $3n$ poles and $2n$ zeros) by manipulating only $3n$ control parameters. Unfortunately, because zeros significantly affect transient behavior [1, 16], their placement can distort the response even if the poles are well-chosen.

In the described PID control system, the two undesirably placed zeros [1, 6] often degrade transient behavior. A pole-placement approach alone cannot account for the resulting shape of the response, since the zeros also influence it [16]. Consequently, if one focuses solely on placing poles first, the zeros remain uncontrolled, making it difficult to predict the ultimate system response. Control engineers must therefore balance stability and regulation performance against transient behavior [20] by reassigning both poles and zeros, a process that demands skill and experience [6, 21].

The connection between pole–zero structure and the transient response of the system has been studied for decades. In "modern" control theory, especially with state-space methods, significant progress has been made over the last thirty years. Since the transient response depends critically on the eigenstructure (eigenvalues and eigenvectors) [22], modifying the transient behavior of the system necessitates changes not just to the eigenvalues (poles) but also to the eigenvectors. The poles control the rate at which the response grows or decays, while the eigenvectors govern the shape of the response.

For the plant in Eq. (7.13), an alternative form of integral compensation—implemented as a two-degrees-of-freedom controller [23] with the capability to assign poles and zeros arbitrarily—will be explored in the following subsection. This approach aims to provide the flexibility lacking in conventional PID methods for precisely shaping both the stability and transient characteristics of the system.

7.3.2 A Two-Degrees-of-Freedom Scheme with Integral Compensation

Being aware of the limitations inherent in the conventional PID control from Eq. (7.14), a modified control with integral compensation is introduced here. This approach is built upon the PD control in Eq. (7.4) while adding an integral term, leading to

$$\boldsymbol{u} := \left(\boldsymbol{r} - \mathbf{K}_v \frac{\mathrm{d}\boldsymbol{q}}{\mathrm{d}t} - \mathbf{K}_p \boldsymbol{q}\right) + \mathbf{K}_i \int \boldsymbol{e} \mathrm{d}t, \tag{7.18}$$

where $\boldsymbol{r}$ is a feedforward control to be determined, and

$$\boldsymbol{e} := \boldsymbol{q}_d - \boldsymbol{q}, \tag{7.19}$$

is the position error vector. With this modified law, the resulting command response becomes

$$\boldsymbol{Q} = \left(\mathbf{I}s^3 + \mathbf{K}_v s^2 + \mathbf{K}_p s + \mathbf{K}_i\right)^{-1}\left(\mathbf{I}s\boldsymbol{R} + \mathbf{K}_i \boldsymbol{Q}_d\right), \tag{7.20}$$

where $\boldsymbol{Q}_d$ is the Laplace transform of $\boldsymbol{q}_d$.. The corresponding disturbance transfer function remains the same as in Eq. (7.17), meaning that the disturbance-attenuation capability of this modified control matches that of the conventional PID scheme. However, since an additional feedforward term $\boldsymbol{r}$ appears in Eq. (7.18), the resulting control system takes the form of a two-degrees-of-freedom controller.

For set-point motion, the feedforward input $\boldsymbol{r}$ in Eq. (7.18) admits two possibilities. One option is to set

$$\boldsymbol{R} := 0, \tag{7.21}$$

yielding

$$\boldsymbol{Q} = \left[\left(\mathbf{I}s^3 + \mathbf{K}_v s^2 + \mathbf{K}_p s + \mathbf{K}_i\right)^{-1}\mathbf{K}_i\right]\boldsymbol{Q}_d, \tag{7.22}$$

so that the command transfer function contains only $3n$ poles, all freely assignable. Alternatively, if $\boldsymbol{R}$ is chosen as

$$\boldsymbol{R} := \mathbf{K}\boldsymbol{Q}_d, \tag{7.23}$$

with $\mathbf{K} \in \mathbb{R}^{n\times n}$ diagonal, the command transfer function becomes

$$\boldsymbol{Q} = \left[\left(\mathbf{I}s^3 + \mathbf{K}_v s^2 + \mathbf{K}_p s + \mathbf{K}_i\right)^{-1}(\mathbf{K}s + \mathbf{K}_i)\right]\boldsymbol{Q}_d, \tag{7.24}$$

allowing all $3n$ poles and n zeros to be placed arbitrarily. In particular, one can reduce the system to order $2n$ by canceling n of the $3n$ poles with the n arbitrarily assignable zeros. A similar feedforward design can be used for trajectory tracking, though this discussion is omitted here.

Thus, the control in Eq. (7.18) constitutes a two-degrees-of-freedom method whose key strength lies in enabling pole/zero assignment independently of each other, in contrast to conventional PID control. In the standard PID framework, the coefficients become interlinked, as indicated by Eq. (7.16).

To illustrate the performance of various outer-loop controllers, we consider a single-link manipulator (Sect. 2.4.1) with either negligible motor dynamics, pre-compensated via Algorithm A.1, or with significant motor dynamics, pre-compensated via Algorithm 6.2. The outer-loop controllers tested in this comparative study are as follows:

C1 PD control in Eq. (7.4) with $k_v = 20$ and $k_p = 100$, placing both poles at -10.
C2 PID control in Eq. (7.14) with $k_v = 30$, $k_p = 300$ and $k_i = 1000$., placing all three poles at -10 while leaving the two zeros at large.
C3 Modified control in Eq. (7.18) with $k_v = 30$, $k_p = 300$, $k_i = 1000$, and $R = 0$, placing all three poles at -10 and no zero.
C4 Modified control in Eq. (7.18) with $k_v = 30$, $k_p = 300$, $k_i = 1000$, and $R = kQ_d$ where $k = 100$. All three poles are at -10, and the single zero is also assigned to -10 to cancel one of the three poles.

In each case, the gains are designed via pole/zero assignment, ensuring critical damping of the assigned poles. The manipulator executes a set-point motion from $q_i = 0$ to $q_f = 1.0$ rad. Consequently, these examples enable direct performance comparisons among PD, PID, and two variants of the modified integral-compensation control.

Example 7.2 In this example, the manipulator parameters are exact, and no uncertainties or disturbances arise in the system. Figure 7.2 {Solid line: **C1** and **C4**. Dashed line: **C2**. Dash-dot line: **C3**} shows the step responses under the different outer-loop controllers. Controller **C1** exhibits no overshoot because it is critically damped, while **C2** experiences a large overshoot despite placing all poles at -10. The excessive overshoot in **C2** stems from the two undesirable zeros embedded in its design. Attempting to retune the gains of **C2** to reduce overshoot would undermine the robustness of the controller against uncertainties and compromise its ability to attenuate disturbances.

Meanwhile, the response of **C4** overlaps with the response of **C1**, indicating that the command-to-output transfer function of **C4** is effectively the same as the transfer function of **C1**. This outcome arises from the pole–zero cancelation in **C4**, which reduces its third-order dynamics to a second-order system matching **C1**. As a result, **C4** achieves a faster response than **C3**, whose additional pole leads to a slight delay while maintaining zero overshoot.

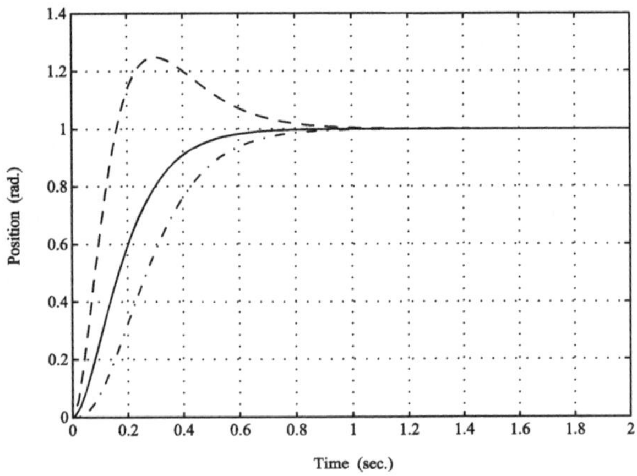

Fig. 7.2 Integral responses without uncertainty

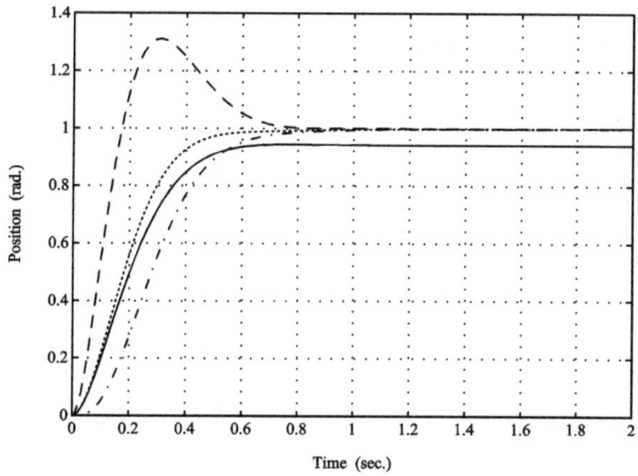

Fig. 7.3 Integral responses with payload variation

Example 7.3 In this example, a 5 kg payload ($m_p = 5$ kg) is added to the upper end of the manipulator, yet the control torque calculation [i.e., Eq. (A.2)] does not account for the added mass. Consequently, this payload variation introduces modeling uncertainty and associated nonlinear effects. Figure 7.3 {Solid line: **C1**. Dashed line: **C2**. Dash-dot line: **C3**. Dotted line: **C4**} shows that the increased payload induces some steady-state error in the response of **C1**, while the overshoot in **C2** grows even larger than what was observed in Fig. 7.2. In contrast, **C4** and **C3** continue to exhibit satisfactory performance.

Example 7.4 In this example, an external step-disturbing torque $\tau_D = 20$ Nm is applied to the manipulator joint at $t = 1$ s. No additional uncertainties exist in the system. As shown in Fig. 7.4 {**C1**: Solid line. **C2**: Dashed line. **C3**: Dash-dot line. **C4**: Solid line $t \in [0, 1]$., Broken line $t \in [1, 2]$}, the responses from $t \in [0, 1]$ match those depicted in Fig. 7.2. However, within $t \in [1, 2]$, the disturbance induces a significant steady-state error in the response of **C1**. By contrast, the overlapping disturbance responses (broken line) for **C2**, **C4**, and **C3** remain satisfactory, exhibiting only minor transient deviations and no steady-state error in reaction to the torque step disturbance.

It is important to note that the step responses of the conventional PID control system (**C2**) differ from those predicted by the idealized transfer function in Eq. (7.16). This discrepancy arises because pure differential control is physically unrealizable. In practical applications—particularly in computer-controlled environments—the discrete, saturated impulse response of differential control under a step input can deviate substantially from the theoretical ideal. Consequently, the results shown for the conventional PID control in this section incorporate these nonlinear effects, and the same behavior is confirmed by the experimental data presented subsequently.

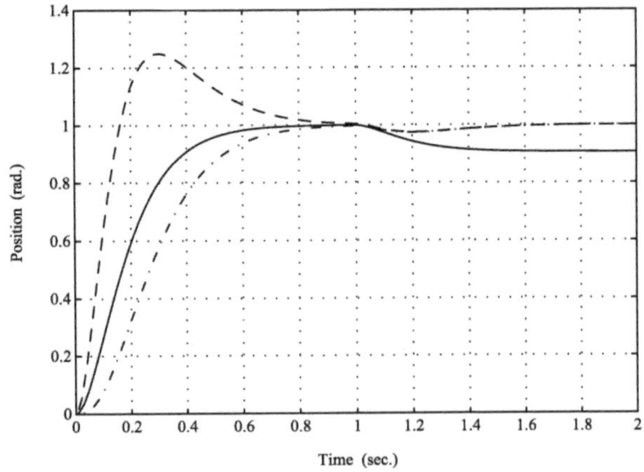

Fig. 7.4 Integral responses with external disturbance

Experiment 7.1

The previously discussed controllers were further evaluated on the physical testbed detailed in Appendix D. In this experiment, the manipulator moved from its initial lower vertical position $q_i = -\pi/2$ to the final set-point $q_f = -3\pi/8$. Figure 7.5 {Solid line: **C1**. Dashed line: **C2**. Dash-dot line: **C3**. Dotted line: **C4**} displays the normalized step-response measurements. During testing, joint friction was intentionally left uncompensated, and gravity torque compensation was deliberately imprecise, effectively modeling external disturbances and payload variations. As a result, the PD controller (solid line in Fig. 7.5) exhibits a large steady-state error. These experimental findings closely match the simulation results, confirming the consistent impact of unmodeled effects on system performance.

Comparative case studies reveal that the modified integral control (**C4**) significantly outperforms the traditional PD (**C1**) and PID (**C2**) schemes. Specifically,

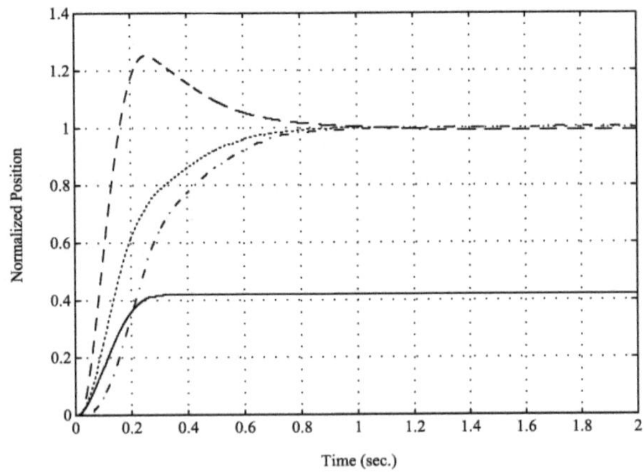

Fig. 7.5 Experimental results

C4 preserves the fast response characteristic of **C1** and the disturbance-attenuation capability of **C2**, yet avoids the steady-state error inherent in **C1** and the excessive overshoot often seen with **C2**.

7.4 Summary

The design of outer-loop controllers for serially pre-compensated mass–stiffness–damping systems based on decoupling and linearization has been presented in this chapter. To simplify the design procedure, a pole-placement approach is employed for feedback-gain selection.

To mitigate steady-state errors in scenarios with payload variations, friction, or constant disturbances, a modified integral control scheme featuring an arbitrarily assignable pole–zero structure is introduced. Detailed analysis demonstrates that this approach not only overcomes and eliminates the shortcomings of conventional PD or PID controllers, but also provides substantially increased flexibility in control implementation. As a result, the method effectively meets the dual goals of high precision and rapid response.

With the capability to freely assign both poles and zeros, the eigenstructure (eigenvalues and eigenvectors) of a system can be tailored more comprehensively. Such eigenstructure assignment has proven highly effective for designing robust control strategies, and has been extensively studied over the past three decades [22, 24, 25]. Appropriately choosing eigenvalues and eigenvectors often fulfills both robustness and performance demands [26]. Although detailed procedures for eigenstructure assignment lie beyond the scope of this discussion, they remain a promising direction for industrial practices.

In contrast, conventional PID control does not afford the same versatility in eigenstructure assignment within the eigenspace of the system. Because zeros, as well as poles (eigenvalues), play a critical role in defining the response [1, 16], their positioning becomes vital for achieving complex performance goals [14]. In [26], a general class of generic algorithms for eigenstructure assignment offers a universally applicable and numerically stable approach to these fundamental problems, enabling direct engagement of time-domain performance metrics, such as integral square error (ISE). This line of research continues to advance the methodologies for pole–zero and eigenstructure assignment in modern control system design.

The two-degrees-of-freedom nature of the presented controllers, consisting of a feedback controller and a feedforward pre-filter, is also model-based. That is, both components rely on the same nominal dynamics assumed for Eq. (7.1) or Eq. (7.2). Hence, uncertainty or parameter mismatch can still affect system performance. Although robust feedback designs have not been discussed, relevant methods [10–12] can deliver strong performance under high-gain conditions [8, 13]. Practical strategies for enhancing robustness in model-based control will be explored in the later sections of this book.

Although integrators positioned in the forward path (between the control input and the disturbance point) can attenuate certain steady-state disturbances, the transient performance may remain unsatisfactory. This was exemplified in Example 7.4, illustrating that passive measures alone do not guarantee optimal disturbance rejection. Active strategies incorporating disturbance observation and feedforward compensation will be introduced later in the book.

References

1. Van de Vegte J (1993) Feedback control systems. Prentice-Hall
2. Luo GL, Saridis GN (1985) L-Q design of PID controllers for robot arms. IEEE J Robot Autom 1(3):152–159
3. J.H. Wei, M.C. Shih, C.P. Chiu, Model-following PID control using hyperstability theory. JSME Int J Ser 3 Vibr Control Eng Eng Ind, 1991, Vol.34, No.3, pp.377–381.
4. Isermann R (1991) Digital control systems, stochastic control, multivariable control, adaptive control, applications, vol 2. Springer, Berlin
5. Åström KJ, Wittenmark B (1997) Computer controlled systems: theory and design. Prentice-Hall
6. Hang CC (1989) The choice of controller zeros. IEEE Control Syst Mag 9(1):72–75
7. Spong MW (1989) On the force control problem for flexible joint manipulators. IEEE Trans Autom Control 34(1):107–111
8. Dawson DM, Qu Z, Carroll JJ (1992) Tracking control of rigid-link electrical-ly-driven robot manipulators. Int J Control 56(5):991–1006
9. Spong MW, Vidyasagar M (1989) Robot dynamics and control. Wiley
10. Leitmann G (1981) On the efficacy of nonlinear control in uncertain linear systems. Trans ASME J Dyn Syst Measur Control 103(2):95–102
11. Corless MJ, Leitmann G (1981) Continuous state feedback guaranteeing uniform ultimate boundedness for uncertain dynamic systems. IEEE Trans Autom Control 26(5):1139–1144
12. Corless M (1989) Tracking controllers for uncertain systems: application to a manutec R3 robot. Trans ASME J Dyn Syst Measur Control 111(4):609–618
13. Spong MW (1992) On the robust control of robot manipulators. IEEE Trans Autom Control 37(11):1782–1786
14. Kailath T (1980) Linear systems. Prentice Hall
15. Sastry S, Bodson M (1989) Adaptive control: stability, convergence, and robustness. Prentice-Hall
16. Chen CT (1993) Analog and digital control system design: transfer-function, state-space, and algebraic methods. Saunders College Publishing, HBJ International
17. Schilling RJ (1990) Fundamentals of robotics: analysis and control. Prentice-Hall
18. Craig JJ (2022) Introduction to robotics: mechanics and control. Pearson Education
19. Shoji Y, Inaba M, Fukuda T (1992) Stable positioning control of 3-dimensional multi-link manipulator for contact tasks. In: Proceedings of the 1992 IEEE/RSJ international conference on intelligent robots and systems, vol 2, pp 983–989
20. Chen Y (1989) Replacing a PID controller by a lag-lead compensator for a robot: a frequency response approach. IEEE Trans Robot Autom 5(2):174–182
21. Litt J (1991) An expert system to perform on-line controller tuning. IEEE Control Syst Mag 11(3):18–23
22. Andry AN, Shapiro EY, Chung JC (1983) Eigenstructure assignment for linear systems. IEEE Trans Aerospace Electr Syst 19(5):711–729
23. Golnaraghi F, Kuo BC (2017) Automatic control systems. McGraw-Hill

24. Broussard JR, Berry PW (1980) The relationship between implicit model following and eigenvalue eigenvector placement. IEEE Trans Autom Control 25(3):591–594
25. Wilson RF, Cloutier JR, Yedavalli RK (1992) Control design for robust eigenstructure assignment in linear uncertain systems. IEEE Control Syst Mag 12(5):29–34
26. Porter B, Borairi M (1992) Genetic design of linear multivariable feedback control systems using eigenstructure assignment. Int J Syst Sci 23(8):1387–1390

Part III
Enhanced Control

The Road Map: Enhanced Control

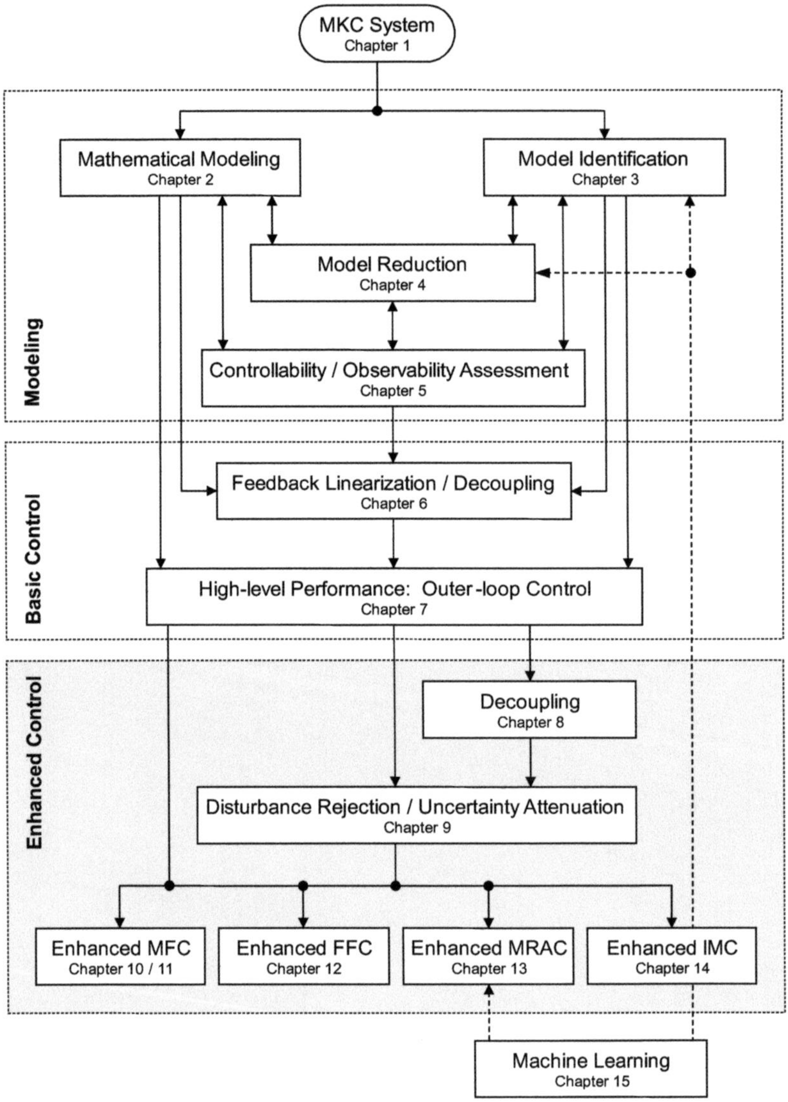

Chapter 8
Model-Based Decoupling

A critical feature of multivariable mass–stiffness–damping systems is the dynamic coupling among their variables and subsystems. Multi-joint robotic manipulators, large rotating machineries, and large-scale structures serve as prime examples: they are inherently large and complex dynamic systems with significant cross-coupling across their degrees-of-freedom, and they often exhibit nonlinear, time-varying, and multivariable behavior. Most precise decoupling control methods [1–4] presume a fixed linear dynamic structure, making them cumbersome to implement on real-world mass–stiffness–damping systems operating under varying conditions.

In this chapter, a linear model-based control strategy is adopted to construct a robust scheme suitable for engineering applications. This approach provides decoupling control that mitigates internal cross-couplings and external disturbances, and it maintains strong performance even in the presence of nonlinear parameter perturbations.

8.1 Introduction

Multivariable dynamic systems often exhibit significant cross-couplings among their internal variables, rendering decoupling a key objective in effective control design [5, 6]. Many conventional decoupling techniques, including similarity transformation and equivalence transformation, require advanced mathematical tools that may not be straightforward to implement. Modal analysis is a more classical methodology commonly used in engineering practices to decouple linear systems. It has been extended, via phase synchronization, to handle virtually all linear systems, even those with non-symmetric coefficient matrices, by compensating for the time drifts introduced by viscous damping and external excitation [7].

A typical linear dynamical mass–stiffness–damping system can be represented by a set of second-order ordinary differential equations. The classical decoupling

H. Zhu, *Model-Based Control of Mass–Stiffness–Damping Systems*, Advances in Industrial Control, https://doi.org/10.1007/978-3-031-97592-9_8

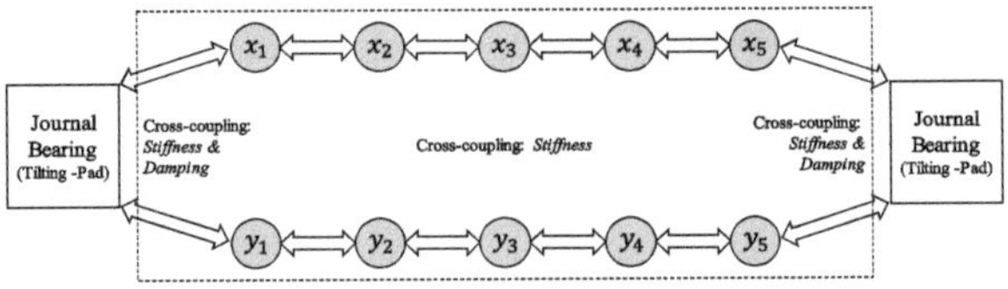

Fig. 8.1 Coupling situation of a rotor-bearing system

problem centers on converting these equations into a diagonalized (or block diagonal) form, thereby isolating each degree-of-freedom. However, real-world mass–stiffness–damping systems—including robotic systems that exhibit strong intrinsic nonlinearity—often demand specialized nonlinear decoupling methods [8, 9].

One notable example of a multivariable, large-scale mass–stiffness–damping system is a multi-degrees-of-freedom (m-DoF) rotor-bearing assembly. Such systems frequently have severe cross-coupling between their degrees-of-freedom, especially at bearing supports that introduce cross-coupled stiffness and damping. Figure 8.1 presents a simplified structural diagram for the low-pressure cylinder of a large compressor, illustrating the primary rotor section (enclosed by dashed lines) and its interconnections. Each circle denotes a degree-of-freedom, and the arrows represent key coupling paths.

In many rotor-bearing systems, the rotor section itself typically has no stiffness coupling between the x- and y-directions, but the bearing supports can introduce strong cross-coupling terms. Additionally, for both x- and y-directions, there may be direct coupling between non-adjacent variables (e.g., x_1 and x_3, x_4, x_5), though Fig. 8.1 omits these for simplicity.

This chapter introduces a model-based control scheme [10] that exploits the structural patterns found in cross-coupled mass–stiffness–damping systems—such as rotor-bearing assemblies and other large-scale, high-dimensional applications—to achieve decoupled control. Crucially, it avoids the heavy reliance on complex transformations or extensive computations found in many existing decoupling approaches, making it more accessible to practical engineering scenarios.

8.2 Model-Based Decoupling Control

In multi-degrees-of-freedom (m-DoF) mass–stiffness–damping systems, each controlled degree-of-freedom is influenced by internal cross-coupling from other subsystems as well as external disturbances. Figure 8.2 illustrates the open-loop dynamic structure for one such degree-of-freedom. If decoupling control and disturbance rejection are not properly implemented, the combined influence of control actions and external perturbations can lead to sustained oscillations that propagate between subsystems, degrading control accuracy and potentially destabilizing the closed-loop system.

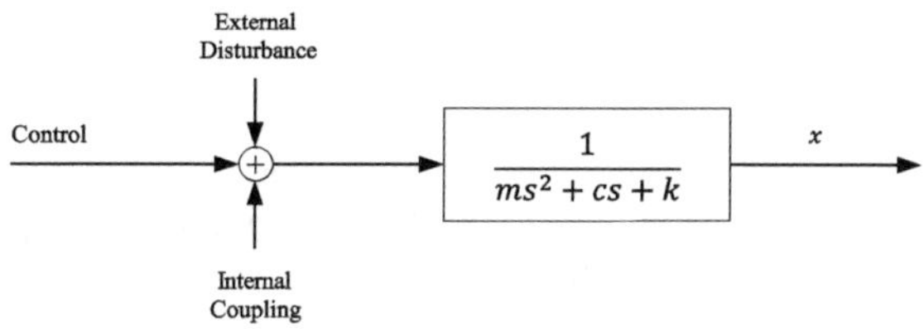

Fig. 8.2 Dynamic structural diagram of a controlled degree-of-freedom

To address these challenges, a linear model-following (MF) robust control scheme for both decoupling and disturbance rejection is introduced in this chapter; its closed-loop dynamic structure appears in Fig. 8.3. The controlled object, denoted by P, encompasses the actuation mechanisms and the targeted degrees-of-freedom. F_B is the closed-loop feedback gain, and C is the regulator that governs the overall control law. The reference model P_m serves as an ideal or desired performance benchmark, often selected to reflect the best operating point or an idealized configuration of the system. W is a dynamic compensator, while R is the set-point input. The combined control signal U enforces the closed-loop behavior. All internal cross-coupling and external disturbances are collectively represented by D. The state X_p corresponds to the displacement of the controlled degree-of-freedom, and X_m is the analogous variable within the reference model P_m.

For a single-degree-of-freedom in a mass–stiffness–damping system, the open-loop transfer function can often be approximated by

$$P = \frac{K_a}{m_p s^2 + c_p s + k_p}, \tag{8.1}$$

where K_a is the equivalent gain representing the amplification and actuation capability, and m_p, c_p, and k_p are the mass, damping, and stiffness parameters of the subsystem, respectively. An associated reference model that captures the ideal or nominal dynamics is given by

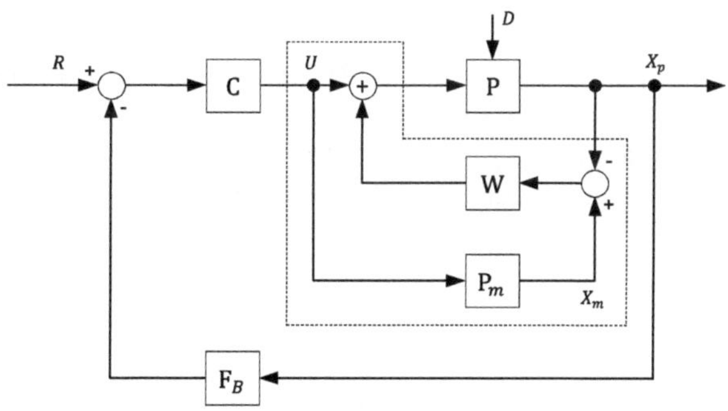

Fig. 8.3 Closed-loop system of MF control

$$\mathrm{P}_m = \frac{\mathrm{K}_a}{m_m s^2 + c_m s + k_m}, \tag{8.2}$$

where m_m, c_m, and k_m represent the mass, damping, and stiffness in the model space. For simplicity, P_m is assumed to have the same amplification and actuation gain as P. However, it may be assigned a different gain, denoted as K_{am}, if needed. Defining

$$\mathrm{W} = \mathrm{K}(c_m s + k_m), \tag{8.3}$$

we can interpret W as a PD compensator scaled by a gain K. Physically, W can be viewed as an "adjustment" to the nominal damping and stiffness (c_m and k_m) of the reference model, thereby allowing for flexible compensation around the set-point or operational regime of interest. The compensator W can also take the more general form:

$$\mathrm{W} = \mathrm{K}(c_c s + k_c),$$

where K is a scaling gain, c_c represents a damping-like coefficient, and k_c is a stiffness-like term. This form allows for additional flexibility in shaping the dynamic response of the system.

Building on this formulation, the chapter examines the robustness of the resulting closed-loop control laws to parameter variations and inter-subsystem couplings. For conceptual clarity, robustness to parameter variations and cross-coupling effects is addressed separately, though in practice they often co-occur. The analysis clarifies how the presented MF control structure sustains acceptable performance, even when real-world mass–stiffness–damping systems deviate from their nominal models or experience dynamic cross-couplings that fall outside the idealized assumptions.

8.2.1 *Robustness Against Nonlinear Perturbations of Parameters*

Based on Eqs. (8.1) and (8.2), let us assume that the deviations of the actual system from the nominal model arise from parameter perturbations Δc and Δk from the nominal damping c_m and stiffness k_m, while K_a and m_p remain identical to those in the nominal model. This can be expressed as

$$\begin{aligned} \mathrm{P} &= \frac{\mathrm{K}_a}{m_m s^2 + (c_m + \Delta c)s + (k_m + \Delta k)} \\ &= \frac{\mathrm{K}_a}{\left(m_m s^2 + c_m s + k_m\right) + (\Delta c s + \Delta k)}. \end{aligned} \tag{8.4}$$

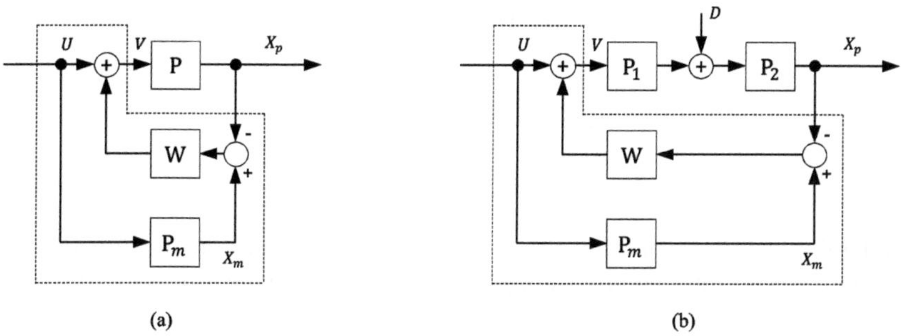

Fig. 8.4 Robustness analysis

Figure 8.4a shows that, for the open-loop system, the transfer function from the control input V to the output X_p is

$$\frac{X_p}{V} = \mathrm{P}. \tag{8.5}$$

When the MF compensator enclosed by the dashed lines is incorporated, the resulting transfer function from the control U to the system output X_p (accounting for the reference model P_m and a compensator W) becomes

$$\begin{aligned}\frac{X_p}{U} &= \frac{\mathrm{P}(1+\mathrm{WP}_m)}{1+\mathrm{PW}} = \frac{\mathrm{P}(1+\mathrm{WP}_m)}{(1+\mathrm{PW})\mathrm{P}_m}\mathrm{P}_m \\ &= \frac{1}{1+\frac{\Delta cs+\Delta k}{m_m s^2+(1+\mathrm{K}_a\mathrm{K})c_m s+(1+\mathrm{K}_a\mathrm{K})k_m}}\mathrm{P}_m.\end{aligned} \tag{8.6}$$

It can be seen that if the deviations Δc and Δk are relatively small, or if K is sufficiently large (while still ensuring closed-loop stability), then

$$\frac{X_p}{U} \approx \mathrm{P}_m, \tag{8.7}$$

meaning that the external behavior of the compensated system closely approximates that of the reference model P_m. Intuitively, this condition indicates robust tracking of the desired dynamics, despite the presence of moderate parametric uncertainty in the plant.

8.2.2 *Robustness Against Cross-Coupling and Disturbances*

At this stage, assume no nonlinear variations have entered the system, allowing us to set $\mathrm{P} = \mathrm{P}_m$. As illustrated in Fig. 8.4b, we define

$$\mathrm{P}_1 = \mathrm{K}_a,$$

and

$$\mathrm{P}_2 = \frac{1}{m_m s^2 + c_m s + k_m}.$$

Under these assumptions, the open-loop transfer function from the disturbance D (representing both internal cross-coupling and external disturbances) to the output X_p is

$$\frac{X_p}{D} = \mathrm{P}_2. \tag{8.8}$$

Once the MF elements enclosed by the dashed lines in Fig. 8.4b are introduced, the resulting disturbance-to-output path becomes

$$\begin{aligned} \frac{X_p}{D} &= \frac{\mathrm{P}_2}{1 + \mathrm{P}_2\mathrm{P}_1\mathrm{W}} \\ &= \frac{1}{m_m s^2 + (1 + \mathrm{K}_a\mathrm{K})c_m s + (1 + \mathrm{K}_a\mathrm{K})k_m}, \end{aligned} \tag{8.9}$$

where W is a compensator, commonly implemented to modify the effective damping and stiffness of the reference model, and K is its primary gain.

By comparing Eq. (8.9) with Eq. (8.8), it is evident that, as the gain K grows larger (within stable limits), the effect of internal cross-coupling and external disturbances on X_p diminishes considerably. Physically, this reflects how an increased gain in the model-based feedback pathway enhances decoupling and disturbance rejection, preventing unwanted oscillations from propagating between subsystems and reinforcing the overall stability of the mass–stiffness–damping system.

8.3 Decoupling Control of MKC Systems

In the previous section, we presented the linear MF robust control scheme in a frequency-domain form for illustrative simplicity. For practical engineering design and implementation, however, a time-domain state-space formulation is often more useful, particularly in modern control systems that leverage state-feedback and observer-based approaches. Figure 8.5 depicts the closed-loop structural diagram used in this time-domain analysis.

Consider a single-degree-of-freedom (1-DoF) in a mass–stiffness–damping system, with the open-loop state-space equations given by

Fig. 8.5 Closed-loop control

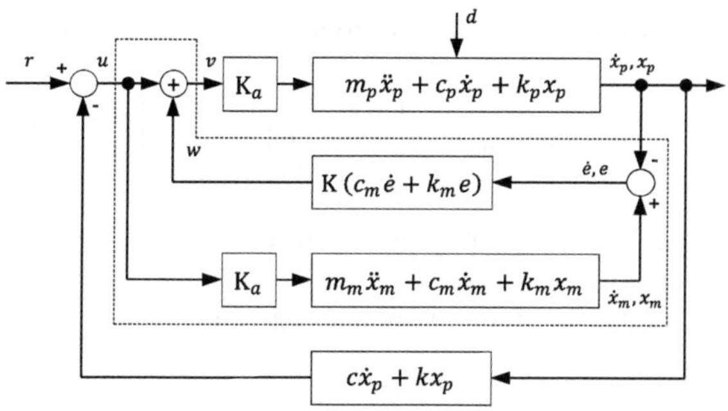

$$\begin{bmatrix} \vdots \\ \dot{x}_p \\ \ddot{x}_p \\ \vdots \end{bmatrix} = \begin{bmatrix} \ddots & \vdots & \vdots & \iddots \\ \cdots & 0 & 1 & \cdots \\ \cdots & -\frac{k_p}{m_p} & -\frac{c_p}{m_p} & \cdots \\ \iddots & \vdots & \vdots & \ddots \end{bmatrix} \begin{bmatrix} \vdots \\ x_p \\ \dot{x}_p \\ \vdots \end{bmatrix} + \begin{bmatrix} \vdots \\ 0 \\ \frac{\mathrm{K}_a}{m_p} \\ \vdots \end{bmatrix} u, \tag{8.10}$$

where x_p denotes the displacement of the controlled degree-of-freedom in the plant, and $\dot{x}_p$ represents the corresponding velocity. m_p, k_p, and c_p are the effective mass, stiffness, and damping of the DoF, respectively, and K_a is the equivalent actuator and amplifier gain.

The closed-loop control u is designed as a PD-type regulator plus an input term r, such that

$$u = \begin{bmatrix} -k & -c \end{bmatrix} \begin{bmatrix} x_p \\ \dot{x}_p \end{bmatrix} + r, \tag{8.11}$$

where $-k$ and $-c$ denote the proportional and derivative gains, respectively. In many rotor-bearing or other high-speed rotating systems, velocity sensors can be impractical due to contact-based designs. Consequently, velocity signals are often derived from displacement measurements via a velocity observer or differentiation algorithm.

A parallel reference model is introduced to represent the ideal or desired dynamics:

$$\begin{bmatrix} \dot{x}_m \\ \ddot{x}_m \end{bmatrix} = \begin{bmatrix} 0 & 1 \\ -\frac{k_m}{m_m} & -\frac{c_m}{m_m} \end{bmatrix} \begin{bmatrix} x_m \\ \dot{x}_m \end{bmatrix} + \begin{bmatrix} 0 \\ \frac{\mathrm{K}_a}{m_m} \end{bmatrix} u, \tag{8.12}$$

where $(x_m, \dot{x}_m)$ characterizes the target trajectory for displacement and velocity, while m_m, k_m, and c_m define the nominal mass, stiffness, and damping in the model space.

To align the plant output x_p with the reference-model output x_m, an error variable is introduced:

$$e = x_m - x_p, \tag{8.13}$$

along with an auxiliary compensator w, defined as

$$w = \mathrm{K}\begin{bmatrix} k_m & c_m \end{bmatrix}\begin{bmatrix} e \\ \dot{e} \end{bmatrix} = \mathrm{K}\begin{bmatrix} k_m & c_m \end{bmatrix}\begin{bmatrix} x_m \\ \dot{x}_m \end{bmatrix} - \mathrm{K}\begin{bmatrix} k_m & c_m \end{bmatrix}\begin{bmatrix} x_p \\ \dot{x}_p \end{bmatrix}, \tag{8.14}$$

where K is a compensator gain that effectively scales the nominal damping and stiffness of the system.

Combining these elements yields the final control signal v:

$$v = u + w = \begin{bmatrix} \mathrm{K}k_m & \mathrm{K}c_m \end{bmatrix}\begin{bmatrix} x_m \\ \dot{x}_m \end{bmatrix} - \begin{bmatrix} (k + \mathrm{K}k_m) & (c + \mathrm{K}c_m) \end{bmatrix}\begin{bmatrix} x_p \\ \dot{x}_p \end{bmatrix} + r. \tag{8.15}$$

Putting Eqs. (8.10)–(8.15) together, one can derive the state-space representation for the entire closed-loop system:

$$\begin{bmatrix} \vdots \\ \dot{x}_p \\ \ddot{x}_p \\ \vdots \\ \dot{x}_m \\ \ddot{x}_m \\ \vdots \end{bmatrix} = \begin{bmatrix} \ddots & \vdots & \vdots & \vdots & \vdots & \vdots & ⋰ \\ \cdots & 0 & 1 & \cdots & 0 & 0 & \cdots \\ \cdots & \frac{-[k_p + \mathrm{K}_a(k + \mathrm{K}k_m)]}{m_p} & \frac{-[c_p + \mathrm{K}_a(c + \mathrm{K}c_m)]}{m_p} & \cdots & \frac{\mathrm{K}_a\mathrm{K}k_m}{m_p} & \frac{\mathrm{K}_a\mathrm{K}c_m}{m_p} & \cdots \\ \cdots & \vdots & \vdots & \ddots & \vdots & \vdots & \cdots \\ \cdots & 0 & 0 & \cdots & 0 & 1 & \cdots \\ \cdots & -\frac{k}{m_m} & -\frac{c}{m_m} & \cdots & -\frac{k_m}{m_m} & -\frac{c_m}{m_m} & \cdots \\ ⋰ & \vdots & \vdots & \vdots & \vdots & \vdots & \ddots \end{bmatrix} \begin{bmatrix} \vdots \\ x_p \\ \dot{x}_p \\ \vdots \\ x_m \\ \dot{x}_m \\ \vdots \end{bmatrix} + \begin{bmatrix} \vdots \\ 0 \\ \frac{\mathrm{K}_a}{m_p} \\ \vdots \\ 0 \\ \frac{\mathrm{K}_a}{m_m} \\ \vdots \end{bmatrix} r, \tag{8.16}$$

where $(x_p, \dot{x}_p, x_m, \dot{x}_m)$ collectively define the augmented state vector. This representation highlights the interplay between the plant and the reference model, facilitated by the compensator w. Adjusting the gains k, c, K, and K_a then enables decoupling and robustness objectives, even when facing moderate parametric uncertainties, unmodeled modes, or other real-world complexities.

8.4 Case Analysis

As an example of evaluating decoupling and disturbance rejection robustness using the MF approach described in this chapter, we consider a rotor-bearing system at two distinct rotational speeds. At $n_1 \approx 10,600$ rpm (i.e., $\omega_1 \approx 1110$ rad/s), the system remains stable; however, at $n_2 \approx 12,720$ rpm (i.e., $\omega_2 \approx 1332$ rad/s), the open-loop system becomes unstable. We aim to demonstrate how the introduced MFcompensator maintains robust decoupling control under these conditions and how it enhances stability even when the rotational speed increases.

For convenience, a nondimensional model is also employed in the analysis. At the lower speed n_1, the nondimensional system matrix **A** of the open-loop system is shown in Fig. 4.2. When the rotational speed increases to n_2, the corresponding non-dimensional system matrix **A** is given in Fig. 8.6.

In mass–stiffness–damping systems, it is common to remove dimensional units by scaling the physical parameters with characteristic frequencies and forces. To handle speed variations in a dimensionless context, we adopt the conversion formulas below, where overbarred quantities (e.g., $\overline{m}, \overline{k}, \overline{c}$) represent nondimensional parameters. The formulas typically used for nondimensionalization are:

$$\overline{m} = m\omega^2 f_m, \tag{8.17a}$$

$$\overline{k} = kf_m, \tag{8.17b}$$

$$\overline{c} = c\omega f_m. \tag{8.17c}$$

Here, ω is the operating angular speed, and f_m is a chosen scaling factor. Substituting Eq. (8.17) into the ratios k/m, c/m, and $1/m$, we get:

$$\frac{k}{m} = \omega^2 \frac{\overline{k}}{\overline{m}}, \tag{8.18a}$$

$$\frac{c}{m} = \omega \frac{\overline{c}}{\overline{m}}, \tag{8.18b}$$

$$\frac{1}{m} = \omega^2 f_m \frac{1}{\overline{m}}. \tag{8.18c}$$

Thus, when comparing speeds ω_1 and ω_2, the nondimensional parameters must satisfy:

$$\frac{\omega_2^2 \left(\frac{\overline{k}}{\overline{m}}\right)_2}{\omega_1^2 \left(\frac{\overline{k}}{\overline{m}}\right)_1} = 1, \tag{8.19a}$$

$$\frac{\omega_2 \left(\frac{\overline{c}}{\overline{m}}\right)_2}{\omega_1 \left(\frac{\overline{c}}{\overline{m}}\right)_1} = 1, \tag{8.19b}$$

$$\frac{\omega_2^2 \left(\frac{1}{\overline{m}}\right)_2}{\omega_1^2 \left(\frac{1}{\overline{m}}\right)_1} = 1, \tag{8.19c}$$

leading to:

$$
\mathbf{A} = \left[\begin{array}{cccccccccccccccccccccccccc}
0 & 0 & 0 & 0 & 0 & 0 & 0 & 0 & 0 & 0 & 1 & 0 & 0 & 0 & 0 & 0 & 0 & 0 & 0 & 0 & 0 & 0 & 0 & 0 & 0 & 0 \\
0 & 0 & 0 & 0 & 0 & 0 & 0 & 0 & 0 & 0 & 0 & 1 & 0 & 0 & 0 & 0 & 0 & 0 & 0 & 0 & 0 & 0 & 0 & 0 & 0 & 0 \\
0 & 0 & 0 & 0 & 0 & 0 & 0 & 0 & 0 & 0 & 0 & 0 & 1 & 0 & 0 & 0 & 0 & 0 & 0 & 0 & 0 & 0 & 0 & 0 & 0 & 0 \\
0 & 0 & 0 & 0 & 0 & 0 & 0 & 0 & 0 & 0 & 0 & 0 & 0 & 1 & 0 & 0 & 0 & 0 & 0 & 0 & 0 & 0 & 0 & 0 & 0 & 0 \\
0 & 0 & 0 & 0 & 0 & 0 & 0 & 0 & 0 & 0 & 0 & 0 & 0 & 0 & 1 & 0 & 0 & 0 & 0 & 0 & 0 & 0 & 0 & 0 & 0 & 0 \\
0 & 0 & 0 & 0 & 0 & 0 & 0 & 0 & 0 & 0 & 0 & 0 & 0 & 0 & 0 & 1 & 0 & 0 & 0 & 0 & 0 & 0 & 0 & 0 & 0 & 0 \\
0 & 0 & 0 & 0 & 0 & 0 & 0 & 0 & 0 & 0 & 0 & 0 & 0 & 0 & 0 & 0 & 1 & 0 & 0 & 0 & 0 & 0 & 0 & 0 & 0 & 0 \\
0 & 0 & 0 & 0 & 0 & 0 & 0 & 0 & 0 & 0 & 0 & 0 & 0 & 0 & 0 & 0 & 0 & 1 & 0 & 0 & 0 & 0 & 0 & 0 & 0 & 0 \\
0 & 0 & 0 & 0 & 0 & 0 & 0 & 0 & 0 & 0 & 0 & 0 & 0 & 0 & 0 & 0 & 0 & 0 & 1 & 0 & 0 & 0 & 0 & 0 & 0 & 0 \\
0 & 0 & 0 & 0 & 0 & 0 & 0 & 0 & 0 & 0 & 0 & 0 & 0 & 0 & 0 & 0 & 0 & 0 & 0 & 1 & 0 & 0 & 0 & 0 & 0 & 0 \\
-1.0580 & 2.0393 & -1.6478 & 0.2594 & -0.0406 & 1.1349 & 0 & 0 & 0 & 0 & -2.2967 & 0 & 0 & 0 & 0 & -0.0003 & 0 & 0 & 0 & 0 & 0.0002 & -0.2982 & 0.2985 & 0 & 0 & 0 \\
1.0850 & -5.1050 & 5.6684 & -1.9534 & 0.3054 & 0 & 0 & 0 & 0 & 0 & 0 & 0 & 0 & 0 & 0 & 0 & 0 & 0 & 0 & 0 & 0 & 0 & 0 & 0 & 0 & 0 \\
-0.9822 & 6.3503 & -8.9227 & 4.6649 & -1.1091 & 0 & 0 & 0 & 0 & 0 & 0 & 0 & 0 & 0 & 0 & 0 & 0 & 0 & 0 & 0 & 0 & 0 & 0 & 0 & 0 & 0 \\
0.1247 & -1.7649 & 3.7623 & -3.2228 & 1.1008 & 0 & 0 & 0 & 0 & 0 & 0 & 0 & 0 & 0 & 0 & 0 & 0 & 0 & 0 & 0 & 0 & 0 & 0 & 0 & 0 & 0 \\
-0.0507 & 0.7181 & -2.3277 & 2.8645 & -1.7481 & 0 & 0 & 0 & 0 & 1.4315 & 0 & 0 & 0 & 0 & -2.8962 & 0 & 0 & 0 & 0 & -0.0001 & 0 & 0 & 0 & 0.0002 & -0.3763 & 0.3764 \\
-1.1321 & 0 & 0 & 0 & 0 & -1.4801 & 2.0393 & -1.6478 & 0.2594 & -0.0406 & -0.0003 & 0 & 0 & 0 & 0 & -2.3453 & 0 & 0 & 0 & 0 & 0.5060 & 0.0970 & 0.0969 & 0 & 0 & 0 \\
0 & 0 & 0 & 0 & 0 & 1.0850 & -5.1050 & 5.6684 & -1.9534 & 0.3054 & 0 & 0 & 0 & 0 & 0 & 0 & 0 & 0 & 0 & 0 & 0 & 0 & 0 & 0 & 0 & 0 \\
0 & 0 & 0 & 0 & 0 & -0.9822 & 6.3503 & -8.9227 & 4.6649 & -1.1091 & 0 & 0 & 0 & 0 & 0 & 0 & 0 & 0 & 0 & 0 & 0 & 0 & 0 & 0 & 0 & 0 \\
0 & 0 & 0 & 0 & 0 & 0.1247 & -1.7649 & 3.7623 & -3.2228 & 1.1008 & 0 & 0 & 0 & 0 & 0 & 0 & 0 & 0 & 0 & 0 & 0 & 0 & 0 & 0 & 0 & 0 \\
0 & 0 & 0 & 0 & -1.3724 & -0.0507 & 0.7181 & -2.3277 & 2.8645 & -2.1922 & 0 & 0 & 0 & 0 & -0.0001 & 0 & 0 & 0 & 0 & -2.8328 & 0 & 0 & 0 & 0.6104 & 0.1224 & 0.1223 \\
2.0193 & 0 & 0 & 0 & 0 & -0.7320 & 0 & 0 & 0 & 0 & 2.0000 & 0 & 0 & 0 & 0 & 2.0797 & 0 & 0 & 0 & 0 & -1.0096 & 0 & 0 & 0 & 0 & 0 \\
1.2042 & 0 & 0 & 0 & 0 & 0.7016 & 0 & 0 & 0 & 0 & -0.3795 & 0 & 0 & 0 & 0 & 2.2258 & 0 & 0 & 0 & 0 & 0 & -0.5198 & 0 & 0 & 0 & 0 \\
-0.5591 & 0 & 0 & 0 & 0 & -1.2792 & 0 & 0 & 0 & 0 & 1.6197 & 0 & 0 & 0 & 0 & -1.5766 & 0 & 0 & 0 & 0 & 0 & 0 & -0.5221 & 0 & 0 & 0 \\
0 & 0 & 0 & 0 & 1.9237 & 0 & 0 & 0 & 0 & -0.7629 & 0 & 0 & 0 & 0 & 2.0000 & 0 & 0 & 0 & 0 & 1.9748 & 0 & 0 & 0 & -0.9619 & 0 & 0 \\
0 & 0 & 0 & 0 & 1.1998 & 0 & 0 & 0 & 0 & 0.6716 & 0 & 0 & 0 & 0 & -0.3486 & 0 & 0 & 0 & 0 & 2.2162 & 0 & 0 & 0 & 0 & -0.5047 & 0 \\
0 & 0 & 0 & 0 & -0.5737 & 0 & 0 & 0 & 0 & -1.2507 & 0 & 0 & 0 & 0 & 1.5876 & 0 & 0 & 0 & 0 & -1.5876 & 0 & 0 & 0 & 0 & 0 & -0.5060
\end{array}\right]
$$

Fig. 8.6 Nondimensional system matrix **A** (unstable)

$$\left(\frac{\overline{k}}{\overline{m}}\right)_2 = \left(\frac{\omega_1}{\omega_2}\right)^2 \left(\frac{\overline{k}}{\overline{m}}\right)_1, \tag{8.20a}$$

$$\left(\frac{\overline{c}}{\overline{m}}\right)_2 = \left(\frac{\omega_1}{\omega_2}\right) \left(\frac{\overline{c}}{\overline{m}}\right)_1, \tag{8.20b}$$

$$\left(\frac{1}{\overline{m}}\right)_2 = \left(\frac{\omega_1}{\omega_2}\right)^2 \left(\frac{1}{\overline{m}}\right)_1. \tag{8.20c}$$

These relationships ensure proper scaling of the inertia, damping, and stiffness terms across different speeds when analyzing stability and robust decoupling in a dimensionless framework.

By employing these nondimensional parameters, we can compare the decoupling performance of the system at ω_1 and ω_2. At ω_1, the rotor-bearing system remains stable in open loop, and the MF compensator is primarily tested for decoupling and disturbance rejection. At ω_2, where the open-loop system would otherwise be unstable, the approach demonstrates its capacity to maintain or restore stability via MF control laws.

This chapter explores how the presented scheme accommodates changes in rotational speed without substantial redesign of the controller. The nondimensional analysis highlights the physical meaning of the control gains and parameter relationships, ensuring that insights gained at one speed can be extrapolated to others with minimal recalibration.

8.4.1 *Decoupling Control Under Stable Conditions*

At the lower rotational speed n_1, the open-loop rotor-bearing system remains stable. The discussion below demonstrates how the system responds to impulse disturbances in specific degrees-of-freedom and how state feedback plus MF compensation can improve decoupling and disturbance rejection.

8.4.1.1 x_3 of the Rotor Under Unit Impulse Disturbance

(1) Open-loop scenario:

A unit impulse disturbance is applied to x_3 of the rotor. Figure 8.7a–d and g shows the resulting responses in x_3, x_2, x_1, and y_1, demonstrating intense vibrations across all degrees-of-freedom connected to x_3. Because y_1 initially diverged, Fig. 8.7g extends the time axis to capture its behavior until convergence begins. This behavior highlights the strong coupling that propagates the disturbance from one DoF to the rest of the system.

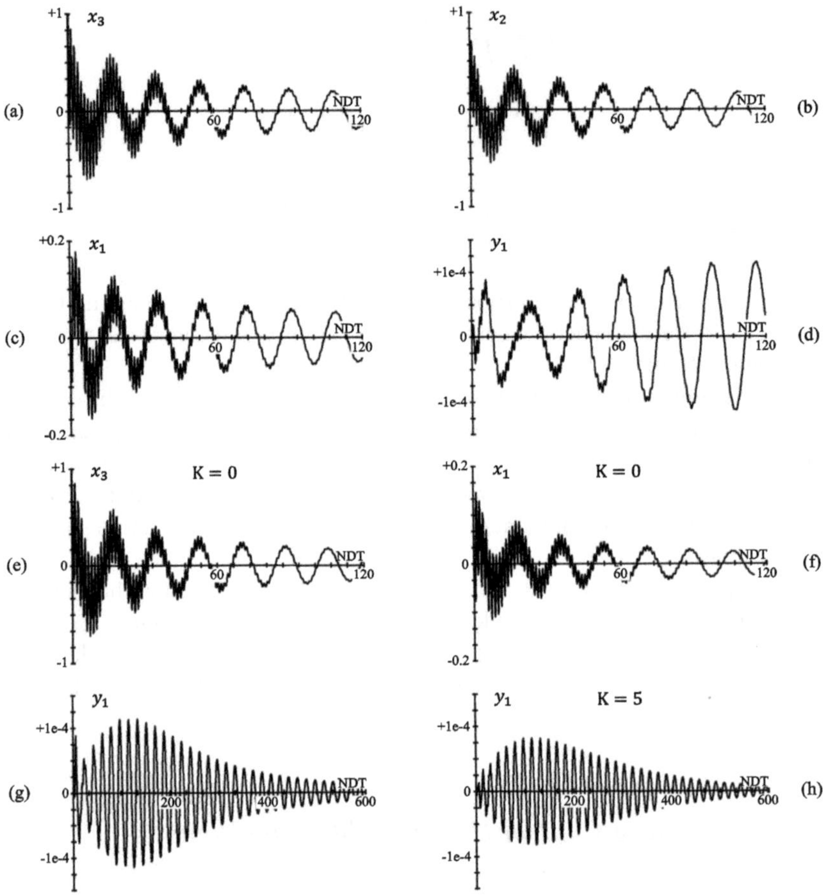

Fig. 8.7 x_3 of the rotor under unit impulse disturbance (at stable speed)

(2) x_1 under closed-loop state-feedback control:

To mitigate vibration in x_1, apply the state feedback:

$$u_1 = \begin{bmatrix} -1 & -1 \end{bmatrix} \begin{bmatrix} x_1 \\ \dot{x}_1 \end{bmatrix}. \tag{8.21}$$

As shown in Fig. 8.7f, this control partially suppresses the vibrations of x_1, illustrating the local effect of applying feedback to a single DoF. Figure 8.7e also indicates a small change in x_3 due to subtle feedback cross-couplings.

(3) x_1 under dual state-feedback and MF control:

Next, MF compensation is added to improve decoupling. From Eq. (8.10) and the system matrix **A** in Fig. 4.2, the subsystem for x_1 becomes

$$\begin{bmatrix} \dot{x}_1 \\ \ddot{x}_1 \end{bmatrix} = \begin{bmatrix} 0 & 1 \\ -1.4855 & -2.8528 \end{bmatrix} \begin{bmatrix} x_1 \\ \dot{x}_1 \end{bmatrix} + \begin{bmatrix} 0 \\ \times \end{bmatrix} u_1, \tag{8.22}$$

where the reference model is

$$\begin{bmatrix} \dot{x}_m \\ \ddot{x}_m \end{bmatrix} = \begin{bmatrix} 0 & 1 \\ -1.4855 & -2.8528 \end{bmatrix} \begin{bmatrix} x_m \\ \dot{x}_m \end{bmatrix} + \begin{bmatrix} 0 \\ \times \end{bmatrix} u_1. \tag{8.23}$$

A PD compensator is defined as

$$w = \mathrm{K}\begin{bmatrix} 1.4855 & 2.8528 \end{bmatrix} \begin{bmatrix} e \\ \dot{e} \end{bmatrix}, \tag{8.24}$$

where $e = x_m - x_1$.

Figures 8.8 and 8.9 illustrate how increasing K (i.e., 1, 2, 3, 4, 5, 10) affects x_3 and x_1 responses after a unit impulse disturbance in x_3. Compared with Fig. 8.7a and c, x_1 vibrations decrease as K grows, revealing robust decoupling and disturbance rejection properties; x_3 remains largely unaffected except for a phase shift.

Because only x_1 is in the closed control loop, other DoFs (e.g., x_3, y_1) are not actively controlled; therefore, their vibrations persist. Figure 8.7g and h confirms

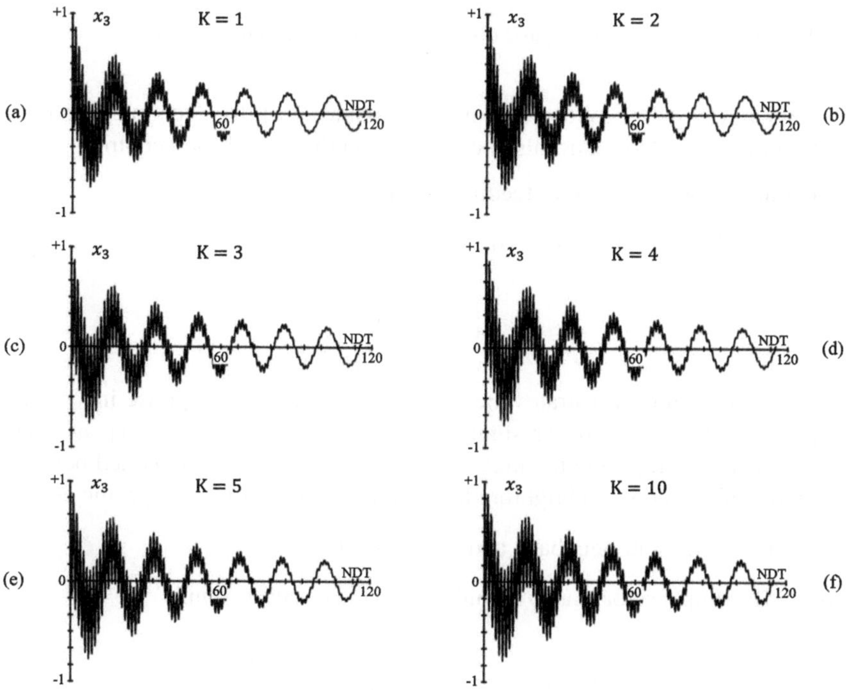

Fig. 8.8 x_3 responses: x_3 under impulse disturbance and x_1 under dual control

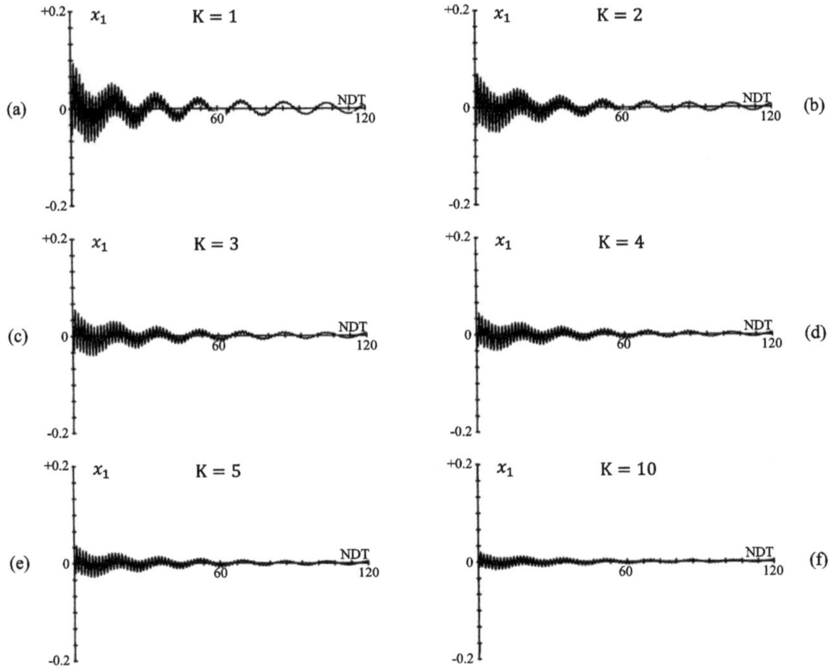

Fig. 8.9 x_1 responses: x_3 under impulse disturbance and x_1 under dual control

that the behavior of y_1 follows a similar rationale. To fully decouple or suppress vibration in x_3, that DoF must also be included in the closed-loop control.

(4) x_3 under closed-loop state-feedback control:

When the feedback is applied to x_3:

$$u_1 = \begin{bmatrix} -1 & -1 \end{bmatrix} \begin{bmatrix} x_3 \\ \dot{x}_3 \end{bmatrix}. \tag{8.25}$$

Figure 8.10a and b illustrates the improved disturbance response in x_3 and x_1. Compared to the open-loop or single-axis control applied only to x_1, as seen in Fig. 8.7a, c, e, and f, providing direct feedback on x_3 produces a marked benefit for that degree-of-freedom, highlighting the localized impact of DoF-specific control.

(5) x_3 under dual state-feedback and MF control:

Next, a MF compensator is also included for x_3. The subsystem is

$$\begin{bmatrix} \dot{x}_3 \\ \ddot{x}_3 \end{bmatrix} = \begin{bmatrix} 1 & 0 \\ -12.8571 & 0 \end{bmatrix} \begin{bmatrix} x_3 \\ \dot{x}_3 \end{bmatrix} + \begin{bmatrix} 0 \\ \times \end{bmatrix} u_1. \tag{8.26}$$

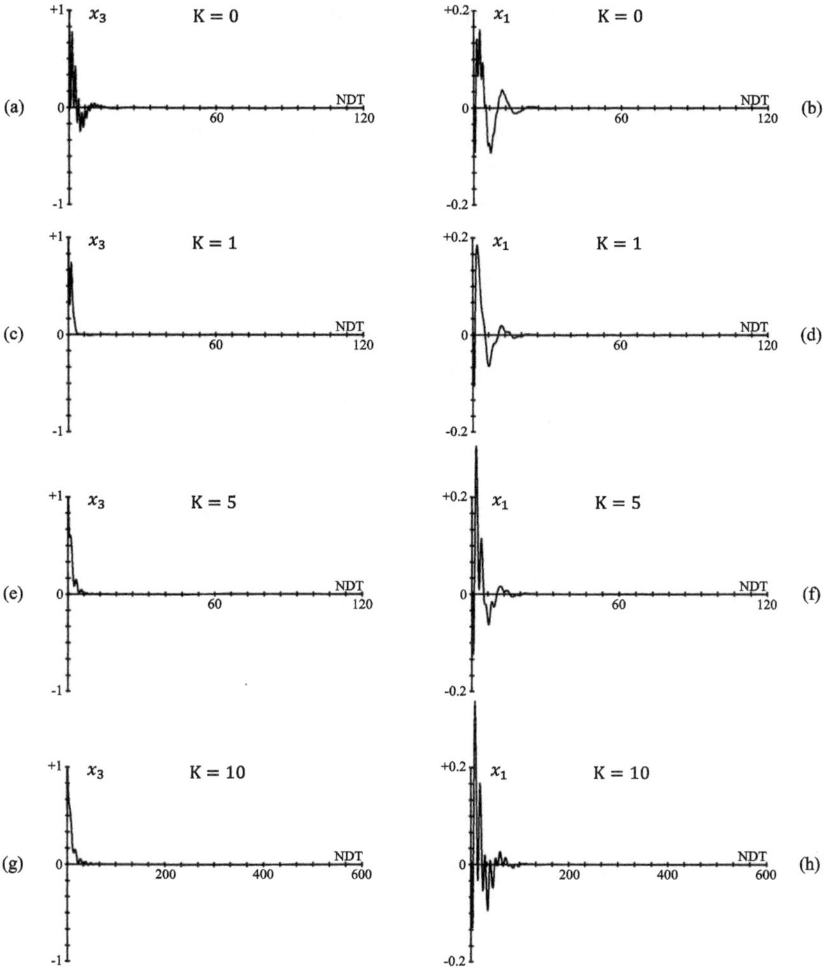

Fig. 8.10 x_3 under unit impulse disturbance with various control schemes

The reference model and PD compensator remain as in Eqs. (8.23) and (8.24), respectively. Figure 8.10c–h illustrates the impulse responses of x_3 and x_1 for K = 1, 5, 10. As K increases, x_3 benefits from greater vibration suppression, while x_1, being uncontrolled, experiences forced vibration driven by cross-coupling from x_3.

8.4.1.2 x_1 of the Rotor Subjected to a Step input

(1) Open-loop case:

Figure 8.11a and e shows the open-loop step responses of x_1 and x_3 when x_1 receives a step input $r_1(t) = 1$. Cross-coupling creates sustained coherent oscillations in both DoFs. For reference, Fig. 8.11a also plots x_m, the open-loop step response of the reference model.

(2) x_1 under closed-loop state feedback:

The control specified by Eq. (8.21) again applies here. Figure 8.11b and f shows reduced vibration amplitudes for x_1 and x_3, though residual oscillations remain due to partial coupling. For comparison, Fig. 8.11b also includes x_m under the same state feedback:

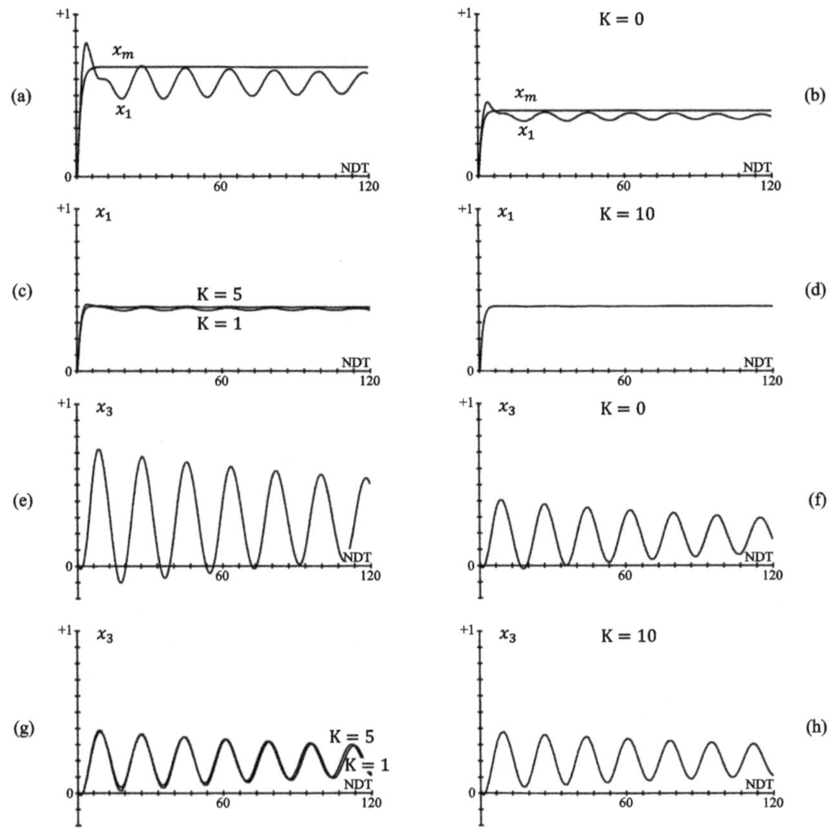

Fig. 8.11 x_1 of the rotor subjected to a step input (at stable speed)

$$u_{1m} = \begin{bmatrix} -1 & -1 \end{bmatrix} \begin{bmatrix} x_m \\ \dot{x}_m \end{bmatrix}. \tag{8.27}$$

(3) x_1 under dual state-feedback and MF control:

Implementing Eqs. (8.21)–(8.24), Fig. 8.11c, d, g, and h shows the step responses of x_1 and x_3 for K = 1, 5, 10. As K increases, x_1 approaches the reference model x_m more closely, illustrating that the closed-loop characteristics of the degree-of-freedom can be shaped by selecting a suitable reference model and adjusting K.

The persistent oscillations of x_3 in Fig. 8.11 arise partly from it being outside the control loop and partly from the PD compensator W not accounting for acceleration terms. Although acceleration compensation theoretically provides complete dynamic decoupling, practical implementation can be hindered by sensor noise, observer lag, and the inherent unreliability of physical differentiators.

(4) x_3 under dual state-feedback and MF control:

The same approach, in Eqs. (8.23)–(8.26), is applied to x_3. Figure 8.12 illustrates the step responses of x_1 and x_3 for K = 1 and K = 5. Controlling x_3 directly reduces its vibration amplitude, though a steady-state error remains due to the lack of integral action in the regulator. Whether integral control is beneficial for eliminating offset depends on the need for additional energy inputs and the overall stability requirements of the larger rotor-bearing system.

From Fig. 8.11a, it is clear that decoupling also benefits x_1, as controlling x_3 or vice versa reduces cross-coupled oscillations. In practice, combining multiple DoFs within closed-loop control is often essential for large rotor-bearing assemblies, where full multi-DoF decoupling is the key to robust, high-performance operation.

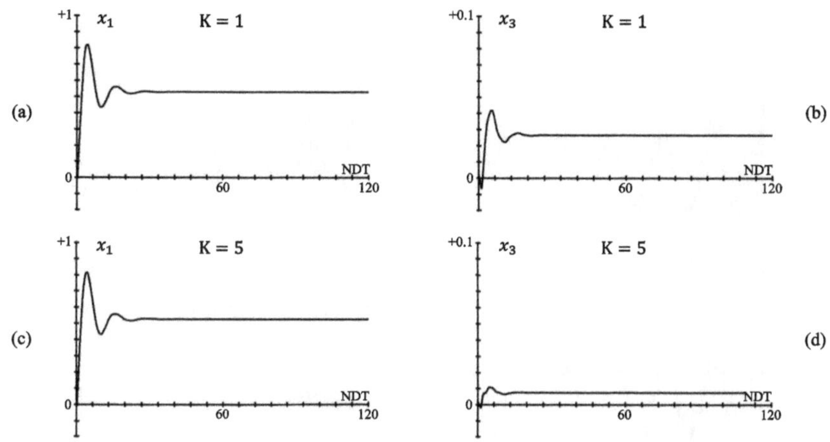

Fig. 8.12 x_1 subjected to step input with x_3 under dual control

8.4.2 Decoupling Control Under Unstable Conditions

At the higher rotational speed n_2, the open-loop rotor-bearing system becomes unstable. The following examples demonstrate how the MF compensation approach can suppress vibrations and stabilize specific degrees-of-freedom, even under changing system conditions.

8.4.2.1 x_3 of the Rotor Impacted by a Unit Impulse Disturbance

(1) Open-loop scenario:

Figure 8.13a and b depicts the responses of x_3 and x_1 subjected to a unit impulse at x_3. Compared to Fig. 8.7a and c (which correspond to n_1), the stability of the system has deteriorated significantly, reflecting the open-loop instability at n_2.

(2) x_1 under dual state-feedback and MF control:

By referring to the system matrix **A** in Fig. 8.6 (valid at speed n_2), the subsystem for x_1 is represented by

$$\begin{bmatrix} \dot{x}_1 \\ \ddot{x}_1 \end{bmatrix} = \begin{bmatrix} 0 & 1 \\ -1.0580 & -2.2967 \end{bmatrix} \begin{bmatrix} x_1 \\ \dot{x}_1 \end{bmatrix} + \begin{bmatrix} 0 \\ \times \end{bmatrix} u_2. \tag{8.28}$$

From the non-dimensional relationships (Sect. 8.4) and the updated parameters, the closed-loop control, reference model, and PD compensator become:

$$u_2 = \begin{bmatrix} -0.6944 & -0.8333 \end{bmatrix} \begin{bmatrix} x_1 \\ \dot{x}_1 \end{bmatrix}, \tag{8.29}$$

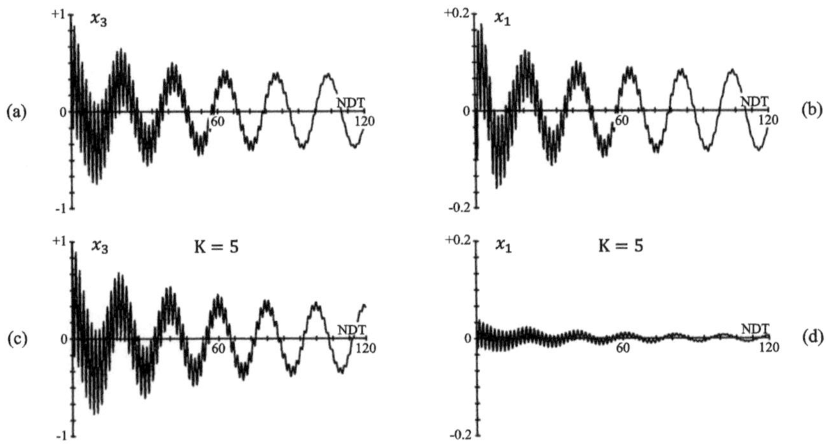

Fig. 8.13 x_3 of the rotor under unit impulse disturbance (at unstable speed)

$$\begin{bmatrix} \dot{x}_m \\ \ddot{x}_m \end{bmatrix} = \begin{bmatrix} 0 & 1 \\ -1.0315 & -2.3772 \end{bmatrix} \begin{bmatrix} x_m \\ \dot{x}_m \end{bmatrix} + \begin{bmatrix} 0 \\ \times \end{bmatrix} u_2, \tag{8.30}$$

and

$$w = \mathrm{K}\begin{bmatrix} 1.0315 & 2.3772 \end{bmatrix} \begin{bmatrix} e \\ \dot{e} \end{bmatrix}. \tag{8.31}$$

Figure 8.13c and d shows the responses of x_3 and x_1 with $\mathrm{K} = 5$. Despite the open-loop instability of the system at n_2, the vibrations of x_1 are significantly suppressed. This outcome highlights the robustness of the introduced control scheme, which adapts effectively to parameter variations and shifting operating conditions.

8.4.2.2 x_1 of the Rotor Subjected to a Step input

(1) Open-loop scenario:

Figure 8.14a and e displays the open-loop step responses of x_1 and x_3, respectively, when x_1 undergoes a step input $r_2(t) = 0.6944$. Both responses begin diverging, underscoring the loss of stability at speed n_2.

(2) x_1 under closed-loop state feedback:

Applying the same control as in Eq. (8.29), Fig. 8.14b and f illustrates marginally improved stability and reduced vibration amplitude for x_1 and x_3. Although the system is no longer unbounded, oscillations remain due to partial coupling and the inherent complexity of the rotor-bearing dynamics.

(3) x_1 under dual state-feedback and MF control:

Combining Eqs. (8.29)–(8.31) leads to the same MF compensation approach. Figure 8.14c, d, g, and h illustrates the step responses for $\mathrm{K} = 1$ and $\mathrm{K} = 5$. As K increases, x_1 more closely follows the trajectory of the reference model, mirroring the success seen at n_1. This aligns with the stabilizing influence of the MF compensator, even though the open-loop system at n_2 is already unstable. Meanwhile, x_3 also experiences improved stability compared to its open-loop behavior.

Note that previous examples focused on directly controlling x_3. In practical rotor-bearing assemblies, the rotor itself is often not directly accessible. Nonetheless, from controllability and observability theory for mass–stiffness–damping systems, a quasi-closed-loop scheme using appropriate state observers can achieve similar results in controlling rotor DoFs. This typically involves designing state observers (for velocity and possibly acceleration) to replace direct sensors, at the expense of additional system complexity.

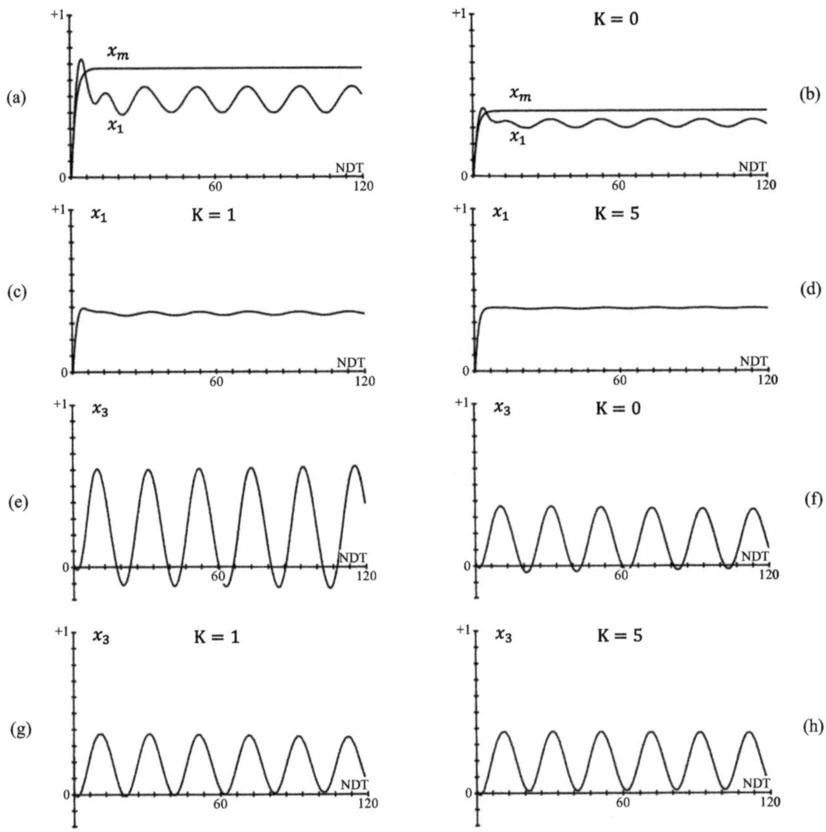

Fig. 8.14 x_1 of the rotor subjected to a step input (at unstable speed)

8.5 Summary

This chapter introduced a model-based decoupling control strategy for mass–stiffness–damping systems, using rotor-bearing systems as a representative example. By leveraging the inherent structural characteristics of these complex, multi-degrees-of-freedom (m-DoF) systems, the scheme was analyzed in three scenarios: open-loop, closed-loop state feedback, and state feedback enhanced with MF compensation. The key insights are as follows:

- *Decoupling and Disturbance Rejection:* For degrees-of-freedom included in the control loop, cross-coupling effects and external disturbances can be effectively mitigated. This highlights the ability of the method to suppress interfering vibrations and achieve localized stabilization.

- *Robustness to Parameter Variations:* The MF compensation scheme maintains favorable dynamic characteristics when the parameters of the plant change under varying operating conditions. Its sensitivity to parametric uncertainties remains low, reflecting robust performance.
- *Stabilization of Uncontrolled DoFs:* Even in the presence of uncontrolled subsystems, the feedback applied to a specific degree-of-freedom can indirectly stabilize other interconnected DoFs through reduced cross-coupling.

Overall, this model-based decoupling control approach proves both robust and practical, relying on relatively straightforward implementations and design procedures. It effectively accomplishes decoupling and stabilization of multi-degrees-of-freedom (m-DoF) systems while retaining an engineering-oriented outlook. Additionally, for strongly nonlinear mass–stiffness–damping systems with highly variable parameters, adaptive mechanisms [11–13] can be integrated into the framework to achieve further improvements in control performance.

References

1. Liu CH (1983) General decoupling theory of multivariable process control systems. Springer, Berlin
2. Vukobratović MK, Stokić SM (1980) Choice of decoupled control law of large-scale mechanical systems. IFAC Proceed 13(6):181–192
3. Zhang SY, Chen CT (1985) Design of compensators for robust tracking and disturbance rejection. IEEE Trans Autom Control 30(7):684–686
4. Ozguler A, Eldem V (1985) Disturbance decoupling problems via dynamic output feedback. IEEE Trans Autom Control 30(8):756–764
5. Albertos P, Sala A (2004) Multivariable control systems: an engineering approach. Springer, Berlin
6. Kučera V (2021) Decoupling with stability of linear systems by static-state feedback. IEEE Trans Autom Control 66(10):4684–4699
7. Salsa RG, Ma F (2021) Advances in the theory of system decoupling. Springer, Berlin
8. Arakelian V (2018) Dynamic decoupling of robot manipulators. Springer, Berlin
9. Moallem M, Patel RV, Khorasani K (2020) Flexible-link robot manipulators: control techniques and structural design. Springer, Berlin
10. Zhu HA (1988) Multivariable decentralized adaptive control of rotor-bearing systems. In: Proceedings of the 1988 IEEE international conference on systems, man, and cybernetics, vol 1, pp 215–218
11. Landau YD (1979) Adaptive control: the model reference approach. Marcel Dekker
12. Chen ZJ, Cook PA (1984) Robustness of model-reference adaptive control systems with unmodelled dynamics. Int J Control 39(1):201–214
13. Kidd PT (1985) Comparison of the performance of a model-reference adaptive system and a classical linear control system under non-ideal conditions. Int J Control 42(3):671–694

Chapter 9
Model-Based Disturbance Rejection and Uncertainty Attenuation

In Chap. 7, we established that employing pure integrators in the forward path—specifically, between the control input and the disturbance location—can attenuate steady-state errors induced by certain disturbances and parametric uncertainties. This approach, however, remains a passive technique, which can compromise transient performance. As a proactive alternative, this chapter introduces a model-based active disturbance rejection strategy that incorporates disturbance observation and feedforward compensation. By actively identifying and counteracting disturbance sources, it aims to achieve quicker adaptation to changing system conditions and more precise control of mass–stiffness–damping systems, ultimately enhancing both transient and steady-state behavior.

9.1 Introduction

In Chap. 6, the derivations of Algorithms 6.1 and 6.2 assumed no external disturbances $\boldsymbol{\tau}_D$ or other uncertainties. In real-world applications, however, such non-ideal effects can significantly degrade the performance of model-based control systems designed under idealized assumptions. Integral compensation, as discussed in Chap. 7, can mitigate the steady-state impact of constant disturbances or time-invariant uncertainties. Yet, this integral action operates passively, since the disturbance must manifest at the system output—such as the end-effector position of a robotic manipulator—before the integral compensation responds [1]. The transient error plots in Fig. 7.4 (Example 7.4) illustrate this effect, where corrective action only begins after the disturbance shifts the output away from the desired set-point or trajectory.

In control theory, unmeasurable disturbances with known signal characteristics can often be compensated asymptotically in the controlled variables. Generally, two approaches exist for compensator design [2]. The first uses a disturbance observer

H. Zhu, *Model-Based Control of Mass–Stiffness–Damping Systems*, Advances in Industrial Control, https://doi.org/10.1007/978-3-031-97592-9_9

[3–6], reconstructing the internal states of the disturbance and feeding them back so the disturbance is counteracted in the output of the plant. If tracking of a reference trajectory is also required, a reference observer can be integrated in parallel. The second approach, known as the Internal Model Principle [7–10], generalizes classical integral control to multivariable (MIMO) systems and more complex signal processes. It ensures robust disturbance rejection and robust reference accommodation by incorporating joint models of the assumed reference and disturbance signals into the augmented plant, driven by tracking errors. Subsequent observer design in this augmented setting remains as straightforward as in conventional state-feedback control [11, 12].

Robotic manipulators—a prime example of complex mass–stiffness–damping systems—commonly encounter external disturbances and internal parametric uncertainties. In this chapter, we focus on disturbance rejection and uncertainty attenuation in manipulators through disturbance observation and feedforward compensation. Here, any uncertainty that can effectively be treated as a disturbance also falls under our framework. The algorithms presented build on Chap. 6 and demonstrate a simple yet effective method of enhancing both modeling and control. Encouragingly, this approach substantially simplifies the overall control system design for complex, high-dimensional mass–stiffness–damping dynamics—ensuring robust performance even under non-ideal and changing operating conditions.

9.2 Disturbance Observation

A feedforward control term can significantly enhance performance in systems with known or estimable disturbances by effectively counteracting the disturbance input [13]. In essence, the disturbance is measured or estimated and then "canceled out" via an additional control signal. However, for nonlinear, cross-coupled mass–stiffness–damping systems (e.g., robotic manipulators), disturbances and parametric uncertainties are often unknown and unmeasurable. Consequently, an equivalent disturbance must first be determined, ideally using indirect methods that infer disturbances from measurable outputs.

Figure 6.2 illustrates a mass–stiffness–damping system that incorporates actuator dynamics, where the output $\boldsymbol{q}$ can be viewed as a function of time t, control torque $\boldsymbol{\tau}_C$, and exogenous disturbance $\boldsymbol{\tau}_D$. Formally, we may write:

$$\boldsymbol{q} = \mathcal{Q}(\boldsymbol{\tau}_C, \boldsymbol{\tau}_D, t). \tag{9.1}$$

This implies that an inverse relationship exists for $\boldsymbol{\tau}_D$ as:

$$\boldsymbol{\tau}_D = \mathcal{F}(\boldsymbol{q}, \dot{\boldsymbol{q}}, \ddot{\boldsymbol{q}}, \boldsymbol{\tau}_C, t). \tag{9.2}$$

In motor-actuated robotic manipulators, the control torque $\boldsymbol{\tau}_C$ is typically not measured directly. Instead, the armature current $\boldsymbol{i}_a$ is often available, and the two are related by Eq. (6.8). Substituting this relationship into Eq. (9.2) yields an alternative formulation for the disturbance term, denoted here as

$$\boldsymbol{\tau}_D = \mathcal{T}_D(\boldsymbol{q}, \dot{\boldsymbol{q}}, \ddot{\boldsymbol{q}}, \boldsymbol{i}_a, t). \tag{9.3}$$

Hence, if the joint positions $\boldsymbol{q}$, their derivatives $\dot{\boldsymbol{q}}$ and $\ddot{\boldsymbol{q}}$, and the armature current $\boldsymbol{i}_a$ are known or estimable, then $\boldsymbol{\tau}_D$ can be indirectly identified. Concretely, from Eq. (6.1) we obtain:

$$\hat{\boldsymbol{\tau}}_D = \left(\mathbf{N}\hat{\mathbf{K}}_T\right)\boldsymbol{i}_a - \left[\hat{\mathbf{D}}(\boldsymbol{q})\hat{\ddot{\boldsymbol{q}}} + \hat{\boldsymbol{h}}(\dot{\boldsymbol{q}}, \boldsymbol{q})\right], \tag{9.4}$$

implying that $\boldsymbol{\tau}_D$ is a function of time, armature currents, and the joint states. Once $\hat{\boldsymbol{\tau}}_D$ is estimated, a feedforward compensation signal can be introduced into the pre-compensating algorithms developed in Chap. 6, thereby actively rejecting disturbances and uncertainties. This approach effectively closes the loop on disturbance estimation and compensation, mitigating the passive drawbacks of pure integral action in highly coupled, nonlinear mass–stiffness–damping systems.

9.3 Feedforward Compensation

Building on the pre-compensating algorithms from Chap. 6, the forward-path signal flow in the control system can be illustrated as in Fig. 9.1. In this representation, $\mathbf{G}$ captures the equivalent dynamics from the control input $\boldsymbol{u}_c$ to the control torque $\boldsymbol{\tau}_C$, while $\boldsymbol{u}_f$ is an additional feedforward control designed to offset the impact of disturbance $\boldsymbol{\tau}_D$. If the resulting torque $\boldsymbol{\tau}$ is kept free of disturbance influences, then any subsequent states (e.g., the position output) remain unaffected by $\boldsymbol{\tau}_D$.

From Fig. 9.1, it follows that the forward-path output torque is

$$\boldsymbol{\tau} = \mathbf{G}\left(\boldsymbol{u}_c + \boldsymbol{u}_f\right) - \boldsymbol{\tau}_D, \tag{9.5}$$

indicating that $\boldsymbol{\tau}$ depends on the original control $\boldsymbol{u}_c$, the feedforward term $\boldsymbol{u}_f$, and the disturbance $\boldsymbol{\tau}_D$. In an ideal, disturbance-free scenario ($\boldsymbol{\tau}_D = 0, \boldsymbol{u}_f = 0$), one obtains

$$\boldsymbol{\tau} = \mathbf{G}\boldsymbol{u}_c \;\text{ for }\; \boldsymbol{\tau}_D = 0 \;\text{ and }\; \boldsymbol{u}_f = 0, \tag{9.6}$$

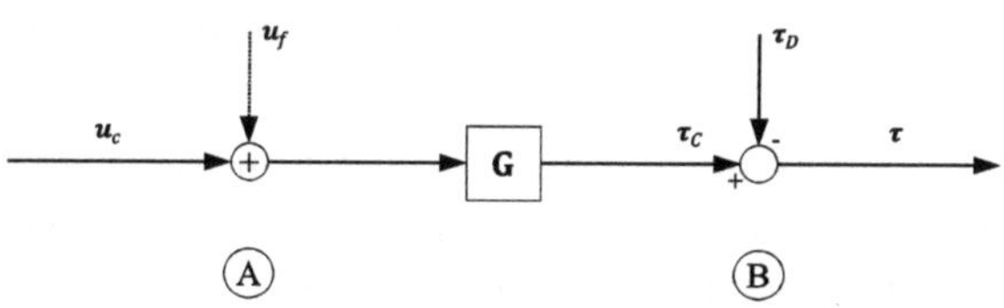

Fig. 9.1 Signal flow in the pre-compensated systems

meaning the torque is controlled solely by $\boldsymbol{u}_c$. Notably, if $\boldsymbol{u}_f$ is chosen such that

$$\mathbf{G}\boldsymbol{u}_f = \boldsymbol{\tau}_D, \tag{9.7}$$

then $\boldsymbol{\tau}_D$ is effectively canceled out at the torque output $\boldsymbol{\tau}$, equivalently eliminating its effect on the rest of the system.

In what follows, this disturbance-rejection concept is integrated into the pre-compensating algorithms from Chap. 6. For clarity, it is assumed that the parameters of the mass–stiffness–damping system used in control design are sufficiently accurate; otherwise, large parameter mismatches may degrade performance. Uncertainties considered here are those which can be modeled or treated as equivalent disturbances. Under these conditions, feedforward compensation complements the existing pre-compensated structure, enhancing the ability of the control system to reject disturbances and maintain stable, high-performance operation.

9.3.1 Feedback Linearization with Disturbance Rejection

Disturbance observation and feedforward compensation have been employed in control systems for decades [14]. Nevertheless, as will be shown, the control algorithms introduced in this chapter not only enable disturbance rejection but also significantly simplify both modeling and control in complex mass–stiffness–damping systems.

Consider a system pre-compensated with Algorithm 6.1. From Algorithm 6.1, it follows that the dynamic transfer function $\mathbf{G}$ from the control input $\boldsymbol{u}_c$ to the control torque $\boldsymbol{\tau}_C$ is

$$\mathbf{G} = \mathbf{I}s^{-1}, \tag{9.8}$$

which represents a set of first-order pure integrators, indicating

$$\boldsymbol{\tau}_C = \int \boldsymbol{u}_c \mathrm{d}t. \tag{9.9}$$

Based on Eqs. (9.7) and (9.8), any feedforward control $\boldsymbol{u}_f$ designed to neutralize the disturbance $\boldsymbol{\tau}_D$ must satisfy:

$$\boldsymbol{u}_f := \hat{\dot{\boldsymbol{\tau}}}_D = \left(\mathbf{N}\hat{\mathbf{K}}_T\right)\hat{\dot{\boldsymbol{i}}}_a - \left\{\hat{\mathbf{D}}(\boldsymbol{q})\hat{\dddot{\boldsymbol{q}}} + \left[\hat{\dot{\mathbf{D}}}(\boldsymbol{q})\hat{\ddot{\boldsymbol{q}}} + \hat{\dot{\boldsymbol{h}}}(\dot{\boldsymbol{q}}, \boldsymbol{q})\right]\right\}. \tag{9.10}$$

Here, $\hat{\dot{\boldsymbol{\tau}}}_D$ represents an estimated disturbance, while $\hat{\dot{\boldsymbol{i}}}_a$ and $\hat{\dddot{\boldsymbol{q}}}$ indicate derivative estimates of the motor armature current $\boldsymbol{i}_a$ and joint acceleration $\ddot{\boldsymbol{q}}$, respectively. Substituting $\boldsymbol{u}_f$ directly into the "A" node in Fig. 9.1 cancels $\boldsymbol{\tau}_D$ at the torque output $\boldsymbol{\tau}$, thus rejecting the disturbance before it affects the system.

Moreover, by separating the terms in Eq. (9.10) and merging them with the original control inputs $\boldsymbol{u}_{bc}$ and $\boldsymbol{u}_c$ from Algorithm 6.1, one can streamline the computations. Specifically:

$$\boldsymbol{u}_{bc} := \left(\mathbf{N}\hat{\mathbf{K}}_T\right)^{-1}\boldsymbol{u}_c + \hat{\dot{\boldsymbol{i}}}_a, \tag{9.11}$$

and

$$\boldsymbol{u}_c := \hat{\mathbf{D}}(\boldsymbol{q})\big[\boldsymbol{u} - \hat{\dddot{\boldsymbol{q}}}\big]. \tag{9.12}$$

This reallocation of feedforward compensation implicitly achieves disturbance rejection while dramatically reducing the complexity of $\boldsymbol{u}_c$. The updated version of Algorithm 6.1—augmented with disturbance rejection—is summarized as Algorithm 9.1. It retains the same pre-compensating structure while incorporating a feedforward disturbance-cancelation mechanism that effectively neutralizes $\boldsymbol{\tau}_D$ at the source, thereby yielding improved control performance in mass–stiffness–damping systems.

Algorithm 9.1 By applying the following compensating control laws:

$$\boldsymbol{u}_a := \boldsymbol{u}_b + \big(\hat{\mathbf{K}}_E\mathbf{N}\big)\dot{\boldsymbol{q}},$$

$$\boldsymbol{u}_b := \hat{\mathbf{L}}_a\boldsymbol{u}_{bc} + \hat{\boldsymbol{R}}_a\boldsymbol{i}_a,$$

$$\boldsymbol{u}_{bc} := \left(\mathbf{N}\hat{\mathbf{K}}_T\right)^{-1}\boldsymbol{u}_c + \hat{\dot{\boldsymbol{i}}}_a,$$

$$\boldsymbol{u}_c := \hat{\mathbf{D}}(\boldsymbol{q})\big[\boldsymbol{u} - \hat{\dddot{\boldsymbol{q}}}\big],$$

the electromechanical dynamics of the mass–stiffness–damping system that incorporates actuator dynamics, as depicted in Fig. 6.2, can be effectively decoupled and linearized with robust disturbance rejection. This control strategy cancels the undesired nonlinear cross-couplings and compensates for external disturbances, resulting in a significantly simplified and more reliable overall control scheme.

In the derivation above, some estimated variables—such as the estimated third derivative of $\boldsymbol{q}$ ($\hat{\dddot{\boldsymbol{q}}}$) and the estimated first derivative of the armature current $\boldsymbol{i}_a$ ($\hat{\dot{\boldsymbol{i}}}_a$)—are used in place of the actual signals. This substitution is necessary because these high-order signals are typically not directly measurable in electromechanical systems.

To obtain the unmeasurable states, state observers [11, 12, 15] or alternative estimation methods [16–20] must be implemented. A major challenge in this context is that the separation principle—which allows for the independent design of observer and controller loops—is not generally applicable for nonlinear systems. Consequently, a comprehensive stability analysis that accounts for both the control law and the observer dynamics is required, and any inaccuracies in model parameters further complicate this task [21].

For simplicity, an approximate estimator for the velocity $\dot{\boldsymbol{x}}$ from the displacement $\boldsymbol{x}$ is employed, as given by [22]:

$$\hat{\dot{\boldsymbol{x}}} : \mathbf{T}_{pf}\frac{\mathrm{d}\hat{\dot{\boldsymbol{x}}}}{\mathrm{d}t} + \hat{\dot{\boldsymbol{x}}} = \frac{\mathrm{d}\boldsymbol{x}}{\mathrm{d}t}. \tag{9.13}$$

Here, $\hat{\dot{\boldsymbol{x}}} \in \mathbb{R}^n$ represents the estimated velocity, and $\mathbf{T}_{pf} \in \mathbb{R}^{n\times n}$ is a diagonal matrix of pre-filter time constants.

With this estimator, the derivative of the armature current, $\hat{\dot{\boldsymbol{i}}}_a$, can be obtained from $\boldsymbol{i}_a$ as well; similarly, the estimated third derivative $\hat{\dddot{\boldsymbol{q}}}$ can be obtained sequentially from $\hat{\dot{\boldsymbol{q}}}$ and $\hat{\ddot{\boldsymbol{q}}}$. In general, the time constants in $\mathbf{T}_{pf}$ must be chosen sufficiently small to minimize time delay and distortion in the state estimates; however, if they are too small, they may compromise system stability. In practice, this approach has been shown to perform effectively [22].

The performance and effectiveness of Algorithm 9.1 are demonstrated in the example that follows.

Example 9.1 In this example, we compare the performance of the new Algorithm 9.1, which incorporates active disturbance rejection, with that of Algorithm 6.1, which lacks such a mechanism. The evaluation is based on the trajectory-tracking performance of a two-link manipulator, as described in Sect. 2.4.2, with the motion task detailed in Appendix C.2.

The study considers three sources of disturbance: payload variation, external impulse disturbances, and external step disturbances. Payload variation is specified in Sect. 2.4.2, while the disturbances are comprehensively described in Appendix C.3. The resulting tracking errors are illustrated in Fig. 9.2 {a, c, e Algorithm 6.1. b, d, f, Algorithm 9.1. Dotted line: Upper link. Solid line: Lower link}, where charts (a) and (b) correspond to payload variation, (c) and (d) to impulse disturbance, and (e) and (f) to step disturbance.

The results clearly indicate that Algorithm 9.1 significantly attenuates the adverse effects of both payload variations and external disturbances through its disturbance rejection mechanism. Although a small residual error remains in Fig. 9.2b due to inaccuracies in state estimation, this error also highlights that payload variation is not merely an external disturbance; it acts as an internal perturbation that modifies the effective gain in the forward path of the system. This observation underscores the need for adaptive compensation strategies to address such internal uncertainties.

In multivariable control systems, disturbances may arise from both external sources and internal parameter variations. Conventional designs often assume disturbances to be purely external; however, payload variations in robotic systems can alter system dynamics by changing inertia and stiffness properties. The integration of a disturbance observer and feedforward compensation, as employed in Algorithm 9.1, is essential for mitigating these effects and achieving robust tracking performance.

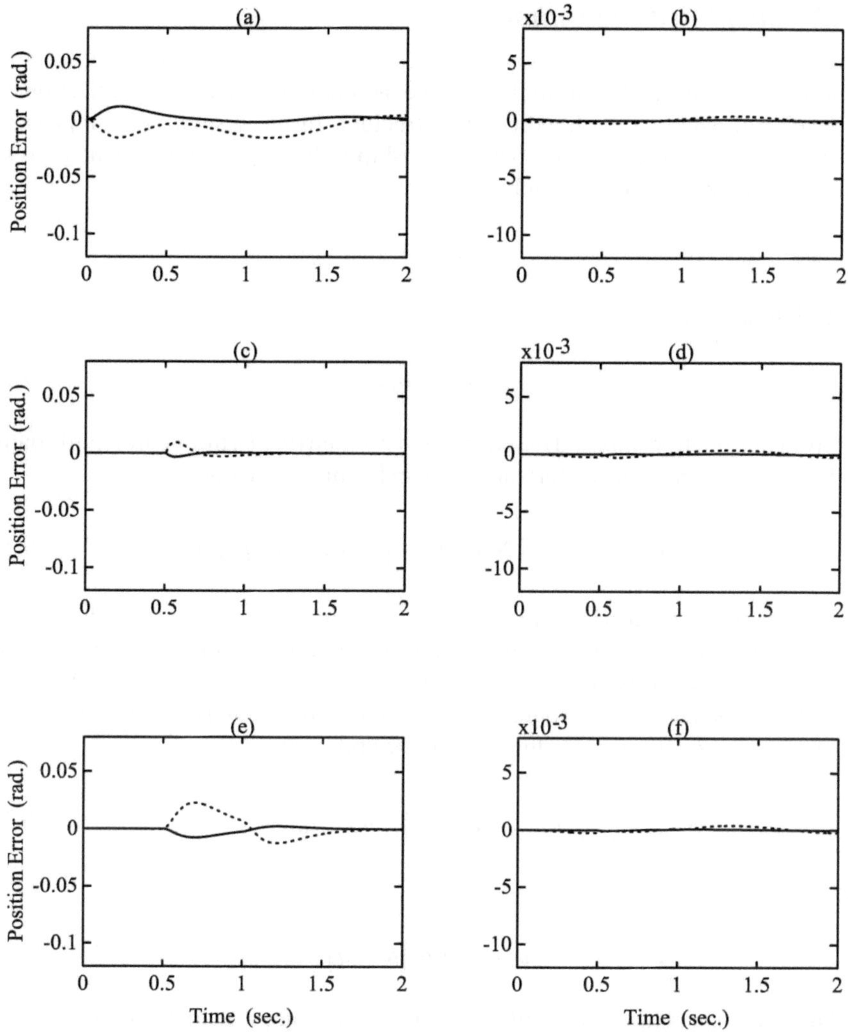

Fig. 9.2 Performance of Algorithm 9.1

Overall, the enhanced performance of Algorithm 9.1 demonstrates that an active disturbance rejection scheme can effectively improve the robustness and precision of model-based control systems for complex mass–stiffness–damping systems.

9.3.2 Simplified Algorithm

Building on Algorithm 6.2, a similar scheme incorporating disturbance rejection can be developed using a simplified control structure. In the system pre-compensated via Algorithm 6.2, the input–output relationship $\mathbf{G}$ from the control input $\boldsymbol{u}_c$ to the control torque $\boldsymbol{\tau}_C$ is approximately

$$\mathbf{G} \approx \mathbf{I}, \tag{9.14}$$

which implies that

$$\boldsymbol{\tau}_C \approx \boldsymbol{u}_c. \tag{9.15}$$

Given this relationship, and based on Eqs. (9.7) and (9.14), the feedforward control $\boldsymbol{u}_f$ designed to cancel the disturbance $\boldsymbol{\tau}_D$ can be directly defined as

$$\boldsymbol{u}_f := \hat{\boldsymbol{\tau}}_D = \left(\mathbf{N}\hat{\mathbf{K}}_T\right)\boldsymbol{i}_a - \left[\hat{\mathbf{D}}(\boldsymbol{q})\hat{\ddot{\boldsymbol{q}}} + \hat{\boldsymbol{h}}(\dot{\boldsymbol{q}}, \boldsymbol{q})\right]. \tag{9.16}$$

This feedforward term is critical because, in practical control systems, disturbance compensation relies on measurable or estimable signals rather than on idealized models. In many cases, a direct cancelation approach would involve splitting $\boldsymbol{u}_f$ into two components for computational simplicity. Accordingly, the original control signals $\boldsymbol{u}_{bc}$ and $\boldsymbol{u}_c$ from Algorithm 6.2 can be updated as follows:

$$\boldsymbol{u}_{bc} := \left(\mathbf{N}\hat{\mathbf{K}}_T\right)^{-1}\boldsymbol{u}_c + \boldsymbol{i}_a, \tag{9.17}$$

and

$$\boldsymbol{u}_c := \hat{\mathbf{D}}(\boldsymbol{q})\left[\boldsymbol{u} - \hat{\ddot{\boldsymbol{q}}}\right]. \tag{9.18}$$

The new control laws are significantly simplified compared to the original expressions—most notably, the complex term $\hat{\boldsymbol{h}}(\dot{\boldsymbol{q}}, \boldsymbol{q})$ is eliminated from the updated $\boldsymbol{u}_c$. This reduction in complexity decreases the computational burden while still effectively compensating for disturbances.

In many control systems, feedforward compensation is crucial for mitigating disturbances before they propagate through the closed loop. Designing an effective feedforward controller generally requires a precise understanding of the plant dynamics. In this context, the approximation $\mathbf{G} \approx \mathbf{I}$ indicates that the dynamics from $\boldsymbol{u}_c$ to $\boldsymbol{\tau}_C$ can be modeled as pure unity-gains, which greatly simplifies the feedforward design. This is particularly beneficial when disturbances arise both externally and internally due to parametric uncertainties. By decoupling these effects, the presented approach provides a streamlined and computationally efficient solution for robust disturbance rejection in mass–stiffness–damping systems that incorporate actuator dynamics.

In summary, by incorporating the compensating controls from Eqs. (9.14) through (9.18), the enhanced version of Algorithm 6.2 achieves effective decoupling, linearization, and disturbance rejection. The resulting algorithm is not only simpler in form but also maintains the robustness required for practical implementation in complex electromechanical systems.

Algorithm 9.2 By implementing the following compensating control laws:

$$\boldsymbol{u}_a := \boldsymbol{u}_b + \left(\hat{\mathbf{K}}_E\mathbf{N}\right)\dot{\boldsymbol{q}},$$

$$\boldsymbol{u}_b : \mathbf{T}_{pf}\frac{\mathrm{d}\boldsymbol{u}_b}{\mathrm{d}t} + \boldsymbol{u}_b = \hat{\mathbf{L}}_a\frac{\mathrm{d}\boldsymbol{u}_{bc}}{\mathrm{d}t} + \hat{\mathbf{R}}_a\boldsymbol{u}_{bc},$$

$$\boldsymbol{u}_{bc} := \left(\mathbf{N}\hat{\mathbf{K}}_T\right)^{-1}\boldsymbol{u}_c + \boldsymbol{i}_a,$$

$$\boldsymbol{u}_c := \hat{\mathbf{D}}(\boldsymbol{q})\left[\boldsymbol{u} - \hat{\ddot{\boldsymbol{q}}}\right],$$

the mass–stiffness–damping system that incorporates actuator dynamics, as depicted in Fig. 6.2, achieves approximate decoupling and linearization with effective disturbance rejection.

This strategy relies on accurate estimation of key state derivatives and leverages the unity-gain property of the pre-compensated system dynamics to simplify the feedforward compensation. As a result, the undesirable effects of disturbances and internal nonlinearities are largely canceled before they can propagate through the closed-loop system, yielding improved performance and robustness.

In order to validate the effectiveness of Algorithm 9.2, a representative example is presented below.

Example 9.2 In this example, the performance of the new Algorithm 9.2—which incorporates active disturbance rejection—is compared with that of Algorithm 6. 2, which lacks this mechanism. The evaluation is conducted under the same plant system, motion task, and non-ideal conditions as those described in Example 9.1. The tracking errors, displayed in Fig. 9.3 {a, c, e: Algorithm 6.2. b, d, f: Algorithm 9.2. Dotted line: Upper link. Solid line: Lower link} across different scales, correspond to three types of disturbances: payload variation [(a) and (b)], impulse disturbances [(c) and (d)], and step disturbances [(e) and (f)].

The results clearly demonstrate that Algorithm 9.2 effectively attenuates the adverse effects of payload variation and external disturbances. In particular, the active disturbance rejection mechanism in Algorithm 9.2 significantly improves transient and steady-state performance compared to Algorithm 6.2. Moreover, a comparison between Examples 9.1 and 9.2 reveals that, in the presence of disturbances, the

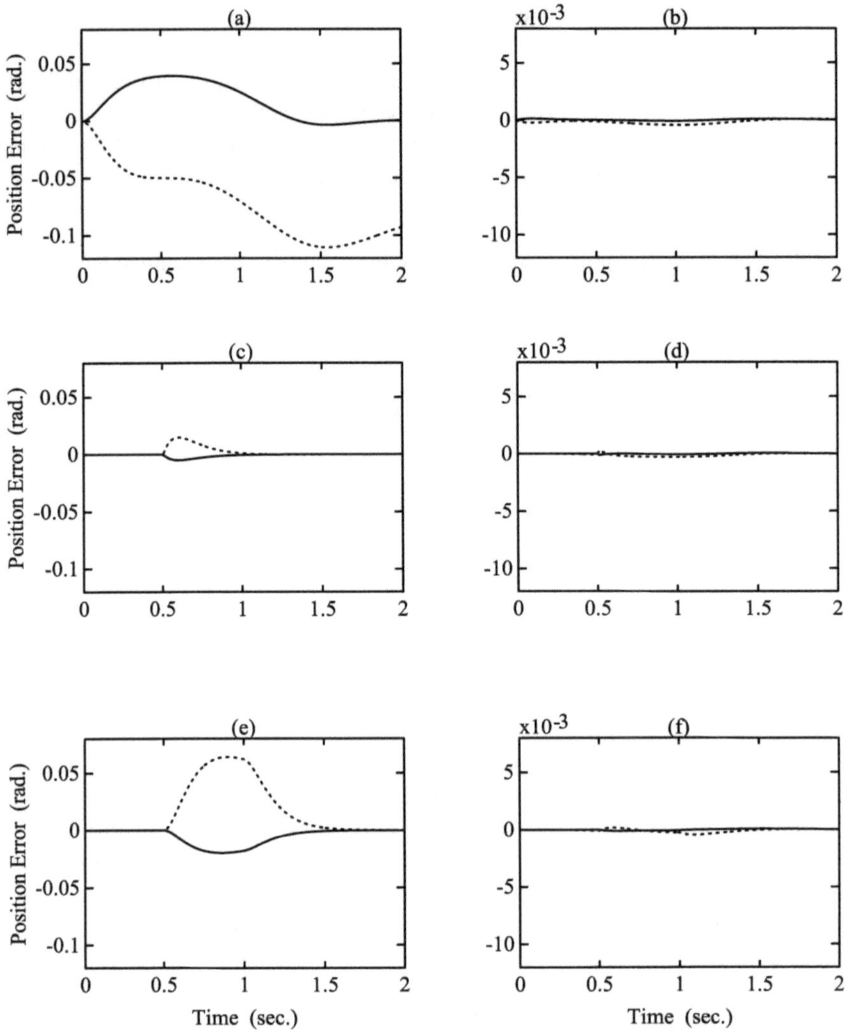

Fig. 9.3 Performance of Algorithm 9.2

performance of Algorithm 6.1 is superior to that of Algorithm 6.2, while Algorithm 9.2 achieves results comparable to those of Algorithm 9.1.

This outcome underscores the importance of integrating disturbance rejection mechanisms into control algorithms for complex mass–stiffness–damping systems. By actively compensating for disturbances, the new algorithm not only enhances robustness and stability but also reduces the overall sensitivity to parameter variations and external perturbations. Such improvements are critical in practical applications where even minor deviations from ideal conditions can significantly impact system performance.

Figure 9.4a {Solid line: Algorithm 6.1. Dashed line: Algorithm 9.1} and b {Solid line: Algorithm 6.2. Dashed line: Algorithm 9.2} shows the control torques $\boldsymbol{\tau}_C$ at the upper joint of the manipulator for step-disturbance rejection, as implemented in Examples 9.1 and 9.2, respectively. Analyzing these torque responses reveals that the improved disturbance rejection performance of the new algorithms is primarily due to the rapid responsiveness of the control torques. Specifically, when a disturbance occurs, the control torques adjust almost immediately, with increments nearly equal in magnitude to the disturbance. This fast response effectively cancels the effect of the disturbance on the system.

Such prompt torque adjustment is critical in applications where transient disturbances can adversely affect performance. The underlying principle is that a high-speed feedforward compensation mechanism, integrated with state feedback, minimizes the time delay between the onset of a disturbance and the corrective control action. This enhanced control performance not only stabilizes the system during

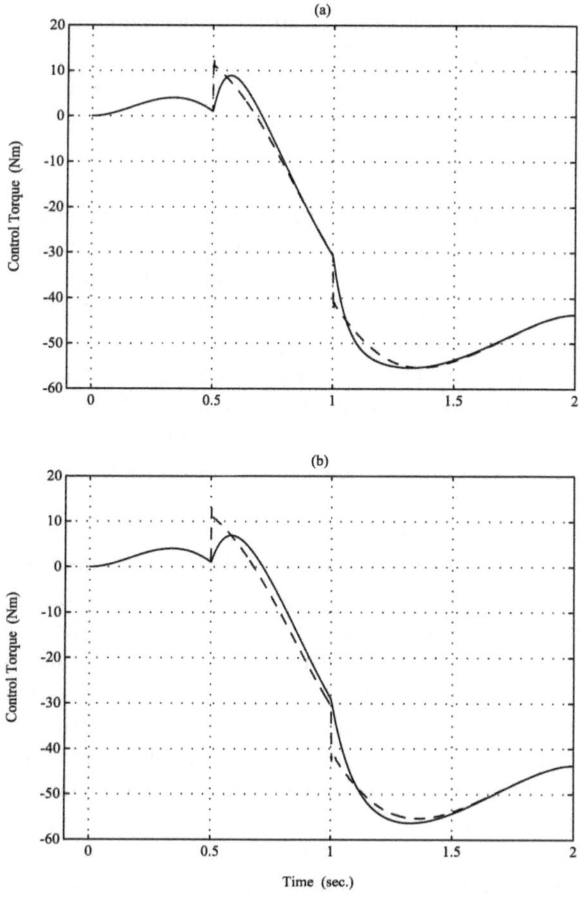

Fig. 9.4 Control torques for disturbance rejection

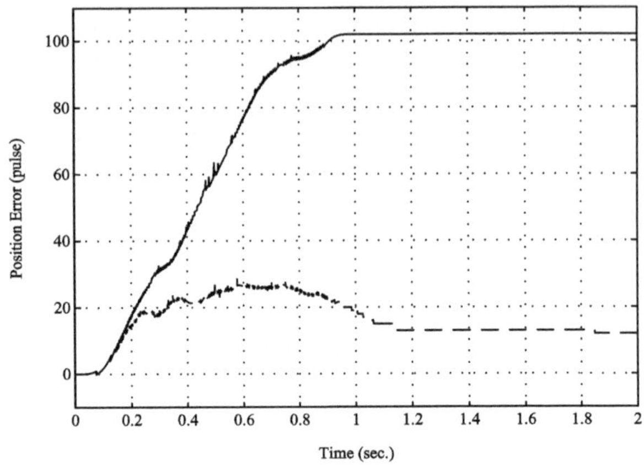

Fig. 9.5 Experimental results of disturbance rejection

disturbances but also contributes to the overall robustness of the model-based control strategy in complex mass–stiffness–damping systems.

To validate the performance of the disturbance rejection and uncertainty attenuation algorithms introduced in this chapter, a series of experiments are conducted on the testbed described in Appendix D.

Experiment 9.1

In this experiment, the manipulator is commanded to move from its initial lower vertical position $q_0 = -\pi/2$ to its final horizontal position $q_f = 0$ over a time interval of $\left[0, t_f\right] = [0, 1.0]$ seconds, following the trajectory specified in Eq. (C.1). The experimental results are shown in Fig. 9.5 {Solid line: Algorithm 6.2. Dashed line: Algorithm 9.2}.

For this experiment, joint friction is deliberately left uncompensated and gravity compensation is intentionally imprecise. These non-ideal effects—particularly the gravity torque—significantly degrade tracking performance and introduce large steady-state errors when using Algorithm 6.2, as indicated by the solid line in Fig. 9.5. By contrast, when Algorithm 9.2 is applied, the undesirable errors are substantially reduced (as seen by the dashed line). Notably, the experimental performance of Algorithm 9.1 is comparable to that of Algorithm 9.2.

The small residual error observed with Algorithm 9.2 can be attributed to modeling inaccuracies and imperfect state estimation. These findings underscore the critical importance of robust disturbance rejection and uncertainty attenuation strategies in model-based control of complex mass–stiffness–damping systems. In subsequent chapters, further enhancements to these algorithms will be explored to improve performance in the presence of practical non-idealities.

In real-world applications, mass–stiffness–damping systems encounter complex nonlinearities, higher-order dynamics, and sensor noise that make exact mathematical modeling impractical. Advanced control methods are thus necessary to meet stringent performance specifications, such as precise command following and effective disturbance rejection. This trade-off between modeling uncertainty and achievable performance drives the need for robust and adaptable controllers.

9.4 Summary

In this chapter, we explore advanced model-based control techniques for disturbance rejection and uncertainty attenuation in complex mass–stiffness–damping systems. Using robotic manipulators as a representative example, we show how integrating disturbance observation and feedforward compensation can effectively decouple the complex nonlinear dynamics of the system from uncertain components—such as friction $\boldsymbol{f}(\dot{\boldsymbol{q}}, \boldsymbol{q})$ and coupling effects in $\boldsymbol{h}(\dot{\boldsymbol{q}}, \boldsymbol{q}) = \left[\mathbf{C}(\dot{\boldsymbol{q}}, \boldsymbol{q})\dot{\boldsymbol{q}} + \boldsymbol{g}(\boldsymbol{q})\right] + \boldsymbol{f}(\dot{\boldsymbol{q}}, \boldsymbol{q})$. This approach not only enhances overall system robustness and performance but also significantly reduces modeling challenges and computational complexity.

In our derivation, the mechanical dynamics of nonlinear, cross-coupled mass–stiffness–damping systems—capturing both their essential dynamic cross-couplings and inherent nonlinearities—are expressed as

$$\mathbf{D}(\boldsymbol{q})\ddot{\boldsymbol{q}} + \boldsymbol{\tau}_d = \boldsymbol{\tau}_C, \tag{9.19}$$

with an equivalent disturbance defined by

$$\boldsymbol{\tau}_d = \boldsymbol{\tau}_D + \boldsymbol{h}(\dot{\boldsymbol{q}}, \boldsymbol{q}), \tag{9.20}$$

where $\boldsymbol{\tau}_D$ represents the uncertain external disturbance and $\boldsymbol{h}(\dot{\boldsymbol{q}}, \boldsymbol{q})$ captures the internal coupling effects

$$\boldsymbol{h}(\dot{\boldsymbol{q}}, \boldsymbol{q}) = \left[\mathbf{C}(\dot{\boldsymbol{q}}, \boldsymbol{q})\dot{\boldsymbol{q}} + \boldsymbol{g}(\boldsymbol{q})\right] + \boldsymbol{f}(\dot{\boldsymbol{q}}, \boldsymbol{q}). \tag{9.21}$$

By employing our disturbance rejection algorithms, these uncertain and complex factors are effectively decoupled from the controller design, resulting in a robust system that is much simpler to model and implement.

There are two main approaches to compensating for disturbances. One is the disturbance observation and feedforward compensation method, which is intuitively straightforward and relies on estimating the disturbance from measurable states. The other, the Internal Model Principle offers asymptotic rejection even in the presence of plant parameter variations but is typically more complex. Our model-based approach, while sensitive to modeling inaccuracies and state estimation dynamics, simplifies implementation and improves robustness. These sensitivities, as well as the residual dynamics from pole–zero cancelation in Algorithm 9.2, will be addressed further in later chapters, with enhanced feedforward compensation strategies discussed in Chap. 12.

In summary, the algorithms introduced in this chapter—rooted in the fundamental physics of cross-coupled electromechanical dynamics—effectively reject external disturbances and attenuate internal uncertainties in complex systems. By integrating an additional control structure specifically designed for disturbance rejection and uncertainty attenuation into conventional model-based control loops, overall system robustness is significantly enhanced. Furthermore, recent advances in robust control theory, including $\mathcal{H}^{\infty}$ and eigenstructure assignment techniques, demonstrate that effectively managing modeling uncertainties is an achievable goal. This underscores the importance of continued development in robust and adaptive control strategies for practical applications.

References

1. LeBlanc SE, Coughanowr DR (2009) Process systems analysis and control. McGraw-Hill
2. Tomizuka M, Chew KK, Yang WC (1990) Disturbance rejection through an external model. Trans ASME J Dyn Syst Measur Control 112(4):559–564
3. Johnson CD (1971) Accommodation of external disturbances in linear regulator and servomechanism problems. IEEE Trans Autom Control 16(6):635–644
4. Komada S, Ohnishi K (1990) Force feedback control of robot manipulator by the acceleration tracing orientation method. IEEE Trans Ind Electr 37(1):6–12
5. Umeno T, Hori Y (1991) Robust speed control of DC servomotors using modern two degrees-of-freedom controller design. IEEE Trans Ind Electr 38(5):363–368
6. Endo S, Kobayashi H, Kempf CJ, Kobayashi S, Tomizuka M, Hori Y (1996) Robust digital tracking controller design for high-speed positioning systems. Control Eng Pract 4(4):527–536
7. Davison EJ, Goldenberg A (1975) Robust control of a general servomechanism problem: the servo compensator. Automatica 11(5):461–471
8. Davison EJ (1976) The robust control of a servomechanism problem for linear time-invariant multivariable systems. IEEE Trans Autom Control 21(1):25–34
9. Hippe P (1992) Design of MIMO compensators for systems with unmeasurable disturbances: the polynomial approach. Automatica 28(5):1003–1009
10. Wonham WM (1985) Linear multivariable control: a geometric approach. Springer, Berlin
11. Luenberger DG (1971) An introduction to observers. IEEE Trans Autom Control 16(6):596–602
12. O'Reilly J (1983) Observers for linear systems. Academic Press
13. Chen CT (1993) Analog and digital control system design: transfer-function, state-space, and algebraic methods. Saunders College Publishing, HBJ International
14. Hashimoto M, Horiuchi T (1992) Advanced motion control implementation for industrial robots. In: Proceedings of the 1992 IEEE/RSJ international conference on intelligent robots and systems, vol 2, pp 943–949
15. Timchenko A, Kircanski N (1992) Control of robots with elastic joints: deterministic observer and Kalman filter approach. In: Proceedings of the 1992 IEEE international conference on robotics and automation, vol 1, pp 722–727
16. Nicosia S, Tomei P (1990) Robot control by using only joint position measurements. IEEE Trans Autom Control 35(9):1058–1061
17. de Wit CC, Fixot N (1991) Adaptive control of robot manipulators via velocity estimated feedback. In: de Wit CC (ed) Advanced robot control. Springer, Berlin, pp 69–82
18. Nicosia S, Tornambe A, Valigi P (1991) Observers in the control of rigid robots. In: de Wit CC (ed) Advanced robot control. Springer, Berlin, pp 273–284

19. Nicosia S, Tomei P (1992) State observers for rigid and elastic joint robots. Robot Comput Integr Manufact 9(2):113–120
20. Belanger PR, Dobrovolny P, Helmy A, Zhang X (1998) Estimation of angular velocity and acceleration from shaft-encoder measurements. Int J Robot Res 17(11):1225–1233
21. Doyle JC, Stein G (1979) Robustness with observers. IEEE Trans Autom Control 24(4):607–611
22. Spong MW, Vidyasagar M (1989) Robot dynamics and control. Wiley

Chapter 10
Enhanced Model-Following Control

Up to this point, the control algorithms have been designed under the ideal assumption that the employed mathematical models perfectly capture the system dynamics. In practice, however, uncertainties—arising from parameter variations, unmodeled dynamics, and external disturbances—are inevitable. Consequently, the performance of any control system depends critically on the robustness of its controller, which must sustain acceptable performance despite these discrepancies.

In this chapter, we shift our focus to enhancing the robustness of model-based controllers. We provide both theoretical insights and practical design guidelines for developing controllers that operate reliably in the face of model mismatches and unmeasurable disturbances. Robust control strategies are indispensable in real-world applications where inherent uncertainties can severely degrade performance. This discussion underscores the critical need for continued innovation in control engineering to develop resilient and adaptable control architectures capable of meeting the demands of modern dynamic systems.

10.1 Introduction

In this chapter, we tackle a fundamental challenge in control system design: achieving accurate output regulation for plants whose dynamics are subject to significant uncertainties. Although a wide range of techniques has been proposed to address plant uncertainty [1–3], our focus is on synthesizing robust controllers with fixed structures and parameters for linear or nearly linearized plants.

Robust control, as defined in this context, refers to the ability of a fixed controller to maintain acceptable performance in the presence of uncertainties—such as parameter variations, unmodeled dynamics, or external disturbances. Conventional robust design typically minimizes the sensitivity of the system to these uncertainties.

H. Zhu, *Model-Based Control of Mass–Stiffness–Damping Systems*, Advances in Industrial Control, https://doi.org/10.1007/978-3-031-97592-9_10

However, if the inherent model mismatch can be actively reduced through an uncertainty reduction mechanism, the overall performance of a model-based controller can be substantially improved. Despite significant advances in robust control theory [4, 5], the toolbox for direct uncertainty reduction remains relatively underdeveloped.

Most controller design methods are fundamentally model-based [6]. Thus, any discrepancy between the actual plant and the model used in the design process can adversely affect performance. To mitigate this issue, we present a straightforward scheme for uncertainty reduction that builds on the conventional model-following control structure [7]. An explicit and compact formulation for reducing modeling uncertainties is provided, showing that with such an uncertainty reducer, model-based controllers can be rendered robust against significant variations.

Subsequently, we demonstrate how this robustness enhancement scheme can be integrated into the model-based controller for robotic manipulators, resulting in a dual-model-based control system. Detailed case studies indicate that the enhanced system maintains robust performance even in the presence of substantial internal uncertainties and large external disturbances.

10.2 Uncertain Dynamic Systems

Robust control theory addresses the challenge of designing control systems that maintain accurate and stable performance despite significant plant uncertainties. To define this problem rigorously, it is essential to establish the class of plant models under consideration, the types of uncertainties affecting system behavior, and the performance measures used for analysis and synthesis.

For linear time-invariant (LTI) systems, three primary modeling frameworks are widely utilized in control engineering: state-variable models, transfer-function matrix models, and matrix-fraction models [8]. The state-space representation provides a dynamic formulation of system evolution through differential equations and is typically expressed as:

$$\dot{\boldsymbol{x}} = \mathbf{A}\boldsymbol{x} + \mathbf{B}\boldsymbol{u}, \tag{10.1}$$

$$\boldsymbol{y} = \mathbf{C}\boldsymbol{x} + \mathbf{C}_o\boldsymbol{u}, \tag{10.2}$$

where $\boldsymbol{x} \in \mathbb{R}^l$ is the state vector, $\boldsymbol{y} \in \mathbb{R}^{l_o}$ is the output vector, and $\boldsymbol{u} \in \mathbb{R}^{l_i}$ is the control input vector [8]. Alternatively, the transfer function matrix model characterizes system behavior in the frequency domain and is given by:

$$\mathbf{P}(s) = \mathbf{C}(s\mathbf{I} - \mathbf{A})^{-1}\mathbf{B} + \mathbf{C}_o, \tag{10.3}$$

where $\mathbf{P}(s)$ is an $(l_o \times l_i)$ matrix of rational functions in the Laplace transform variable s, representing the input–output relationship of the system [8]. Finally, the

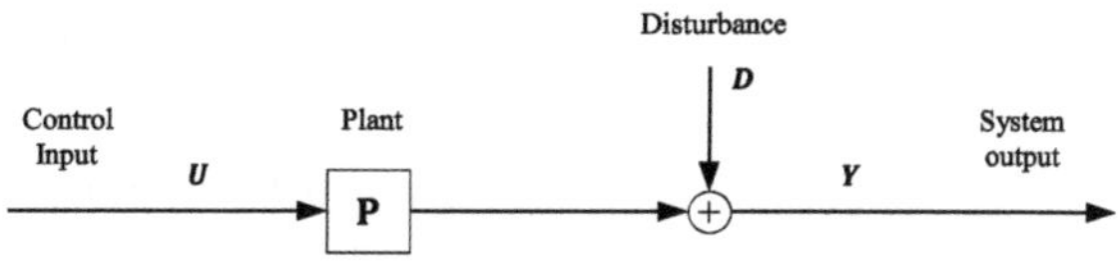

Fig. 10.1 General open-loop system

matrix-fraction representation, often used in robust control, employs right-coprime fraction factorization in the form:

$$\mathbf{P}(s) = \mathbf{N}(s)\mathbf{M}(s)^{-1}, \tag{10.4}$$

where $\mathbf{N}(s)$ and $\mathbf{M}(s)$ are matrices with polynomial or stable rational-function entries. Coprime factorization plays a crucial role in many robust control methodologies, particularly in $\mathcal{H}^{\infty}$ optimization, which is foundational to modern robust control theory [9, 10].

In subsequent discussions, transfer function matrix models will be employed for multi-input, multi-output (MIMO) systems, while transfer-function models will be used for single-input, single-output (SISO) systems. Consider the general control system depicted in Fig. 10.1, where $\mathbf{P}$ represents the dynamics of a stable plant with potential uncertainties, $\boldsymbol{U}$ is the control input, $\boldsymbol{Y}$ is the system output, and $\boldsymbol{D}$ is an equivalent disturbance acting at the output. The system dynamics are defined by:

$$\boldsymbol{Y} = \mathbf{P}\boldsymbol{U} + \boldsymbol{D}. \tag{10.5}$$

For analytical convenience, the plant $\mathbf{P}$ is assumed to be stable. If the system exhibits unstable modes that are controllable, stability can always be ensured through inner-loop pre-stabilization.

The impact of plant uncertainty and external disturbances on system performance is evident from Eq. (10.5). When a control input $\boldsymbol{U}$ is designed based on a nominal model of $\mathbf{P}$, any deviation from this model due to uncertainties or disturbances can prevent the system from achieving its desired performance. Common sources of uncertainty include modeling errors, parameter variations, and unmodeled dynamics, all of which contribute to discrepancies between the actual plant and its model.

Uncertainty can be classified in various ways, including parametric versus nonparametric, structured versus unstructured, and stochastic versus deterministic (set uncertainty). In the transfer function matrix framework, plant uncertainty is often modeled using a combination of multiplicative and additive uncertainty, expressed as follows [10–13]:

$$\mathbf{P} = [\mathbf{I} + \boldsymbol{\Delta}_m]\mathbf{P}_m + \boldsymbol{\Delta}_a, \tag{10.6}$$

where $\mathbf{P}_m$ represents the nominal model, $\boldsymbol{\Delta}_m$ accounts for multiplicative uncertainty, and $\boldsymbol{\Delta}_a$ represents additive uncertainty. In the frequency domain, these uncertainty terms are typically bounded by:

$$\|\mathbf{\Delta}_m(j\omega)\| < \delta_m(\omega), \tag{10.7}$$

$$\|\mathbf{\Delta}_a(j\omega)\| < \delta_a(\omega), \tag{10.8}$$

where $\|\bullet\|$ denotes a matrix norm. Generally, plant models exhibit higher accuracy at low frequencies, while uncertainty increases at higher frequencies due to unmodeled dynamics [14]. Structured uncertainty arises when only specific elements of $\mathbf{\Delta}_m$ and $\mathbf{\Delta}_a$ vary, whereas nonparametric models allow for the treatment of unmodeled system dynamics, providing greater flexibility in robust control design.

Stability is a fundamental requirement for control systems and is often treated as an implicit performance measure in control design. While ensuring stability, additional performance criteria such as tracking accuracy, disturbance rejection, and noise suppression must also be considered [15, 16]. To streamline the discussion, this book emphasizes performance comparison rather than explicit performance metrics.

The robust control problem considered here can now be precisely formulated. Given a linear or approximately linearized and stable plant with bounded uncertainties and a model-based linear controller designed for the nominal plant dynamics, the objective is to develop a controller that ensures robustness, maintaining the performance specified by the original model-based controller despite uncertainties.

Conventional robust control methodologies address uncertainty by designing controllers that are inherently resilient to plant variations. Classical techniques such as $\mathcal{H}^\infty$ control, adaptive control, and sliding-mode control embed robustness directly into the primary controller. However, an alternative strategy is introduced in this discussion: instead of incorporating robustness into the main controller, an additional control component actively mitigates uncertainty.

This alternative framework decouples performance optimization from uncertainty management. The primary controller retains its original performance objectives, while an auxiliary uncertainty-reduction mechanism operates in parallel to counteract disturbances and plant variations. This approach offers several advantages, including improved adaptability to time-varying uncertainties, reduced computational complexity compared to optimization-based robust control methods, and enhanced overall system robustness.

By actively reducing uncertainty rather than embedding it within the main controller, this approach enables computationally efficient robust control without requiring a complete redesign of the existing control structure. Even when system parameters deviate significantly from their nominal values, the closed-loop system remains stable and achieves the desired performance. The following chapters will explore the mathematical foundations, practical implementation details, and case studies demonstrating the effectiveness of this methodology. By shifting the focus from robust control design to active uncertainty reduction, this approach provides a streamlined yet highly effective strategy for managing uncertainties in control systems.

10.3 A Scheme for Uncertainty Reduction

To achieve uncertainty reduction, an additional structural component is introduced into the original uncertain system depicted in Fig. 10.1. The modified system, as illustrated in Fig. 10.2, incorporates a model $\mathbf{P}_m$ representing the nominal plant dynamics, a weighting function $\mathbf{W}$, a complementary control signal $\boldsymbol{U}_i$, and an enhanced control signal $\boldsymbol{U}_{ec}$. This architecture follows a model-following or model-reference paradigm, a well-established approach in adaptive and robust control, and is inherently model-based.

The complementary control signal is formulated as:

$$\boldsymbol{U}_i := \mathbf{W}(\mathbf{P}_m\boldsymbol{U} - \boldsymbol{Y}). \tag{10.9}$$

This control augmentation arises from the additional model-following structure enclosed within the dotted box in Fig. 10.2. The complementary control is injected into the original control pathway in the forward path, yielding the enhanced control:

$$\boldsymbol{U}_{ec} := \boldsymbol{U} + \boldsymbol{U}_i. \tag{10.10}$$

This simple yet effective modification improves robustness by actively reducing the impact of system uncertainties. The introduced framework is referred to in this book as the uncertainty reducer or the robustness enhancer, emphasizing its role in mitigating the adverse effects of modeling inaccuracies and external disturbances.

A fundamental requirement of any control system is stability, which must be ensured through appropriate controller design. Stability analysis often focuses on input–output stability, wherein a system is deemed stable if its output remains bounded for any bounded input. However, internal stability provides a stronger guarantee by ensuring that no unbounded internal signals exist within the closed-loop system [10]. Internal stability encompasses not only external disturbances but also the effects of unmodeled dynamics and parametric uncertainties.

The concept of internal stability is more comprehensive than the conventional stability analysis methods found in many control textbooks, which primarily assess stability by examining the roots of the characteristic equation [17]. While root-locus techniques and frequency–response methods such as Nyquist and Bode plots are

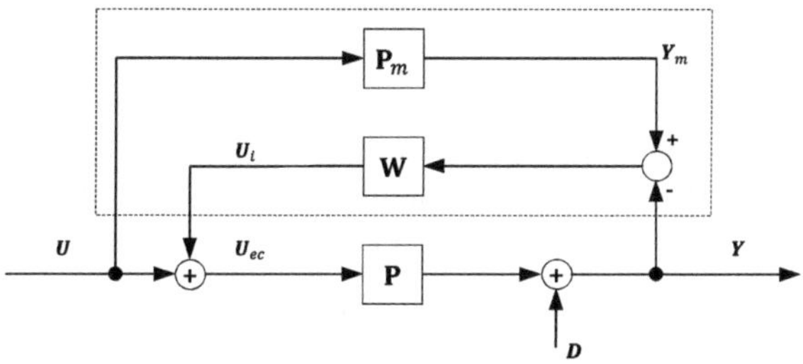

Fig. 10.2 Structure for uncertainty reduction

widely used to determine closed-loop stability, internal stability ensures robustness against disturbances and model variations across all system states. This distinction is particularly critical in model-following control architectures, where mismatches between the nominal model $\mathbf{P}_m$ and the actual plant $\mathbf{P}$ can introduce instability in the absence of careful design.

The inclusion of the uncertainty reducer within a model-following framework enhances stability by dynamically compensating for model mismatches in real time. The weighting function $\mathbf{W}$ plays a crucial role in shaping the dynamic response of the uncertainty compensation mechanism. Proper tuning of $\mathbf{W}$ ensures that the system maintains desirable performance characteristics while suppressing the effects of uncertainties. This structured approach to uncertainty reduction and robustness enhancement demonstrates the effectiveness of augmenting conventional model-based controllers with auxiliary compensation loops.

In the subsequent sections, the stability properties of this modified control structure will be examined in greater detail. Specific conditions for ensuring robust internal stability, as well as guidelines for selecting appropriate weighting functions and controller gains, will be discussed. By leveraging model-following principles and real-time uncertainty compensation, this approach offers a systematic and computationally efficient means of achieving robust control in uncertain environments.

Property 10.1 (Internal Stability) *The structure depicted in* Fig. 10.2 *is internally stable if its components* $\mathbf{P}$, $\mathbf{P}_m$, *and* $\mathbf{W}$, *along with the factors* $(\mathbf{I} + \mathbf{W}\mathbf{P}_m)$ *and* $(\mathbf{I} + \mathbf{P}\mathbf{W})^{-1}$, *are stable.*

Proof From Fig. 10.2, the system dynamics can be expressed as:

$$\boldsymbol{Y} = \mathbf{P}(\boldsymbol{U} + \boldsymbol{U}_i) + \boldsymbol{D}, \tag{10.11}$$

$$\boldsymbol{U}_i = \mathbf{W}(\boldsymbol{Y}_m - \boldsymbol{Y}), \tag{10.12}$$

$$\boldsymbol{Y}_m = \mathbf{P}_m \boldsymbol{U}, \tag{10.13}$$

which leads to:

$$\boldsymbol{Y} = (\mathbf{I} + \mathbf{P}\mathbf{W})^{-1}\mathbf{P}(\mathbf{I} + \mathbf{W}\mathbf{P}_m)\boldsymbol{U} + (\mathbf{I} + \mathbf{P}\mathbf{W})^{-1}\boldsymbol{D}, \tag{10.14}$$

provided that the inverses exist. Therefore, for bounded exogenous inputs $\boldsymbol{U}$ and $\boldsymbol{D}$, the internal stability of the components $\mathbf{P}$, $\mathbf{P}_m$, $\mathbf{W}$, and the factors $(\mathbf{I} + \mathbf{W}\mathbf{P}_m)$ and $(\mathbf{I} + \mathbf{P}\mathbf{W})^{-1}$ ensure the stability of $\boldsymbol{U}_i$, $\boldsymbol{Y}_m$, and $\boldsymbol{Y}$, thereby guaranteeing the internal stability of the structure in Fig. 10.2. □

The above property provides a useful criterion for assessing the stability of the enhanced control structure. However, since stability analysis is not the primary focus of this book, further discussion on this topic is omitted. For simplicity, it is assumed that all systems considered in this book are internally stable.

The structural characteristics and robustness of the framework in Fig. 10.2, particularly in relation to uncertainties and external disturbances, have been analyzed using sensitivity analysis methods [18, 19], as detailed in [7]. Sensitivity functions are widely used in control theory to quantify how system performance is affected by model uncertainties and external perturbations. These functions provide insight into how disturbances propagate through the system and how changes in plant dynamics impact closed-loop performance.

While traditional sensitivity analysis offers a systematic way to evaluate robustness, a more compact and intuitive formulation will be presented in the following sections. This alternative approach aims to provide direct and insightful results on how the additional structure enhances system robustness, effectively reducing the impact of uncertainties and disturbances on the closed-loop system.

10.3.1 Robustness Against Model Uncertainties

Various control methodologies enhance robustness against uncertainties such as parameter variations, external disturbances, and unmodeled dynamics. Conventional approaches rely on high-gain feedback, adaptive mechanisms, or robust optimization techniques, indirectly compensating for uncertainties rather than actively reducing them. This limitation highlights the need for a systematic framework that directly mitigates uncertainty at its source, improving control accuracy and efficiency.

The enhanced model-following scheme introduced in this chapter provides a direct mechanism for uncertainty reduction. By integrating a model-following structure with an additional compensation loop, it actively counteracts plant uncertainties at their origin. Instead of merely tolerating model discrepancies, the control system dynamically minimizes their impact, leading to greater precision and performance.

The following property demonstrates how this enhanced control structure offers a compact and analytically explicit formulation of uncertainty reduction. It quantifies the degree of uncertainty attenuation and provides a structured approach for designing controllers inherently resilient to plant variations and disturbances. The subsequent results establish a rigorous foundation for uncertainty mitigation, offering a powerful alternative to conventional robustness-driven control strategies.

Property 10.2 (Uncertainty Reduction) *With the structure in* Fig. 10.2, *both uncertainties,* $\mathbf{\Delta}_m$ *and* $\mathbf{\Delta}_a$, *in the plant system* $\mathbf{P}$,

$$\mathbf{P} = [\mathbf{I} + \mathbf{\Delta}_m]\mathbf{P}_m + \mathbf{\Delta}_a,$$

are attenuated by a factor of $(\mathbf{I} + \mathbf{PW})$.

Proof Equation (10.14) can be rewritten as.

$$Y = \mathbf{P}_c U + (\mathbf{I} + \mathbf{PW})^{-1} D, \tag{10.15}$$

where

$$\mathbf{P}_c = (\mathbf{I} + \mathbf{PW})^{-1} \mathbf{P}(\mathbf{I} + \mathbf{WP}_m), \tag{10.16}$$

represents the compensated plant, serving as the equivalent transfer function matrix of the enhanced control system. Expanding $\mathbf{P}_c$, we obtain

$$\begin{aligned}\mathbf{P}_c &= (\mathbf{I} + \mathbf{PW})^{-1}(\mathbf{P} + \mathbf{PWP}_m)\\ &= (\mathbf{I} + \mathbf{PW})^{-1}[(\mathbf{I} + \mathbf{PW})\mathbf{P}_m + \mathbf{P} - \mathbf{P}_m].\end{aligned}$$

Substituting the expression for the original plant $\mathbf{P}$ from Eq. (10.6), yields

$$\begin{aligned}\mathbf{P}_c &= (\mathbf{I} + \mathbf{PW})^{-1}\{(\mathbf{I} + \mathbf{PW})\mathbf{P}_m + [(\mathbf{I} + \boldsymbol{\Delta}_m)\mathbf{P}_m + \boldsymbol{\Delta}_a] - \mathbf{P}_m\}\\ &= (\mathbf{I} + \mathbf{PW})^{-1}\{(\mathbf{I} + \mathbf{PW})\mathbf{P}_m + \boldsymbol{\Delta}_m\mathbf{P}_m + \boldsymbol{\Delta}_a\}\\ &= (\mathbf{I} + \mathbf{PW})^{-1}\{[(\mathbf{I} + \mathbf{PW}) + \boldsymbol{\Delta}_m]\mathbf{P}_m + \boldsymbol{\Delta}_a\}.\end{aligned}$$

From this derivation, it follows that

$$\mathbf{P}_c = \left[\mathbf{I} + (\mathbf{I} + \mathbf{PW})^{-1}\boldsymbol{\Delta}_m\right]\mathbf{P}_m + (\mathbf{I} + \mathbf{PW})^{-1}\boldsymbol{\Delta}_a. \tag{10.17}$$

By comparing Eq. (10.17) with Eq. (10.6), it becomes evident that in the enhanced structure, both uncertainties, $\boldsymbol{\Delta}_m$ and $\boldsymbol{\Delta}_a$, are attenuated by a factor of $(\mathbf{I} + \mathbf{PW})$. □

The parallel model-following structure has long been employed in control theory, primarily to improve tracking performance by ensuring that the system output follows a reference model. However, in this book, it is utilized in a fundamentally different manner—as an uncertainty reducer. By actively modifying the response of the system, this structure forces the resulting compensated dynamics $\mathbf{P}_c$ to approximate the desired nominal dynamics $\mathbf{P}_m$, thereby minimizing the impact of plant uncertainties.

The key insight from the above analysis is that the model-following structure not only enhances tracking but also serves as a direct mechanism for uncertainty attenuation. Unlike conventional robust control approaches that seek to mitigate the effects of uncertainties through indirect measures such as high-gain feedback or conservative controller tuning, the enhanced structure explicitly reduces the plant uncertainties by a factor of $(\mathbf{I} + \mathbf{PW})$. This reduction is analytically compact and systematically quantifiable, making it a powerful tool for robustness enhancement in model-based control.

By leveraging this property, robustness enhancement becomes both structurally straightforward and computationally efficient. Rather than relying on complex adaptive schemes or robust optimization techniques, the introduced approach effectively strengthens the resilience of the system against model mismatches and disturbances with minimal effort. The explicit formulation of uncertainty reduction simplifies controller design while improving performance, ensuring that the compensated system closely adheres to its nominal model even in the presence of significant uncertainties.

This perspective underscores the broader applicability of the enhanced model-following framework. Beyond classical tracking applications, this structure provides a systematic foundation for achieving high-precision control in systems with inherent uncertainties, reinforcing its value as an integral component of modern robust control strategies.

10.3.2 Robustness Against External Disturbances

An immediate benefit of incorporating the enhanced model-following structure is that it not only mitigates internal uncertainties but also attenuates the effects of external disturbances. This disturbance rejection capability arises naturally from the additional feedback loop defined by $\mathbf{W}$, which dynamically adjusts the control input based on deviations between the model-following system and the actual plant.

The feedback action of $\mathbf{W}$ effectively modifies the response of the system to external disturbances, reducing their impact on the output. Specifically, as seen in Eq. (10.14), the disturbance term in the compensated system is scaled by $(\mathbf{I} + \mathbf{PW})^{-1}$, implying that the disturbance effects are inherently suppressed by the same factor that reduces the internal uncertainties. This structural property is particularly valuable in practical applications where systems are subjected to unknown and time-varying disturbances, as it enables enhanced robustness without requiring explicit disturbance estimation or compensation.

In classical robust control, disturbance rejection is typically achieved through either high-gain feedback or feedforward compensation schemes that rely on explicit disturbance models. However, these approaches often introduce design trade-offs, such as increased sensitivity to measurement noise or additional complexity in controller synthesis. The model-following structure, by contrast, provides a more integrated solution that simultaneously reduces uncertainty and disturbance effects without requiring extensive modifications to the original controller.

This result underscores the dual functionality of the enhanced model-following framework. Not only does it ensure that the compensated system maintains the desired nominal dynamics $\mathbf{P}_m$, but it also enhances disturbance attenuation, making the system more resilient to environmental perturbations. The following property formally establishes this effect and provides a foundation for further analysis of disturbance rejection in the enhanced control system.

Property 10.3 (Disturbance Reduction) *With the structure in* Fig. 10.2, *the influence of the external disturbance* $\boldsymbol{D}$ *in*

$$\boldsymbol{Y} = \mathbf{P}\boldsymbol{U} + \boldsymbol{D}$$

is attenuated by a factor of $(\mathbf{I} + \mathbf{PW})$, *reducing its impact on the output of the system.*

Proof For the original system, the output response is given by

$$\boldsymbol{Y} = \mathbf{P}\boldsymbol{U} + \boldsymbol{D}.$$

For the enhanced control structure, the modified system dynamics yield

$$\boldsymbol{Y} = \mathbf{P}_c\boldsymbol{U} + (\mathbf{I} + \mathbf{PW})^{-1}\boldsymbol{D},$$

where $\mathbf{P}_c$ is the compensated plant.

By comparing these expressions, it is evident that in the enhanced structure, the influence of the external disturbance $\boldsymbol{D}$ is attenuated by a factor of $(\mathbf{I} + \mathbf{PW})$, effectively reducing its impact on the system output. □

It is important to note that the attenuation of external disturbances in the enhanced structure is a secondary effect, rather than the primary objective of its design. Unlike traditional feedforward-based disturbance rejection strategies, which proactively compensate for disturbances by directly incorporating disturbance estimates into the control input, the attenuation achieved here is inherently passive. The reduction in disturbance influence arises naturally from the additional feedback loop introduced by the model-following structure, rather than through an explicit disturbance-cancelation mechanism.

However, this structure can still play a valuable role in scenarios where feedforward control is employed but suffers from model uncertainties. In feedforward-based disturbance rejection, the accuracy of disturbance compensation heavily depends on the precision of the system model used to predict and counteract disturbances. If the model exhibits discrepancies from the actual plant dynamics, the effectiveness of the feedforward compensation can be significantly compromised. In such cases, the model-following architecture can serve as a robustness enhancer by mitigating the residual effects of model uncertainty, thereby improving overall disturbance rejection performance.

This interplay between model-following structures and feedforward-based disturbance rejection becomes particularly relevant when dealing with uncertain or time-varying systems. A more detailed discussion on the integration of these approaches for enhanced robustness in disturbance rejection will be presented in Chap. 12.

10.3.3 *Design of the Robustness Enhancer*

Many advanced control methodologies, including the powerful $\mathcal{H}^\infty$ optimization framework, can be employed to design the uncertainty reducer presented above. $\mathcal{H}^\infty$ optimization provides a systematic approach to minimizing the worst-case impact of uncertainties and disturbances by shaping the closed-loop frequency response to achieve robust performance. This method is particularly effective in handling structured and unstructured uncertainties while ensuring stability and performance guarantees.

However, implementing such sophisticated techniques often requires significant computational resources and complex controller synthesis procedures. In practical applications, especially for real-time systems with limited computational capacity, a simpler yet effective approach is often preferable. Therefore, this book introduces a set of streamlined design methodologies for uncertainty reduction that maintain effectiveness while minimizing implementation complexity. These designs leverage fundamental principles of robust control, ensuring that the resulting system retains its resilience against uncertainties while remaining computationally feasible for real-world engineering applications.

Design Based on Inverse Dynamics

The compensated transfer function matrix $\mathbf{P}_c$ in Eq. (10.17) provides a structured approach for achieving robust performance by systematically reducing plant uncertainties. Ideally, perfect compensation should result in the desired system dynamics converging to the nominal model $\mathbf{P}_m$, as expressed by:

$$\mathbf{P}_c \to \mathbf{P}_m. \tag{10.18}$$

To achieve this, the weight $\mathbf{W}$ must be appropriately designed as a compensator. A straightforward approach is to define $\mathbf{W}$ as:

$$\mathbf{W} = \mathbf{P}_m^{-1}\mathbf{K}, \tag{10.19}$$

where $\mathbf{K}$ is a constant gain matrix. By substituting this compensator into the system dynamics, the compensated plant becomes:

$$\begin{aligned}
\mathbf{P}_c &= \left(\mathbf{I} + \mathbf{P}\mathbf{P}_m^{-1}\mathbf{K}\right)^{-1}\mathbf{P}\left(\mathbf{I} + \mathbf{P}_m^{-1}\mathbf{K}\mathbf{P}_m\right) \\
&= \left[\mathbf{P}^{-1}\left(\mathbf{I} + \mathbf{P}\mathbf{P}_m^{-1}\mathbf{K}\right)\right]^{-1}\left(\mathbf{P}_m^{-1} + \mathbf{P}_m^{-1}\mathbf{K}\right)\mathbf{P}_m \\
&= \left(\mathbf{P}^{-1} + \mathbf{P}_m^{-1}\mathbf{K}\right)^{-1}\mathbf{P}_m^{-1}(\mathbf{I} + \mathbf{K})\mathbf{P}_m \\
&= \left[\mathbf{P}_m\left(\mathbf{P}^{-1} + \mathbf{P}_m^{-1}\mathbf{K}\right)\right]^{-1}(\mathbf{I} + \mathbf{K})\mathbf{P}_m \\
&= \left[\left(\mathbf{P}_m\mathbf{P}^{-1} + \mathbf{K}\right)^{-1}(\mathbf{I} + \mathbf{K})\right]\mathbf{P}_m.
\end{aligned}$$

From this formulation, it is evident that if the gain matrix $\mathbf{K}$ is sufficiently large, then the term $\left(\mathbf{P}_m\mathbf{P}^{-1}+\mathbf{K}\right)^{-1}(\mathbf{I}+\mathbf{K})$ approaches identity,

$$\left[\left(\mathbf{P}_m\mathbf{P}^{-1}+\mathbf{K}\right)^{-1}(\mathbf{I}+\mathbf{K})\right] \to \mathbf{I}, \tag{10.20}$$

leading to:

$$\mathbf{P}_c \approx \mathbf{P}_m, \tag{10.21}$$

as an approximate result. Furthermore, if the ratio $\mathbf{P}\mathbf{P}_m^{-1}$ approaches identity, then the factor $(\mathbf{I}+\mathbf{P}\mathbf{W})$ simplifies to:

$$\mathbf{I}+\mathbf{P}\mathbf{W} = \mathbf{I}+\left(\mathbf{P}\mathbf{P}_m^{-1}\right)\mathbf{K} \approx (\mathbf{I}+\mathbf{K}). \tag{10.22}$$

This implies that when $\mathbf{P}\mathbf{P}_m^{-1} \approx \mathbf{I}$, uncertainties and disturbances in the system are attenuated by a factor of $(\mathbf{I}+\mathbf{K})$. This effect will be further substantiated through simulation results and experimental validation. Specifically, under ideal modeling conditions where $\mathbf{P}_m = \mathbf{P}$, the system response simplifies to:

$$\boldsymbol{Y} = \mathbf{P}_m\boldsymbol{U} + (\mathbf{I}+\mathbf{K})^{-1}\boldsymbol{D}, \tag{10.23}$$

which confirms that in the absence of modeling discrepancies, the external disturbance $\boldsymbol{D}$ is mitigated by a factor of $(\mathbf{I}+\mathbf{K})$ through the compensator $\mathbf{W}$ in Eq. (10.19).

Although the theoretical weight $\mathbf{W} = \mathbf{P}_m^{-1}\mathbf{K}$ offers mathematical elegance, it is often physically unrealizable for practical systems. Specifically, if an element $\mathrm{p}_m^{\mathrm{inv}}$ in $\mathbf{P}_m^{-1}$ has a numerator order exceeding its denominator order by n, then the resulting $\mathbf{W}$ would involve n-th order differentiation, which is mathematically valid but physically impractical. To address this, a pre-filter p_f is typically introduced to ensure realizability. A common choice for the pre-filter is:

$$\mathrm{p}_f = \left(T_{pf}s+1\right)^{-n}, \tag{10.24}$$

where T_{pf} is the time constant of the pre-filter and n is the order difference mentioned earlier. As an illustrative example, consider a plant model:

$$\mathrm{P}_m = \frac{c}{s(as+b)}. \tag{10.25}$$

In this case, the practical weight design becomes:

$$\mathrm{W} = \mathrm{P}_m^{-1}\mathrm{P}_f\mathrm{K} = \mathrm{K}\frac{s(as+b)}{c\left(T_{pf1}s+1\right)\left(T_{pf2}s+1\right)}, \tag{10.26}$$

which ensures that differentiation terms are properly conditioned by the pre-filters. More generally, a realizable weight must take the form:

$$\mathbf{W} = \mathbf{P}_m^{-1}\mathbf{P}_f\mathbf{K}, \tag{10.27}$$

where $\mathbf{P}_f$ is the pre-filter matrix. The introduction of pre-filters inevitably delays the response of uncertainty reduction. Thus, the time constants in $\mathbf{P}_f$ should be chosen as small as possible while maintaining system stability.

With the revised weight design, the compensated plant becomes:

$$\mathbf{P}_c = \left[\left(\mathbf{P}_m\mathbf{P}^{-1} + \mathbf{P}_f\mathbf{K}\right)^{-1}\left(\mathbf{I} + \mathbf{P}_f\mathbf{K}\right)\right]\mathbf{P}_m, \tag{10.28}$$

and under perfect modeling conditions, the compensated system response simplifies to:

$$\boldsymbol{Y} = \mathbf{P}_m\boldsymbol{U} + \left(\mathbf{I} + \mathbf{P}_f\mathbf{K}\right)^{-1}\boldsymbol{D}. \tag{10.29}$$

This highlights that while the pre-filter $\mathbf{P}_f$ improves realizability, it introduces a delay in disturbance attenuation.

Design Based on State Error

If full state measurements are available, an alternative uncertainty reduction approach can be formulated using a state feedback-based complementary control:

$$\boldsymbol{u}_i = \mathbf{K}(\boldsymbol{x}_m - \boldsymbol{x}), \tag{10.30}$$

where $\boldsymbol{x}_m$ and $\boldsymbol{x}$ are the state vectors of the model $\mathbf{P}_m$ and the actual plant $\mathbf{P}$, respectively, and $\mathbf{K}$ is a gain matrix. This design simplifies system dynamics compared to the inverse-model-based weight $\mathbf{W}$ in Eq. (10.27).

In this state-error-based design, the weight $\mathbf{W}$ serves as a gain matrix for state error weighting and is expressed as

$$\mathbf{W} = \mathbf{K}(s), \tag{10.31}$$

where $\mathbf{K}(s)$ is a weight matrix with polynomial entries.

More generally, regardless of how $\mathbf{W}$ is designed, the compensated plant can always be rewritten as:

$$\begin{aligned}\mathbf{P}_c &= (\mathbf{I} + \mathbf{P}\mathbf{W})^{-1}\mathbf{P}(\mathbf{I} + \mathbf{W}\mathbf{P}_m) \\ &= \left[\left(\mathbf{P}^{-1} + \mathbf{W}\right)^{-1}\left(\mathbf{P}_m^{-1} + \mathbf{W}\right)\right]\mathbf{P}_m.\end{aligned}$$

This formulation suggests that if the weight $\mathbf{W}$ is sufficiently large, the compensated dynamics satisfy:

$$\mathbf{P}_c \approx \mathbf{P}_m, \tag{10.32}$$

as an approximate result.

The structure presented in Fig. 10.2 serves as a general-purpose framework for uncertainty reduction. By varying the design of the parallel compensation structure, different levels of robustness enhancement can be achieved. While further refinements are possible, the core principle remains unchanged: by dynamically compensating for plant-model discrepancies, the system maintains a closer match to the nominal dynamics while attenuating uncertainties and external disturbances.

This uncertainty reduction scheme is particularly advantageous in practical applications where precise system modeling is infeasible. The combination of model-following control and the parallel compensation structure ensures that the system remains robust against modeling errors and external perturbations. While this book does not explore alternative optimization strategies for designing $\mathbf{W}$, the methodology outlined here provides a solid foundation for developing robust controllers capable of mitigating plant uncertainties and disturbances in complex environments.

10.4 Dual-Model-Based Control

Engineering research is largely application-driven, and this principle is particularly evident in the field of control systems. The challenges of translating theoretical control methodologies into practical implementations have been a major driving force behind the development of robust control techniques. This issue arises because conventional control strategies are predominantly model-based, relying on mathematical representations of plant dynamics that often diverge from real-world system behavior. Discrepancies between theoretical models and actual plant dynamics—caused by unmodeled dynamics, parameter variations, and external disturbances—frequently lead to performance degradation. Consequently, the field of robust control has evolved to bridge this gap by designing control methodologies that maintain stability and performance despite such uncertainties [20]. The strategies developed under this framework have found applications in diverse domains where robustness is crucial, ranging from aerospace systems and industrial automation to biomedical devices and precision manufacturing.

Building on this foundation, this section applies the uncertainty reduction framework introduced earlier to the robust control of complex mass–stiffness–damping systems, with robotic manipulators as a representative example. Robotic manipulators are inherently nonlinear, dynamically cross-coupled, and highly sensitive to parameter variations such as payload fluctuations, joint friction, and actuator dynamics. Conventional control strategies, including computed torque control and other model-based approaches, rely heavily on precise dynamic models to achieve accurate motion control. However, real-world systems inevitably deviate from these models due to unmodeled disturbances, external perturbations, and parameter variations. If these uncertainties are not properly addressed, they can lead to instability, degraded performance, and significant steady-state errors.

As demonstrated in Chap. 6, algorithms for decoupling and linearizing mass–stiffness–damping systems—including actuator dynamics—typically involve model-based serial pre-compensators incorporated into the forward control path. These pre-compensators aim to transform the inherently nonlinear and cross-coupled system dynamics into an approximately linearized and decoupled form, thereby facilitating conventional controller design. However, their effectiveness depends critically on the accuracy of the underlying plant model used for controller synthesis. Any mismatch between the model and the actual system can introduce residual uncertainties, reducing control performance and robustness.

To address these challenges, the model-based parallel compensation structure introduced earlier can be integrated as a post-compensator to enhance robustness. The resulting architecture effectively combines serial and parallel model-based compensation, forming a dual-model-based control system. In this configuration, the serial model-based controller serves as a pre-compensator, primarily aimed at decoupling and linearization, while the parallel compensator operates as a post-compensator to mitigate residual uncertainties and disturbances. This dual-layered approach leverages the strengths of both model-based feedforward control for precise trajectory tracking and model-following feedback for robustness enhancement.

For simplicity, it is assumed that the pre-compensated system is already approximately linearized, decoupled, and internally stable—meaning that residual uncertainties are not dominant but still contribute to non-ideal performance. These residual discrepancies, including parametric uncertainties, unmodeled nonlinearities, and external perturbations, are then treated as modeling uncertainties to be actively reduced by the post-compensator. The combination of these two complementary compensation mechanisms enables significant improvement in system robustness while preserving high control accuracy.

By implementing this dual-model-based strategy, the system benefits from the strengths of both feedforward model-based control and feedback-based robustness enhancement, resulting in improved tracking precision and resilience to uncertainties. In the subsequent sections, the application of this methodology to robotic manipulator control will be analyzed in greater detail, with theoretical formulations and experimental validation demonstrating its effectiveness.

10.4.1 Post-compensation for Uncertainty Reduction

The general structure of a mass–stiffness–damping system, pre-compensated using appropriate algorithms, is illustrated in Fig. 10.3. Since the structural modifications required for robustness enhancement in systems based on Algorithms 6.1, 6.2, and A.1 (computed-torque method) are fundamentally similar, the discussion that follows will focus on systems utilizing Algorithms 6.2 and A.1, which yield second-order

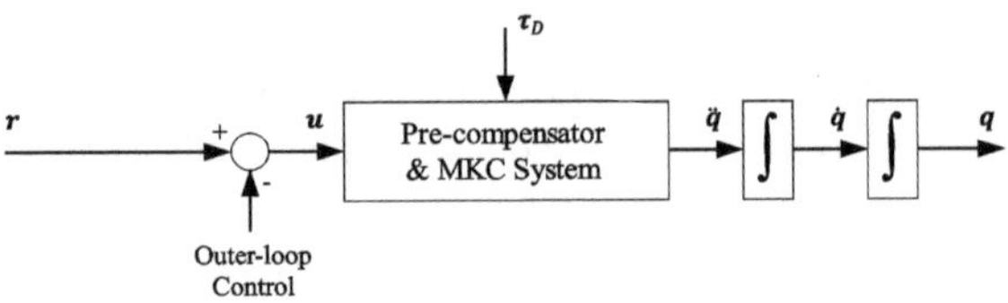

Fig. 10.3 Pre-compensation with outer-loop control

dynamics. This choice is made for simplicity and without loss of generality, as the principles extend to other configurations as well.

When Algorithm 6.2 is applied to pre-compensate for the nonlinear and cross-coupled dynamics of a mass–stiffness–damping system with significant actuator dynamics, or when the computed-torque method (Algorithm A.1) is used for systems with negligible actuator dynamics, the resulting system achieves near-decoupling and linearization. Ideally, such systems exhibit the behavior of double integrators, though they remain susceptible to internal uncertainties and external disturbances.

The ideal inner-loop dynamics of each degree-of-freedom in a serially pre-compensated mass–stiffness–damping system can be expressed as:

$$Q = \frac{1}{s^2} U. \tag{10.33}$$

For the open-loop dynamics described in Eq. (10.33), implementing an outer-loop proportional–derivative (PD) feedback control, as referenced in Eq. (7.4), results in:

$$u = r - \left(k_v \dot{q} + k_p q\right), \tag{10.34}$$

which, in terms of the control input U, is given by:

$$U = R - \left(k_v s + k_p\right) Q. \tag{10.35}$$

By substituting Eq. (10.35) into Eq. (10.33), the closed-loop system dynamics become:

$$\ddot{q} + k_v \dot{q} + k_p q = r, \tag{10.36}$$

or, equivalently, in transfer function form:

$$Q = \frac{1}{s^2 + k_v s + k_p} R. \tag{10.37}$$

The second-order "closed-loop" or "open-loop" dynamics established above provide the foundation for further robustness enhancements. By appending an additional post-compensator to this existing structure, both internal uncertainties and external disturbances can be mitigated. This augmentation effectively strengthens the resilience of the system while maintaining the desired dynamic characteristics. The subsequent analysis will explore the design and implementation of such compensatory mechanisms to further enhance the robustness and disturbance rejection capabilities of mass–stiffness–damping systems.

10.4.2 Possible Configurations

For the system depicted in Fig. 10.3 and the structure outlined in Fig. 10.2, two potential configurations for robustness enhancement are introduced. These configurations aim to reduce internal uncertainties and mitigate external disturbances, thereby improving the overall stability and performance of the mass–stiffness–damping system.

Outer-Loop Post-compensation

One approach to enhancing robustness involves incorporating an additional post-compensator in parallel with the existing outer closed-loop system, as illustrated in Fig. 10.4. This configuration transforms the system into a conventional model-following control structure, which allows for dynamic error correction and improved tracking accuracy in the presence of uncertainties. By leveraging the model-following principle, the post-compensator actively compensates for discrepancies between the actual system dynamics and the nominal model.

From Eq. (10.37), the nominal parallel model P_m for this outer-loop configuration can be defined as:

$$\mathrm{P}_m = \frac{1}{s^2 + k_v s + k_p}. \tag{10.38}$$

This model represents the ideal second-order closed-loop response of the system. To ensure robustness, a compensating weight W_1 is designed to shape the system dynamics and suppress the impact of uncertainties. The compensator can be constructed as:

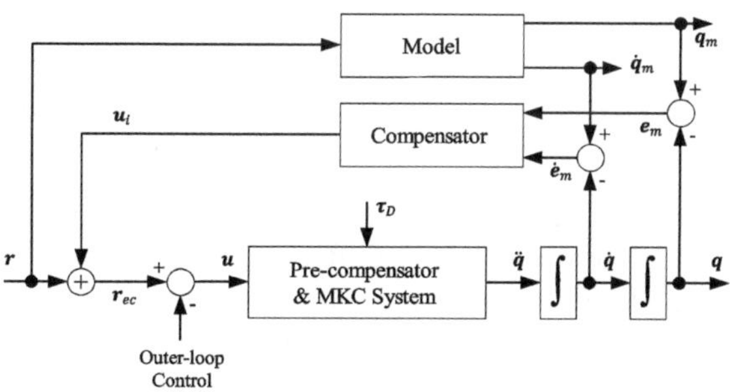

Fig. 10.4 Outer-loop post-compensation

$$W_1 = P_m^{-1} P_f K, \tag{10.39}$$

where P_f is a pre-filter introduced to ensure physical realizability and prevent high-frequency noise amplification. The pre-filter is defined as:

$$P_f = \frac{1}{\left(T_{pf} s + 1\right)^2}, \tag{10.40}$$

where T_{pf} is the time constant of the pre-filter. The introduction of this filtering element ensures that the compensator remains practical by eliminating the need for pure differentiation, which is often physically unrealizable in real-world systems.

Substituting Eq. (10.40) into Eq. (10.39), the compensator weight W_1 simplifies to:

$$W_1 = K \frac{s^2 + k_v s + k_p}{\left(T_{pf} s + 1\right)^2}. \tag{10.41}$$

This formulation ensures that the compensator dynamically adapts to system variations, enhancing robustness while maintaining the desired closed-loop characteristics. The above control methodology can be consolidated into Algorithm 10.1, which outlines the steps for implementing outer-loop post-compensation in serially pre-compensated mass–stiffness–damping systems.

Algorithm 10.1 For the post-compensation of the pre-compensated outer closed-loop system, the enhanced reference input R_{ec} in the control system with post-compensation, as shown in Fig. 10.4, can be designed as:

$$R_{ec} := R + U_i$$

$$U_i := K \frac{s^2 + k_v s + k_p}{\left(T_{pf} s + 1\right)^2} E_m$$

$$E_m := Q_m - Q$$

$$Q_m := \frac{1}{s^2 + k_v s + k_p} R,$$

where R is the original reference input. The additional compensatory signal U_i is introduced to mitigate the effects of uncertainties and disturbances. E_m represents the model-following error, which captures the discrepancy between the desired and actual system responses. Q_m is the ideal response of the parallel reference model. The enhanced reference input R_{ec} is then fed into the closed-loop system, which has been pre-compensated using Algorithm 6.2 or Algorithm A.1.

By incorporating the model-following compensation loop, the system actively reduces the impact of model uncertainties and external disturbances. The weighting function applied to E_m ensures that the control system dynamically adapts to deviations from the ideal response, thereby improving robustness and tracking accuracy.

The introduced approach effectively mitigates the impact of parameter variations, unmodeled dynamics, and external disturbances, offering a structured framework for enhancing system robustness. The effectiveness of Algorithm 10.1 is demonstrated in the following example.

Example 10.1 In this example, the performance of Algorithm 10.1 is compared to that of Algorithm A.1 in the presence of payload variation. For simplicity, trajectory tracking of a single-link manipulator (as described in Sect. 2.4.1) with negligible motor dynamics is considered. The manipulator moves from its initial position $q_0 = 0$ to the final position $q_f = 2$rad. within the time interval $[t_0, t_f] = [0, 2]$, following the trajectory given in Appendix C.1.

The outer-loop PD feedback controllers for both Algorithm 10.1 and Algorithm A.1 are set with gains $k_v = 20$ and $k_p = 100$. For the complementary control U_i in Algorithm 10.1, the gain is set to K $= 10$, and the pre-filter time constant is $T_{pf} = 0.001$.

The tracking errors under payload variation ($m_p = 5$ kg) are shown in Fig. 10.5 {Dashed line: Algorithm 10.1. Solid line: Algorithm A.1}. The solid line represents the error of Algorithm A.1, while the dashed line corresponds to Algorithm 10.1. The results clearly indicate that the robustness enhancement introduced in Algorithm 10.1 significantly reduces the tracking error, demonstrating its improved resilience against payload variations. Since Algorithm 10.1 is built upon Algorithm A.1 (i.e., Algorithm A.1 serves as the pre-compensator), the additional robustness enhancer effectively mitigates the impact of parameter variations.

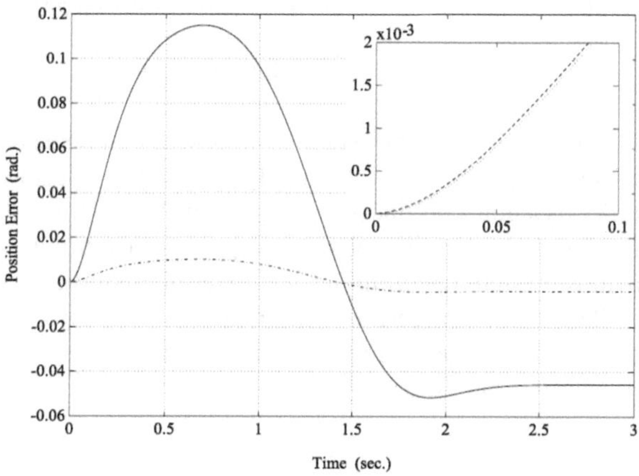

Fig. 10.5 Performance of Algorithm 10.1

A closer inspection of Fig. 10.5 reveals that the dashed line (tracking error of Algorithm 10.1 with $m_p = 5\,\text{kg}$) nearly overlaps with a dotted line representing the tracking error of Algorithm A.1 under a much smaller payload variation $m_p/(1 + \text{K}) = 0.4545455$ kg. This confirms that by applying Algorithm 10.1, the impact of payload variation is reduced by a factor of $(1 + \text{K})$, as predicted by the robustness enhancement analysis.

To further illustrate this effect, an enlarged section of the dash-dot line from Fig. 10.5 is provided in an inset plot. Here, the dashed line and dotted line separate slightly. This deviation arises because payload variation is neither a purely external disturbance nor a linear internal perturbation, but rather a nonlinear internal perturbation. Consequently, it cannot be perfectly attenuated by a factor of $(1 + \text{K})$ using Algorithm 10.1. Additionally, the pre-filtering dynamics P_f also influences the effectiveness of the compensation, further contributing to the observed difference.

The application of Algorithm 10.1 results in a sixth-order system from the reference input R to the position output Q. While this structure effectively enhances robustness, higher-order dynamics can introduce unnecessary complexity, potentially affecting system response characteristics such as transient behavior and computational efficiency. To address this issue and achieve a lower-order control system, a state-error-based compensator is introduced.

To simplify the resulting system, the compensating weight W is designed as a proportional–derivative (PD) state-error-based compensator given by

$$\text{W}_2 = k_1 s + k_2, \tag{10.42}$$

which translates into the following state-error compensation law

$$u_i = k_1(\dot{q}_m - \dot{q}) + k_2(q_m - q), \tag{10.43}$$

where k_1 and k_2 are constant gains. Here, q_m and $\dot{q}_m$ represent the position and velocity of the parallel model, respectively, which follows the reference trajectory dynamics

$$\ddot{q}_m + k_v\dot{q}_m + k_p q_m = r. \tag{10.44}$$

By employing this state-error-based compensator, the order of the resulting closed-loop system is effectively reduced to fourth-order, compared to the sixth-order dynamics produced by Algorithm 10.1. This order reduction simplifies control implementation and improves computational efficiency while maintaining robustness against system uncertainties and external disturbances.

The control algorithm outlined above is now formalized as Algorithm 10.2, offering a structured approach for achieving robust performance with a more compact dynamic formulation.

Algorithm 10.2 For the post-compensation of the pre-compensated outer closed-loop system, the enhanced reference input r_{ec} in the control system with post-compensation (as shown in Fig. 10.4) is designed as follows:

$$r_{ec} := r + u_i$$

$$u_i := k_1 \dot{e}_m + k_2 e_m$$

$$e_m := q_m - q$$

$$q_m : \ddot{q}_m + k_v \dot{q}_m + k_p q_m = r,$$

where u_i is the compensatory control input. e_m is the error signal. q_m represents the position of the parallel model. r is the original reference input. The enhanced reference input r_{ec} is then fed into the closed-loop system based on Algorithm 6.2 or Algorithm A.1, thereby improving robustness and reducing the impact of uncertainties while maintaining the desired dynamic response.

The effectiveness of Algorithm 10.2 is demonstrated in the following example.

Example 10.2 In this example, the performance of Algorithm 10.2 is compared with that of Algorithm A.1 under payload variation. The manipulator and motion task remain the same as those in Example 10.1. For the complementary control u_i in Algorithm 10.2, the gains are set to $k_1 = 80$ and $k_2 = 400$. All other controller parameters for Algorithm 10.2 and Algorithm A.1 are identical to those used in Example 10.1.

The tracking errors for Algorithm 10.2 and Algorithm A.1 with a payload increase of $m_p = 5\,\text{kg}$ are shown in Fig. 10.6 {Dashed line: Algorithm 10.2. Solid line: Algorithm A.1}. The results clearly demonstrate that, in the presence of payload variation, the tracking error of Algorithm 10.2 (dashed line) is significantly smaller than that of Algorithm A.1 (solid line), highlighting the effectiveness of the additional robustness enhancer in mitigating the impact of uncertainty.

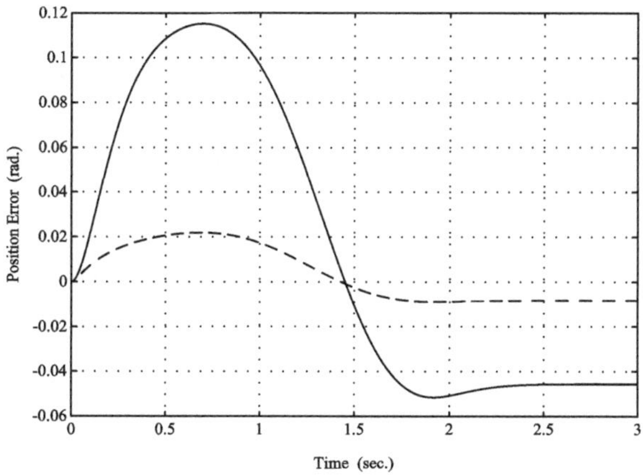

Fig. 10.6 Performance of Algorithm 10.2

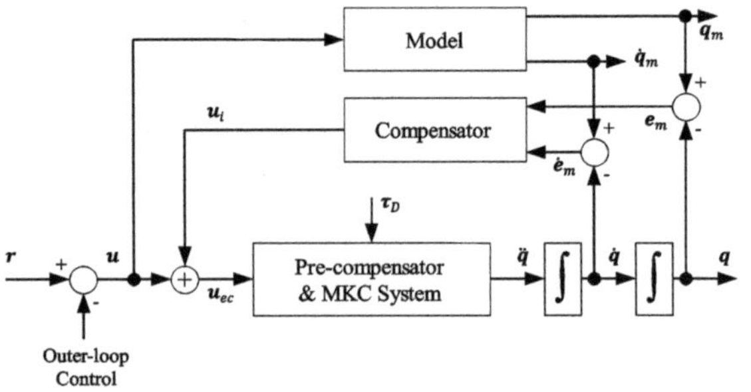

Fig. 10.7 Inner-loop post-compensation

The performance of Algorithm 10.2 is not directly compared with that of Algorithm 10.1, as their respective weights, W_1 and W_2, differ in design and implementation.

Inner-Loop Post-compensation

A modified configuration for enhancing the robustness of pre-compensated mass–stiffness–damping systems involves integrating an additional post-compensator in parallel with the pre-compensated inner open-loop system within the outer control loop, as illustrated in Fig. 10.7. This inner-loop post-compensation approach provides a more direct means of mitigating internal uncertainties and disturbances before they propagate through the system, thereby improving overall robustness and dynamic response.

With this inner-loop configuration, the parallel model can be defined based on Eq. (10.33) as:

$$P_m = \frac{1}{s^2}. \tag{10.45}$$

To compensate for uncertainties and disturbances, the weight W can be designed as:

$$W_3 = P_m^{-1} P_f K, \tag{10.46}$$

where P_f is a pre-filter that stabilizes the compensator and ensures practical realizability:

$$P_f = \frac{1}{\left(T_{pf} s + 1\right)^2}. \tag{10.47}$$

Here, T_{pf} represents the time constant of the pre-filter, and K is a constant gain. Substituting these into Eq. (10.46) results in:

$$W_3 = \mathrm{K}\frac{s^2}{\left(T_{pf}s + 1\right)^2}. \tag{10.48}$$

This formulation parallels the structure of Algorithm 10.1, but with an emphasis on inner-loop compensation to achieve robustness at an earlier stage in the control process. By actively mitigating uncertainties at the inner loop, this approach enhances stability, reduces sensitivity to model discrepancies, and improves disturbance rejection. Based on this formulation, Algorithm 10.3 is introduced as an alternative robust control solution, leveraging inner-loop post-compensation for enhanced system performance.

Algorithm 10.3 For the post-compensation of the pre-compensated inner open-loop system, the enhanced control U_{ec} for the system with post-compensation, as shown in Fig. 10.7, is designed as follows:

$$U_{ec} := U + U_i$$

$$U_i := \mathrm{K}\frac{s^2}{\left(T_{pf}s + 1\right)^2}E_m$$

$$E_m := Q_m - Q$$

$$Q_m := \frac{1}{s^2}U.$$

Here, U_i is the complementary control input, E_m is the model-based error term, Q_m represents the reference-model output, and U denotes the original closed-loop control signal. The enhanced control U_{ec} is applied to the pre-compensated open-loop system, which is based on either Algorithm 6.2 or Algorithm A.1. This additional compensation mechanism actively mitigates internal uncertainties and disturbances at the inner-loop level, improving overall robustness and system performance.

The effectiveness of Algorithm 10.3 is demonstrated through the following example and experiment.

Example 10.3 In this example, the performance of Algorithm 10.3 is compared with that of Algorithm A.1 in the presence of payload variation. The manipulator and its motion task are identical to those in Example 10.1, and the controller parameters for both Algorithm 10.3 and Algorithm A.1 remain the same as in that example.

The tracking errors for Algorithm 10.3 and Algorithm A.1 under a payload variation of $m_p = 5$ kg are shown in the main plot of Fig. 10.8 {Dashed line: Algorithm 10.3. Solid line: Algorithm A.1}. As observed, the tracking performance of Algorithm 10.3 (dashed line) is significantly improved compared to Algorithm A.1 (solid line), highlighting the effectiveness of the additional robustness enhancement.

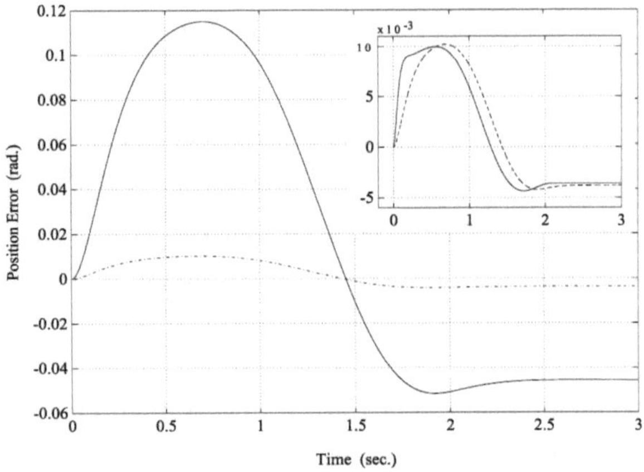

Fig. 10.8 Performance of Algorithm 10.3

Notably, the main figure in Fig. 10.8 reveals that the dashed line nearly overlaps another dotted line, where the dashed line represents the tracking error of Algorithm 10.3 and the dotted line corresponds to the tracking error of Algorithm 10.1. This suggests that, with the given weights W_1 and W_3, both algorithms exhibit identical performance. This property will be further elaborated in the following chapter.

It is well-known that increasing the feedback gains of the outer-loop PD controller in Algorithm A.1 can reduce the tracking error. However, this comes at the cost of increasing the dominant frequency of the closed-loop dynamics, which may compromise system stability in the high-frequency domain. To illustrate this trade-off, the dashed line from the main figure of Fig. 10.8 (representing Algorithm 10.3 with $k_v = 20$, $k_p = 100$) is redrawn in the smaller inset figure. The solid line in the inset represents the performance of Algorithm A.1 with significantly higher feedback gains of $k_v = 68$ and $k_p = 1156$. The comparison clearly demonstrates that achieving similar error levels in Algorithm A.1 requires substantially larger outer-loop gains than in Algorithm 10.3, further emphasizing the advantage of the robustness-enhancing compensator.

Experiment 10.1
To further evaluate the performance of Algorithm 10.3, experiments were conducted using the testbed described in Appendix D, with the results presented in Fig. 10.9 {Solid line: K = 0 (= Algorithm A.1). Dashed line: K = 1. Dash-dot line: K = 3. Dotted line: K = 9}. In this experiment, the manipulator was required to move from its initial lower vertical position at $q_0 = -\pi/2$ to the final horizontal position at $q_f = 0$ within the time interval $\left[0, t_f\right] = [0, 1.0]$ s. The reference trajectory followed Eq. (C.1).

To assess the robustness of the introduced method, joint friction was deliberately left uncompensated, and gravity compensation was intentionally kept inaccurate. These non-ideal factors, particularly the gravitational torque, introduced significant

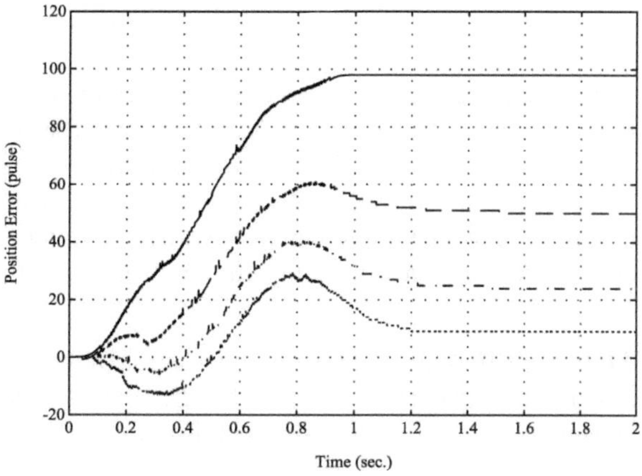

Fig. 10.9 Experimental results of Algorithm 10.3

tracking and steady-state errors in the conventional Algorithm A.1 (solid line in Fig. 10.9).

The experiment examined different values of the compensating gain K in Algorithm 10.3. For K = 0, the algorithm effectively reduces to Algorithm A.1, providing a baseline for comparison. The results in Fig. 10.9 clearly demonstrate that applying Algorithm 10.3 significantly mitigates the undesirable tracking error. Specifically, for K = 1 (dashed line), K = 3 (dash-dot line), and K = 9 (dotted line), the observed errors were approximately reduced by factors of 2, 4, and 10, respectively.

These experimental results closely align with the simulation findings from Examples 10.1 and 10.3, further confirming the effectiveness of Algorithm 10.3 in enhancing robustness against modeling inaccuracies and external disturbances.

To achieve a lower-order control system while maintaining robustness, an alternative state-error-based compensating weight can be employed. This approach simplifies the control structure while still effectively mitigating disturbances and uncertainties. The required compensating weight is defined as

$$\mathrm{W}_4 = k_1 s + k_2, \tag{10.49}$$

where k_1 and k_2 are constant gains. This formulation represents a proportional–derivative (PD) compensator, incorporating both position and velocity feedback to enhance system stability and dynamic response. The corresponding compensatory control input is given by

$$u_i = k_1(\dot{q}_m - \dot{q}) + k_2(q_m - q), \tag{10.50}$$

where q_m and $\dot{q}_m$ denote the position and velocity of the parallel model, ensuring that the compensator actively adjusts for system deviations. The parallel model follows the simplified double-integrator dynamics

$$\ddot{q}_m = u. \tag{10.51}$$

With this state-error-based compensator, the overall system dynamics from the reference input r to the position output q remain fourth-order, reducing computational complexity compared to higher-order compensatory strategies.

This design effectively balances robustness and simplicity, making it particularly useful for real-time implementations where computational efficiency is critical. The control algorithm incorporating this compensator is formally presented as Algorithm 10.4.

Algorithm 10.4 For the post-compensation of the pre-compensated inner open-loop system, the enhanced control u_{ec} for the control system with post-compensation, as shown in Fig. 10.7, is designed as follows:

$$u_{ec} := u + u_i$$

$$u_i := k_1 \dot{e}_m + k_2 e_m$$

$$e_m := q_m - q$$

$$q_m : \ddot{q}_m = u.$$

Here, u_i is the compensatory control input, while e_m denotes the state error. The variable q_m represents the position of the parallel model, which follows a double-integrator dynamic behavior.

In this formulation, u is the original closed-loop control applied to the system before post-compensation. The enhanced control input u_{ec} is used in place of u within the pre-compensated open-loop system, which is based on either Algorithm 6.2 or Algorithm A.1.

By incorporating state-error-based compensation, the system effectively mitigates the impact of model uncertainties and external disturbances, enhancing overall robustness and control precision while preserving computational efficiency. The performance of Algorithm 10.4 is demonstrated in the following example and experiment.

Example 10.4 **(*Payload Variation*)** In this example, the performance of Algorithm 10.4 is compared with that of Algorithm A.1 under payload variation. The manipulator and its motion task remain the same as in Example 10.1, while the control parameters for Algorithm 10.4 and Algorithm A.1 are consistent with those used in Example 10.2.

The tracking errors of both algorithms under a payload increase of $m_p = 5$ kg are illustrated in Fig. 10.10a {Dashed line: Algorithm 10.4. Solid line: Algorithm A.1}. As shown in the figure, the tracking error of Algorithm 10.4 (dashed line) is significantly smaller than that of Algorithm A.1 (solid line), highlighting the effectiveness of the additional robustness compensator. A notable advantage of Algorithm 10.4 over Algorithms 10.1–10.3 is its ability to completely eliminate steady-state error, which is particularly beneficial in precision control applications.

To further illustrate this point, the tracking error of Algorithm 10.4 with $k_v = 20$ and $k_p = 100$ is redrawn in the small inset of Fig. 10.10a, alongside the error of Algorithm A.1 with increased feedback gains, $k_v = 110$ and $k_p = 3025$ (solid line). Although the peak error of the high-gain Algorithm A.1 has been reduced to match that of Algorithm 10.4, its transient and steady-state errors remain noticeably larger, demonstrating the superior robustness of Algorithm 10.4.

Figure 10.10b {Dashed line: Algorithm 10.4. Solid line: Algorithm A.1} presents the corresponding control torques for both algorithms. It is evident that the control torque of Algorithm 10.4 is not larger than that of Algorithm A.1, confirming that its improved performance is not due to excessive control effort but rather to a

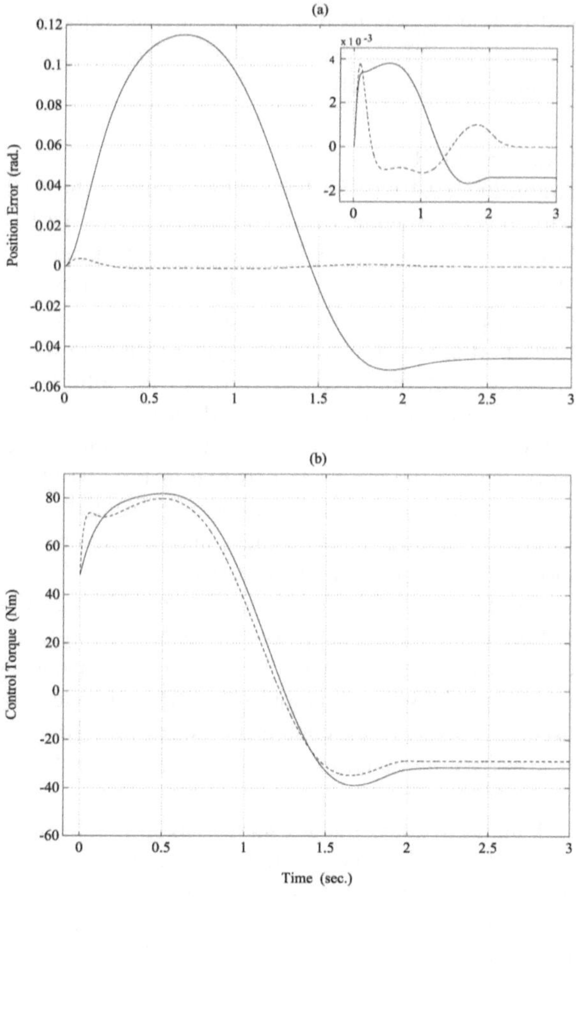

Fig. 10.10 Algorithm 10.4: payload variation

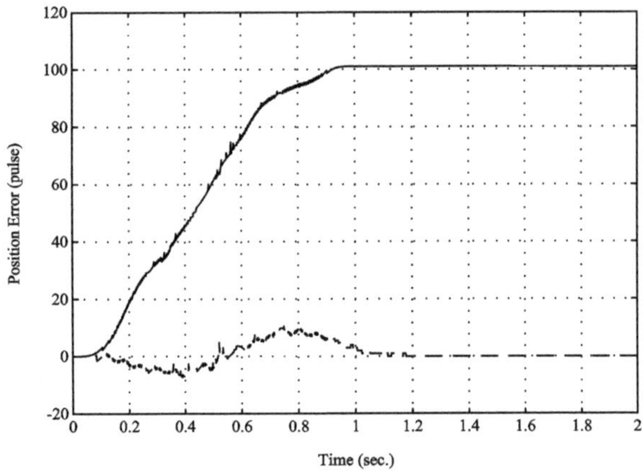

Fig. 10.11 Experimental results of Algorithm 10.4: case 1

fast-responding compensatory action. This rapid response capability enables Algorithm 10.4 to effectively mitigate the impact of uncertainties and disturbances while maintaining control efficiency.

Experiment 10.2

Algorithm 10.4 was further evaluated using the testbed described in Appendix D, with the experimental results presented in Fig. 10.11 {Dashed line: Algorithm 10.4. Solid line: Algorithm A.1}. The motion task remained the same as in Experiment 10.1, and the system continued to operate under non-ideal conditions, including uncompensated joint friction and inaccurate gravity compensation. These imperfections, particularly the gravity torque, resulted in significant tracking and steady-state errors for Algorithm A.1, as shown by the solid line in Fig. 10.11.

As evident from the figure, Algorithm 10.4 effectively mitigates these errors, demonstrating a substantial reduction in undesirable deviations while maintaining zero steady-state error (dashed line in Fig. 10.11). The experimental results align closely with the simulation outcomes presented in Example 10.2 and Example 10.4, further validating the robustness and effectiveness of Algorithm 10.4 in handling model uncertainties and disturbances.

As previously discussed, a system utilizing Algorithm 10.1 with weight W_1 is functionally equivalent to one based on Algorithm 10.3 with weight W_3. This equivalence suggests that both outer-loop and inner-loop post-compensation strategies can achieve comparable robustness enhancement when appropriately configured. However, when comparing Algorithm 10.4 and Algorithm 10.2 under identical weighting conditions ($W_2 = W_4$), Algorithm 10.4 demonstrates significantly superior performance.

One particularly notable advantage of Algorithm 10.4 is its ability to completely eliminate steady-state error, a limitation inherent in the other three algorithms. This is a crucial improvement in precision control applications, where steady-state deviations can compromise accuracy and long-term stability. The superior performance of Algorithm 10.4 stems from its explicit state-error-based compensation, which ensures continuous correction for model mismatches and disturbances, thereby maintaining exact reference tracking.

These properties, along with the theoretical implications of the equivalence between different weighting strategies, will be further analyzed in the next chapter. The discussion will delve deeper into the underlying principles governing the effectiveness of different compensation approaches, providing a more rigorous foundation for selecting optimal control configurations in practical applications.

10.4.3 Performance of Dual-Model-Based Control

The effectiveness of the dual-model-based control scheme will be further examined through detailed case studies. As a representative example, the performance of the control system implemented using Algorithm 10.4 will be analyzed.

Example 10.5 ***(External Disturbances)*** In this example, the performance of Algorithm 10.4 is compared with that of Algorithm A.1 in the presence of external disturbances. The manipulator and its motion task remain the same as those described in Example 10.1. Additionally, the controller parameters for both Algorithm 10.4 and Algorithm A.1 are identical to those used in Example 10.4.

To evaluate robustness against external disturbances, two types of perturbations—impulse disturbance and step disturbance, as described in Appendix C.3—are considered. The resulting tracking errors due to these disturbances are illustrated in Fig. 10.12a {Impulse disturbance. Dashed line: Algorithm 10.4. Solid line: Algorithm A.1}, b { Step disturbance. Dashed line: Algorithm 10.4. Solid line: Algorithm A.1}, respectively. As shown in Fig. 10.12, the implementation of Algorithm 10.4 significantly attenuates the impact of external disturbances, demonstrating its effectiveness in enhancing system robustness.

Experiment 10.3

In this experiment, the manipulator testbed described in Appendix D is initially required to move from its starting lower vertical position, $q_0 = -\pi/2$, to the final horizontal position, $q_f = 0$ over the time interval $\left[0, t_f\right] = [0, 1.0]$ seconds. The trajectory defined in Eq. (C.1) is used to govern this motion. Joint friction remains uncompensated, and gravity compensation is deliberately left inaccurate to introduce additional non-ideal effects.

At $t = 2.0$ s, an additional payload is applied to the end-effector of the manipulator. This added payload introduces both a variation in system dynamics and an external disturbance, making it a suitable test case for evaluating robustness under real-world uncertainties.

The experimental results, presented in Fig. 10.13 {Dashed line: Algorithm 10.4. Solid line: Algorithm A.1}, illustrate the impact of these non-ideal effects, particularly gravity torque, on tracking performance. As seen in Fig. 10.13, the Algorithm

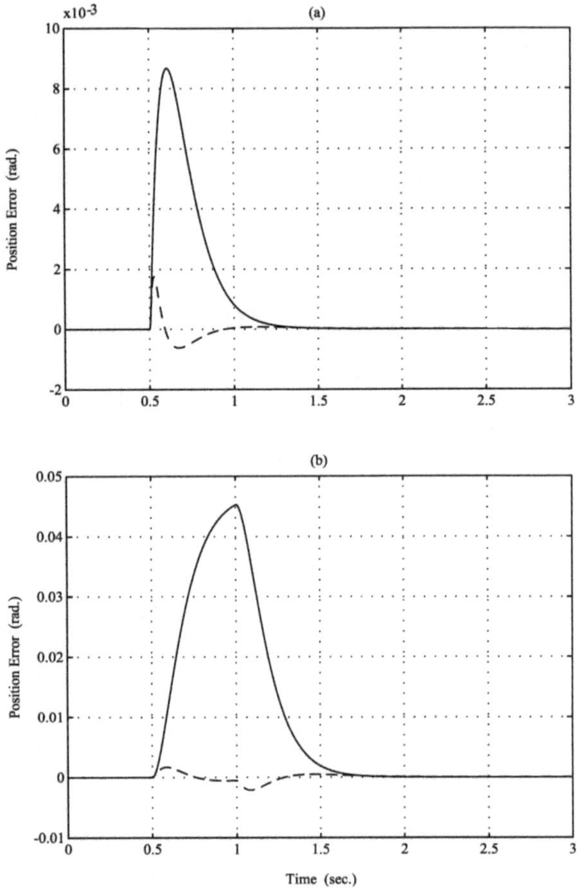

Fig. 10.12 Algorithm 10.4: external disturbances

A.1 (solid line) exhibits significant tracking and steady-state errors, exacerbated by the additional payload.

From Fig. 10.13, it is evident that during the initial time interval $t \in [0, 2]$, the errors closely match those reported in Fig. 10.11. However, in the subsequent interval $t \in [2, 4]$, the tracking error of Algorithm A.1 increases further due to the added payload. In contrast, when employing Algorithm 10.4, the adverse effects of the additional payload are significantly mitigated, with the system maintaining zero steady-state error (dashed line in Fig. 10.13). These experimental findings align well with the simulation results presented in Examples 10.4 and 10.5, confirming the effectiveness of Algorithm 10.4 in enhancing robustness against payload variations and external disturbances.

Example 10.6 (***Unmodeled Damping***) In this example, the performance of Algorithm 10.4 is compared with that of Algorithm A.1 in the presence of unmodeled joint damping. The manipulator and its motion task remain the same as described in

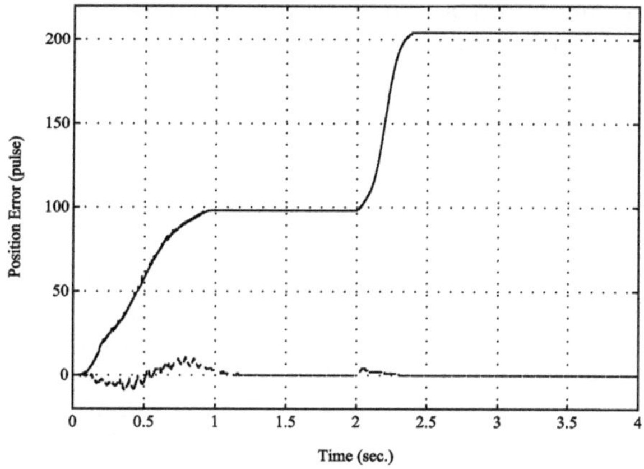

Fig. 10.13 Experimental results of Algorithm 10.4: case 2

Example 10.1, and the controller parameters for both Algorithm 10.4 and Algorithm A.1 are identical to those used in Example 10.4.

To evaluate the impact of unmodeled damping, the actual joint dynamics of the manipulator are modified as follows:

$$m_0 l^2 \ddot{q} + m_0 g l \cos(q) + k_d \dot{q} + \tau_D = \tau_C,$$

where $k_d = 10$ represents the coefficient of the unmodeled damping. This damping effect is not included in the model used for controller design, making it an unmodeled perturbation that challenges the robustness of the control algorithms.

The tracking errors due to the unmodeled damping are illustrated in Fig. 10.14 {Dashed line: Algorithm 10.4. Solid line: Algorithm A.1}. As seen in the figure, the control system based on Algorithm 10.4, which incorporates enhanced robustness, effectively mitigates the impact of the unmodeled damping. In contrast, the system based on Algorithm A.1 exhibits a significantly larger transient error, demonstrating its sensitivity to unmodeled dynamics. These results confirm the robustness of Algorithm 10.4 in handling unmodeled damping effects, ensuring more accurate trajectory tracking and improved system performance.

Example 10.7 **(*Unmodeled Cross-Coupling*)** In this example, Algorithm 10.4 is applied to enhance the performance of a decentralized PD controller for multi-link robotic manipulators with negligible motor dynamics.

The conventional computed-torque control law is given by:

$$\boldsymbol{\tau}_C := \hat{\mathbf{D}}(\boldsymbol{q})\boldsymbol{u} + \hat{\boldsymbol{h}}(\dot{\boldsymbol{q}}, \boldsymbol{q}),$$

where $\hat{\mathbf{D}}(\boldsymbol{q})$ represents the estimated inertia matrix, and $\hat{\boldsymbol{h}}(\dot{\boldsymbol{q}}, \boldsymbol{q})$ accounts for Coriolis, centrifugal, and gravitational effects.

To reduce computational complexity in real-time control, a simplified control law is often used as an alternative:

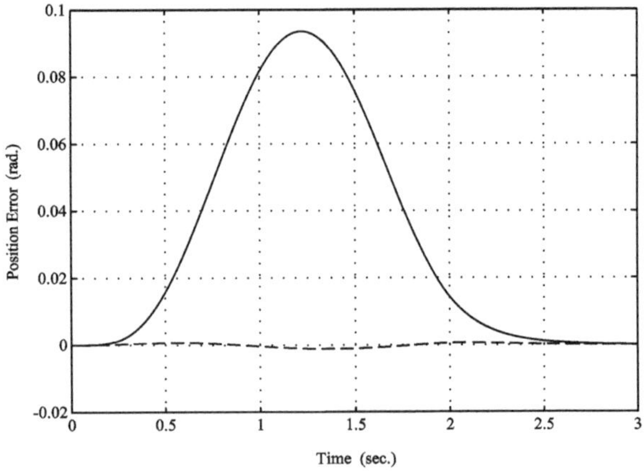

Fig. 10.14 Algorithm 10.4: unmodeled damping

$$\boldsymbol{\tau}_C := \hat{\mathbf{D}}_s \boldsymbol{u},$$

where $\hat{\mathbf{D}}_s$ consists of only the constant (static) terms of the diagonal elements in $\hat{\mathbf{D}}(\boldsymbol{q})$, and $\boldsymbol{u}$ is the conventional PD-based outer-loop feedback control. This simplified approach, known as decentralized PD control, neglects the nonlinear and coupling terms in $\hat{\boldsymbol{h}}(\dot{\boldsymbol{q}}, \boldsymbol{q})$ as well as all off-diagonal and variable elements in $\hat{\mathbf{D}}(\boldsymbol{q})$. While it significantly reduces computational cost, it does so at the expense of control performance. Although other simplifications incorporating additional compensatory elements exist, this particular decentralized scheme is chosen here as a representative example.

The above simplified control is applied to the motion control of a two-link manipulator described in Sect. 2.4.2. For this manipulator, the simplified inertia matrix is:

$$\hat{\mathbf{D}}_s = \begin{bmatrix} M_2 l_2^2 + (m_1 + M_2) l_1^2 & 0 \\ 0 & M_2 l_2^2 \end{bmatrix}.$$

To focus on the impact of control structure, parameter variations are not considered in this case study. Algorithm 10.4 is implemented to enhance the robustness of the simplified control, effectively serving as a post-compensator. The feedback gains for the outer-loop PD controllers in both the simplified control and the enhanced control are set as $\mathbf{K}_v = 20\mathbf{I}$ and $\mathbf{K}_p = 100\mathbf{I}$. The compensator gains in Algorithm 10.4 are chosen as $\mathbf{K}_1 = 80\mathbf{I}$ and $\mathbf{K}_2 = 400\mathbf{I}$. The motion task follows the specifications provided in Appendix C.2.

The tracking errors for the decentralized PD control are shown in Fig. 10.15a {Dotted line: Upper link. Solid line: Lower link}, while the tracking errors for the enhanced control based on Algorithm 10.4 are displayed in Fig. 10.15b {Dotted line: Upper link. Solid line: Lower link}. The steady-state errors observed in the decentralized PD control stem from the omission of the gravitational term $\boldsymbol{g}(\boldsymbol{q})$ in $\hat{\boldsymbol{h}}(\dot{\boldsymbol{q}}, \boldsymbol{q})$.

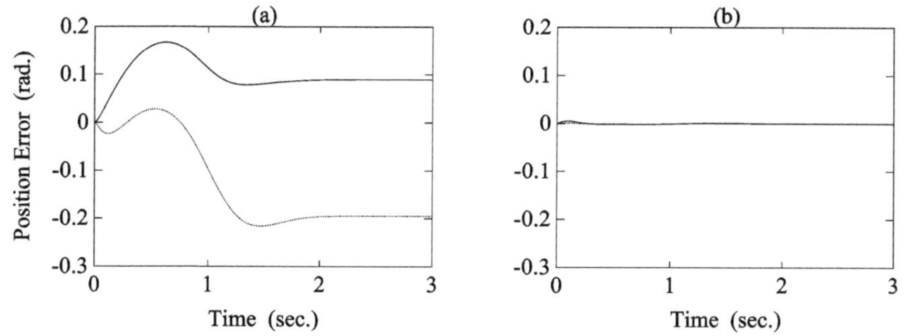

Fig. 10.15 Algorithm 10.4: unmodeled coupling

As seen in Fig. 10.15, the enhanced control significantly improves system performance by mitigating these unmodeled dynamics, demonstrating the effectiveness of Algorithm 10.4 in improving robustness and accuracy.

Example 10.8 (***Unmodeled Motor Dynamics***) In this example, the performance of Algorithm 10.4 is compared with that of Algorithm A.1 in a scenario where motor dynamics are non-negligible but remain unmodeled in the control design. The manipulator and its motion task are identical to those described in Example 10.1. The parameters of the controllers for both Algorithm 10.4 and Algorithm A.1 are also set to the same values as in Example 10.4.

The tracking errors resulting from the unmodeled motor dynamics are shown in Fig. 10.16 {Dashed line: Algorithm 10.4. Solid line: Algorithm A.1}. As observed, the control system based on Algorithm 10.4 exhibits strong robustness against the unmodeled motor dynamics, maintaining superior tracking performance. In contrast, the system based on Algorithm A.1 experiences significant transient errors, highlighting its sensitivity to unmodeled actuator dynamics. This demonstrates that the additional robustness enhancement provided by Algorithm 10.4 effectively mitigates the adverse effects of motor dynamics mismatches, improving overall control performance.

Example 10.9 (***Unmodeled Residual Dynamics***) In this example, the performance of Algorithm 10.4 is compared with that of Algorithm 6.2 in a scenario where the residual dynamics from pole–zero cancelation in Algorithm 6.2 are non-negligible. Algorithm 6.2 is used as the pre-compensator. The manipulator, which includes non-negligible motor dynamics, and its motion task are the same as those described in Example 10.8. The controller parameters for both Algorithm 10.4 and Algorithm 6. 2 are set to the same values as those used in Example 6.2.

The tracking errors resulting from the unmodeled residual dynamics of pole–zero cancelation are shown in Fig. 10.17 {Dashed line: Algorithm 10.4. Solid line: Algorithm A.1}. The results indicate that the control system based on Algorithm 10.4, which incorporates enhanced robustness, effectively mitigates the adverse effects of

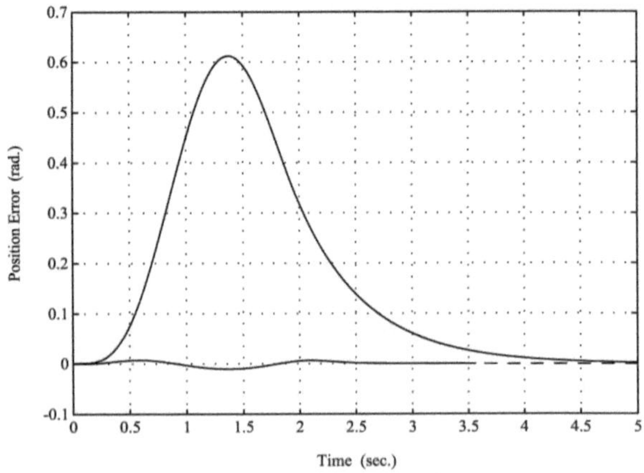

Fig. 10.16 Algorithm 10.4: unmodeled motor-dynamics

unmodeled residual dynamics. In contrast, the system using Algorithm 6.2 experiences noticeable transient errors due to its sensitivity to these dynamics. This demonstrates that Algorithm 10.4 provides superior robustness and improved tracking accuracy by actively compensating for the uncertainties introduced by residual dynamics in pole–zero cancelation.

The structural modification introduced for robustness enhancement in model-based control of mass–stiffness–damping systems, as depicted in Figs. 10.4 and 10.7, is both conceptually simple and practically effective. The compensatory control component is seamlessly integrated into the existing control framework by directly injecting a corrective control signal into the original control flow. This streamlined approach enhances the ability of the system to mitigate uncertainties and external disturbances without introducing significant computational overhead or implementation complexity. By preserving the fundamental structure of the control system

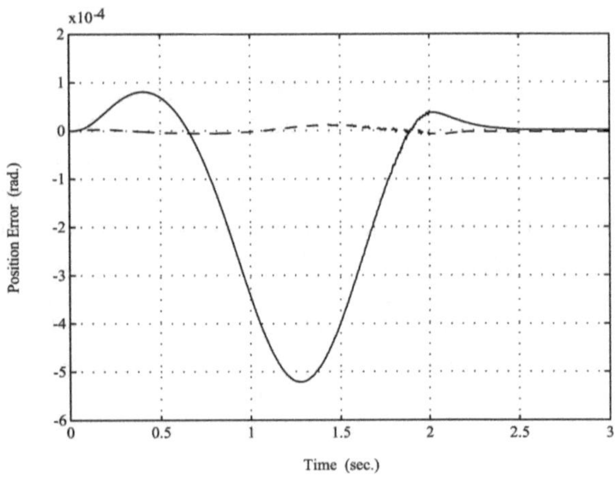

Fig. 10.17 Algorithm 10.4: unmodeled residual dynamics

while improving its resilience, this method offers a practical and efficient solution for achieving robust performance in uncertain environments.

Extensive case studies and detailed examples have validated the effectiveness of this approach. The results consistently show that the introduced robustness enhancement scheme significantly mitigates the effects of unmodeled dynamics, parameter variations, and external disturbances—factors that often degrade the performance of conventional model-based controllers. By integrating an additional parallel compensation structure, the modified system actively minimizes deviations caused by these uncertainties, leading to enhanced tracking accuracy, improved stability, and greater overall system resilience.

Moreover, the introduced method offers an alternative to conventional robust control strategies, which often rely on high-gain feedback, adaptive control, or complex optimization techniques. Instead of passively tolerating uncertainties, this scheme actively reduces their effects at the source, leading to a more predictable and resilient control response. The findings presented in this chapter confirm that even in scenarios with significant model mismatches and perturbations, the presented structural enhancement maintains system performance with minimal additional complexity. These insights highlight the potential of integrating such robustness-enhancing techniques into a wide range of mass–stiffness–damping systems, further extending their applicability in practical engineering environments.

10.5 Loop-Gain Analysis

Robust controllers with fixed gains and structures typically achieve their desired performance through high-gain designs in specific frequency ranges. However, such high-gain configurations may introduce limitations, including increased sensitivity to measurement noise and actuator saturation. In this section, we analyze the loop-gain properties of the introduced robust control structure through comparative studies. As an illustrative example, we examine the control performance of Algorithm 10.4, which incorporates a parallel compensator for robustness enhancement.

Consider a single-degree-of-freedom (1-DoF) mass–stiffness–damping system that has been decoupled and linearized via an appropriate inner-loop compensator (pre-compensator). The resulting stable plant dynamics can be expressed as:

$$\mathrm{P} = \frac{c}{s^2 + as + b}, \tag{10.52}$$

where the parameters a, b, and c may be subject to uncertainties and perturbations. The ideal nominal dynamics of the system, assuming perfect pre-compensation, is given by:

$$\mathrm{P}_m = \frac{1}{s^2}. \tag{10.53}$$

This corresponds to an idealized double integrator, with nominal parameter values: $a_0 = 0$, $b_0 = 0$, and $c_0 = 1$.

To regulate system behavior, a conventional PD (proportional–derivative) feedback controller is applied as the outer-loop controller:

$$\mathrm{F}_B = k_v s + k_p, \tag{10.54}$$

where k_v and k_p are the velocity and position feedback gains, respectively.

For a given reference input R, the ideal closed-loop response (assuming perfect compensation) is:

$$Y_m = (1 + \mathrm{P}_m\mathrm{F}_B)^{-1}\mathrm{P}_m k_p R = \frac{k_p}{s^2 + k_v s + k_p} R. \tag{10.55}$$

However, in the presence of parameter variations (i.e., deviations of a, b, and c from their nominal values), the actual perturbed closed-loop response of the system under conventional PD control is:

$$Y_1 = (1 + \mathrm{P}\mathrm{F}_B)^{-1}\mathrm{P} k_p R = \frac{c k_p}{s^2 + (a + c k_v)s + \left(b + c k_p\right)} R. \tag{10.56}$$

It is evident that the actual system response is highly sensitive to variations in the parameters a, b, and c, leading to degraded performance when significant deviations occur.

To enhance robustness against parameter variations and unmodeled dynamics, we apply the Algorithm 10.4 robustness enhancer. The resulting parallelly compensated system has the closed-loop response:

$$Y_2 = (1 + \mathrm{P}_c\mathrm{F}_B)^{-1}\mathrm{P}_c k_p R, \tag{10.57}$$

where the compensated plant dynamics P_c is defined as:

$$\mathrm{P}_c = (1 + \mathrm{PW})^{-1}\mathrm{P}(1 + \mathrm{W}\mathrm{P}_m). \tag{10.58}$$

Here, W is a parallel PD compensator designed to reduce the impact of uncertainties:

$$\mathrm{W} = k_1 s + k_2. \tag{10.59}$$

By substituting W into the expression for P_c, we derive the modified closed-loop system dynamics:

$$Y_2 = \frac{n_1 s^2 + n_2 s + n_3}{m_1 s^4 + m_2 s^3 + m_3 s^2 + m_4 s + m_5} R, \tag{10.60}$$

where the coefficients are:

$$n_1 = ck_p$$

$$n_2 = ck_1k_p$$

$$n_3 = ck_2k_p$$

$$m_1 = 1$$

$$m_2 = a + c(k_1 + k_v)$$

$$m_3 = b + c\big(k_2 + k_p + k_1k_v\big)$$

$$m_4 = c\big(k_1k_p + k_2k_v\big)$$

$$m_5 = ck_2k_p.$$

With this enhanced control scheme, the desired system performance can be maintained even in the presence of uncertainties. The loop-gain analysis demonstrates that the parallel compensator modifies the frequency response characteristics, effectively increasing robustness without excessively increasing controller gains, thereby avoiding issues such as instability or excessive control effort.

The analysis of loop-gain properties in the introduced robust control structure confirms that the parallel compensator effectively mitigates the impact of parameter variations and enhances system robustness without requiring aggressive high-gain tuning. This provides a systematic approach to improving stability and performance while keeping control effort and noise amplification within reasonable limits. The effectiveness of this methodology is demonstrated in the following numerical examples, which compare conventional PD control with the enhanced robustness scheme under different uncertainty conditions.

Example 10.10 This example investigates the unit-step and frequency responses of three control systems:

1. The ideal closed-loop system (y_m, Y_m), which serves as the reference performance.
2. The perturbed system with a conventional PD controller (y_1, Y_1), which demonstrates the impact of model uncertainties.
3. The perturbed system with an additional robustness enhancer (y_2, Y_2), which illustrates the effectiveness of the introduced method in mitigating uncertainties.

The system parameters are set as follows:

- Plant dynamics: $a = 0$, $b = 10$, $c = 0.5$.
- PD controller gains: $k_v = 20$, $k_p = 100$.
- Robustness enhancement gains: $k_1 = 30$, $k_2 = 225$.

The unit-step responses of the three systems are shown in Fig. 10.18a {Solid line: y_m. Dashed line: y_1. Dotted line: y_2. Dash-dot line: High-gain y_1}.

- The ideal response y_m (solid line) represents the reference trajectory with a steady-state value of:

$$y_{m\infty} = \frac{k_p}{k_p} = 1.$$

- The conventional control system response y_1 (dashed line) exhibits noticeable transient and steady-state errors due to unmodeled dynamics. The steady-state

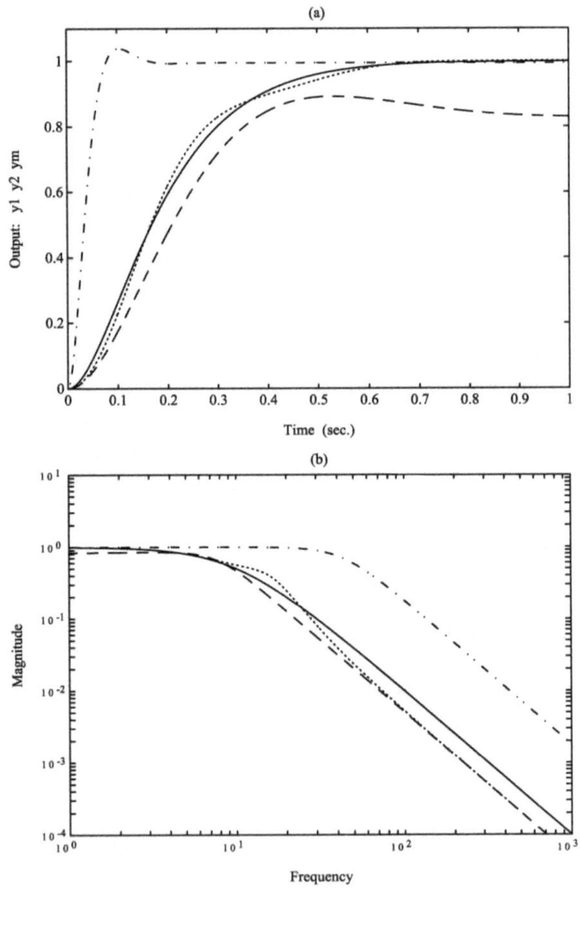

Fig. 10.18 Time and frequency responses

output is:

$$y_{1\infty} = \frac{ck_p}{b + ck_p} = 0.8333.$$

The resulting steady-state error of 0.1667 is clearly visible.

- The enhanced control system response y_2 (dotted line) closely follows the ideal trajectory. The steady-state output of the robustly enhanced system is:

$$y_{2\infty} = \frac{ck_2k_p}{ck_2k_p} = 1,$$

demonstrating zero steady-state error.

To improve the steady-state accuracy of the conventional control system, a high-gain design is applied with increased controller gains: $k_v = 120$, $k_p = 3600$. With this modification, the steady-state output is significantly improved:

$$y_{1\infty} = 0.9945.$$

However, the step response of the high-gain system [dash-dot line in Fig. 10.18a] exhibits undesirable overshoot despite being ideally designed for critical damping. This illustrates a common drawback of high-gain tuning—while it improves steady-state accuracy, it exacerbates transient overshoot and sensitivity to disturbances.

The frequency responses of the three systems are plotted in Fig. 10.18b {Solid line: Y_m. Dashed line: Y_1. Dotted line: Y_2. Dash-dot line: High-gain Y_1}.

- The conventional high-gain system Y_1 (dash-dot line) exhibits a significantly expanded bandwidth. While high-gain tuning can improve robustness against model uncertainties and disturbances, it also increases system bandwidth, which may be undesirable.
- The ideal system Y_m (solid line) provides the target bandwidth, ensuring an optimal balance between response speed and robustness.
- The enhanced system Y_2 (dotted line) maintains a frequency response closely aligned with the ideal system, demonstrating that the robustness enhancer effectively compensates for uncertainties without excessively increasing bandwidth.

In control design, bandwidth plays a critical role in defining system performance:

1. Response Speed
    - Bandwidth is intuitively related to response speed—higher bandwidth often results in faster rise and settling times [21].
    - However, excessive bandwidth can cause undesirable transient effects, including high overshoot and oscillations.

2. Noise Rejection

   - The filtering capability of a system is determined by its bandwidth. Signals with frequency components beyond the bandwidth threshold ($\omega > \omega_b$) are effectively attenuated or eliminated.
   - Excessively high bandwidth allows high-frequency noise to enter the system, potentially leading to control instability and degraded performance.

3. Implementation Costs

   - High-bandwidth designs demand high-performance components with greater precision and faster response times, increasing hardware and implementation costs [22].
   - If the desired bandwidth is set to match that of the ideal system Y_m, the excessive bandwidth expansion of the high-gain system Y_1 is clearly unwanted.

In contrast, the robustness-enhanced system (Y_2, dotted line) achieves improved disturbance rejection and uncertainty mitigation while maintaining a reasonable bandwidth. This demonstrates that the introduced approach effectively enhances robustness without the drawbacks of aggressive high-gain tuning, offering a more balanced and efficient control solution.

The concept of loop gain is fundamental in control design, as high loop gains or "tight" feedback loops generally contribute to improved performance in terms of disturbance rejection and tracking accuracy. However, designing loop gains appropriately requires careful consideration of frequency-dependent trade-offs. This section provides a comparative study of the loop-gain characteristics for the three systems discussed earlier.

The loop gain quantifies the open-loop amplification of system dynamics before feedback correction. The loop gains for the ideal closed-loop system, the conventional control system, and the enhanced control system are defined as follows:

- Ideal closed-loop system:

$$\mathrm{L}_m = \mathrm{P}_m \mathrm{F}_B. \tag{10.61}$$

- Conventional control system:

$$\mathrm{L}_1 = \mathrm{P}\mathrm{F}_B. \tag{10.62}$$

- Enhanced control system:

$$\mathrm{L}_2 = \mathrm{P}[\mathrm{W} + (1 + \mathrm{W}\mathrm{P}_m)\mathrm{F}_B]. \tag{10.63}$$

In these equations:

- P_m represents the idealized plant dynamics.
- P denotes the actual plant dynamics, which may include model uncertainties.
- F_B is the conventional outer-loop PD feedback controller.
- W is the additional compensator introduced for robustness enhancement.

With a properly designed weight W, the loop gain of the enhanced control system is expected to be larger than that of the conventional control system within a specific frequency range, ensuring stronger robustness and improved disturbance rejection:

$$|L_2(j\omega)| > |L_1(j\omega)|, \quad \forall\omega \in \Omega, \tag{10.64}$$

where Ω denotes a critical frequency range where the control system must effectively respond to command inputs while compensating for plant variations and rejecting disturbances.

For SISO systems, this notion is well-understood in terms of magnitude response. In MIMO systems, an equivalent analysis is conducted using matrix norms and singular values [23]. The singular value decomposition (SVD) technique can be employed to assess the frequency-dependent robustness of MIMO systems.

A key measure of loop performance is the return-difference operator $(1 + L)$, which provides insight into how well the feedback system regulates disturbances [24]. In particular, the return-difference magnitudes:

$$(1 + L)$$

approximate the loop gain whenever L is large compared to unity. The inverse of the return-difference operator $(1 + L)^{-1}$ defines the sensitivity function, which quantifies the sensitivity of the system to disturbances and model uncertainties.

The sensitivity functions for the three systems are given below:

- Ideal system sensitivity function

$$S_m = (1 + L_m)^{-1}. \tag{10.65}$$

- Conventional system sensitivity function:

$$S_1 = (1 + L_1)^{-1}. \tag{10.66}$$

- Enhanced system sensitivity function:

$$S_2 = (1 + L_2)^{-1}. \tag{10.67}$$

A well-designed feedback system aims to maximize loop gain (i.e., increase L) while minimizing sensitivity (i.e., reducing S) over the critical frequency range where disturbances and uncertainties have significant effects [11].

While the goal of feedback control is to maximize loop gain, practical design constraints prevent arbitrarily high gains over all frequency ranges. Several factors limit the achievable loop gain:

- *Bandwidth and Stability Trade-off:* Increasing loop gain in low frequencies improves steady-state accuracy and disturbance rejection, but excessive gains at high frequencies can cause instability due to unmodeled high-frequency dynamics.
- *Noise Sensitivity:* While a high loop-gain reduces low-frequency disturbances, it amplifies high-frequency noise. A careful balance must be struck to ensure that sensor noise and actuator bandwidth constraints do not degrade performance.
- *System Bandwidth Limitations:* The physical constraints of actuators and sensors impose practical limits on gain magnitude, especially in high-bandwidth systems.
- *Robustness to Model Uncertainty:* The loop gain must remain stable under plant parameter variations, necessitating careful tuning of controllers and robustness compensators.

Although the fundamental objective of feedback control is to achieve high loop-gain and low sensitivity within critical frequency ranges, the design process is non-trivial due to performance trade-offs and fundamental system limitations [11, 24, 25].

The loop performance characteristics of the ideal system, conventional PD control system, and enhanced control system are quantitatively demonstrated in the following numerical example.

Example 10.11 In this example, the loop performance of the three control systems analyzed in Example 10.10 is further examined. The system parameters and control gains remain identical to those used in Example 10.10.

The frequency responses of the loop gains L_m are depicted in Fig. 10.19a {Solid line: L_m. Dashed line: L_1. Dotted line: L_2. Dash-dot line: High-gain L_1}. It is evident that the loop gain of the conventional high-gain control system (dash-dot line) is significantly larger than that of the system with lower, ideal gains (dashed line). The advantage of the enhanced control system becomes apparent from the distinct shape of its loop-gain frequency response. Specifically, $|L_2(j\omega)|$ (dotted line) exhibits a high magnitude in the low-frequency range but then drops off steeply at higher frequencies. Beyond this range, $|L_2(j\omega)|$ becomes even smaller than $|L_1(j\omega)|$ in the high-gain design. This characteristic suggests that the enhanced control system maintains strong low-frequency robustness while avoiding excessive high-frequency amplification, which is often associated with noise sensitivity and implementation challenges.

The return-difference operators $(1 + L)$, which serve as an alternative measure of loop performance, are shown in the inset of Fig. 10.19a. As expected, the return-difference magnitudes closely approximate the loop gains, particularly in the low-frequency range, further validating the loop characteristics.

Figure 10.19b {Solid line: S_m. Dashed line: S_1. Dotted line: S_2. Dash-dot line: High-gain S_1} presents the frequency responses of the sensitivity functions S for the

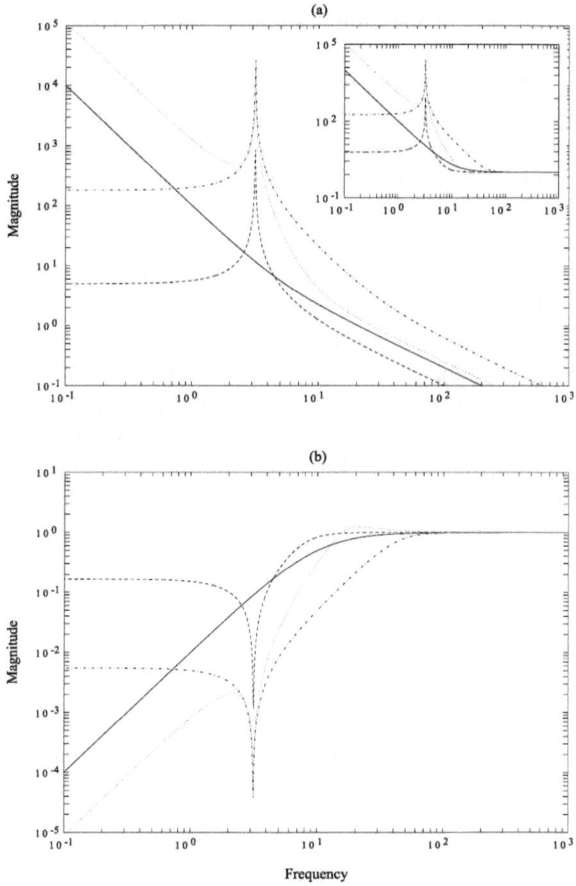

Fig. 10.19 Loop gain and sensitivity function

three systems. The sensitivity function of the conventional high-gain system (dash-dot line) is significantly lower than that of the system with lower gains (dashed line), indicating improved disturbance rejection. However, a key observation is that in the low-frequency range,

$$|S_2(j\omega)| \ll |S_1(j\omega)|,$$

demonstrating that the enhanced control system exhibits superior robustness compared to its conventional counterpart. This suggests that the introduced robustness enhancement effectively mitigates the impact of uncertainties and disturbances while maintaining desirable system dynamics.

Loop shaping is a well-established and powerful graphical technique for controller design, particularly in achieving robust performance in control systems [10, 11, 26, 27]. This method is grounded in frequency-domain analysis, where many key performance and robust stability objectives can be formulated as constraints on the magnitude or maximum singular values of specific closed-loop transfer functions [10].

The fundamental principle behind loop shaping is that the magnitude (or singular values) of these closed-loop transfer functions can be directly influenced by shaping the open-loop transfer function, or loop gain, over the relevant frequency ranges. By appropriately selecting a controller that modifies the open-loop gain characteristics, desired closed-loop specifications can be systematically met. This approach provides a structured methodology for balancing robustness and performance while ensuring stability across various operating conditions.

In the context of the robust control framework discussed in this chapter, improving the reduction of system uncertainties can also be achieved through intentional loop shaping. By designing a compensator that optimally adjusts the loop gain across critical frequency ranges, the sensitivity of the system to parameter variations, disturbances, and unmodeled dynamics can be minimized. This presents a promising avenue for future development, where advanced loop-shaping techniques could be explored to further enhance robustness and performance in practical implementations.

10.6 Summary

This chapter introduces a structured framework for enhancing the robustness of model-based control strategies in uncertain dynamic systems. The core principle of the presented scheme is uncertainty reduction, achieved through a parallel compensation mechanism that aligns system dynamics with an idealized reference model. By actively mitigating deviations between the actual system and its nominal counterpart, the introduced approach significantly improves tracking accuracy, disturbance rejection, and overall stability, particularly in model-based control architectures where robustness is critical.

Structurally, the introduced scheme aligns with conventional model-following (model-reference) control architectures, making it a natural extension of existing methodologies. By reconfiguring the loop structure, it can also be interpreted as a two-degrees-of-freedom control system, a well-established framework that enables independent design of reference tracking and disturbance rejection. For over five decades, two-degrees-of-freedom control has been recognized for its ability to decouple command input response from closed-loop dynamics [28–30]. The presented scheme advances this paradigm by introducing an additional degree-of-freedom dedicated explicitly to uncertainty reduction, significantly enhancing robust controller design flexibility.

The scheme is systematically applied to mass–stiffness–damping systems that are serially pre-compensated using model-based algorithms, such as those in Chap. 6, resulting in a dual-model-based control system. Here, pre-compensation and parallel compensation operate synergistically to improve robustness. Case studies and experimental validation confirm that the dual-model-based framework effectively reduces the impact of large uncertainties and external disturbances, ensuring superior closed-loop performance while preserving computational efficiency.

The analysis in this chapter is intentionally structured to be accessible, avoiding excessive mathematical complexity while maintaining rigor in theoretical foundations. While the synthesis of the control structure is not explored in depth, its design is inherently compatible with advanced robust control techniques, such as $\mathcal{H}^\infty$ optimization, which could further extend its applicability. $\mathcal{H}^\infty$ control, often presented in an abstract mathematical setting, fundamentally revolves around shaping the closed-loop frequency response, aligning well with the objectives of the introduced scheme [31]. Future work could integrate $\mathcal{H}^\infty$ design principles with this structure to optimize robustness across a broader range of system uncertainties.

Beyond its theoretical foundation, the scheme provides a compact and quantitative uncertainty reduction mechanism that improves the robustness of model-based controllers. Its generic and modular nature enables seamless integration into existing control loops with minimal structural modifications and computational overhead, making it a practical and scalable solution. This study establishes a strong foundation for applying the uncertainty reduction framework to a wide range of engineering systems, further broadening the scope of robust control methodologies. The structural properties and broader applications of this framework will be explored in greater depth in subsequent chapters, providing deeper insights into its potential for addressing complex control challenges in real-world applications.

References

1. Dorato P (1987) A historical review of robust control. IEEE Control Syst Mag 7(2):44–47
2. Petersen IR, Tempo R (2014) Robust control of uncertain systems: classical results and recent developments. Automatica 50(5):1315–1335
3. Bhattacharyya SP (2017) Robust control under parametric uncertainty: an overview and recent results. Annu Rev Control 44:45–77
4. Dorato P, Tempo R, Muscato G (1993) Bibliography on robust control. Automatica 29(1):201–213
5. Safonov MG (2012) Origins of robust control: early history and future speculations. Annu Rev Control 36(2):173–181
6. Fisher DG (1991) Process control: an overview and personal perspective. Can J Chem Eng 69(1):5–26
7. Zhu HA, Teo CL, Hong GS, Poo AN (1992) An enhanced scheme for the model-based control of robot manipulators. Int J Control 56(6):1243–1261
8. Kailath T (1980) Linear systems. Prentice Hall, New Jersey
9. Vidyasagar M (2011) Control system synthesis: a factorization approach. Morgan & Claypool, California
10. McFarlane DC, Glover K (1990) Robust controller design using normalized coprime factor plant descriptions. Springer, Berlin
11. Doyle JC, Stein G (1981) Multivariable feedback design: concepts for a classical/modern synthesis. IEEE Trans Autom Control 26(1):4–16
12. Wei K, Yedavalli RK (1989) Robust stabilizability for linear systems with both parameter variation and unstructured uncertainty. IEEE Trans Autom Control 34(2):149–156

13. Dorato P, Fortuna L, Muscato G (1992) Robust control for unstructured perturbations: an introduction. Springer, Berlin
14. Callier FM, Desoer CA (1991) Linear system theory. Springer, Berlin
15. Shinskey FG (1990) How Good are our controllers in absolute performance and robustness? Meas Control 23(4):114–121
16. Jaworska, Finkelstein L (1991) Control systems robustness and its measurement. Trans Inst Meas Control 13(4):170–174
17. Morari M, Zafiriou E (1989) Robust process control. Prentice-Hall, New Jersey
18. Stefani RT, Shahian B, Savant Jr CJ, Hostetter GH (2002) Design of feedback control systems. Oxford University Press, Oxford
19. Isermann R (1989) Digital control systems, vol. 1: fundamentals, deterministic control. Springer, Berlin
20. Derbel N, Nouri AS, Zhu Q (2023) Advances in robust control and applications. Springer Nature, Switzerland
21. Chen CT (1993) Analog and digital control system design: transfer-function, state-space, and algebraic methods. Saunders College Publishing, HBJ International
22. Ogata K (2010) Modern control engineering. Prentice-Hall, New Jersey
23. Stewart GW (1973) Introduction to matrix computation. Academic Press, California
24. Safonov MG, Laub AJ, Hartmann GL (1981) Feedback properties of multivariable systems: the role and use of the return difference matrix. IEEE Trans Autom Control 26(1):47–65
25. Boyd SP, Barratt CH (1991) Linear controller design: limits of performance. Prentice Hall, New Jersey
26. Anderson BDO, Moore JB (1989) Optimal control: linear quadratic methods. Prentice-Hall, New Jersey
27. Doyle JC, Francis BA, Tannenbaum AR (1992) Feedback control theory. Macmillan Publishing Company, New York
28. Truxal JG (1955) Automatic feedback control system synthesis. McGraw-Hill, London
29. Horowitz IM (1963) Synthesis of feedback systems. Academic Press, California
30. Desoer CA, Lin CA (1985) A comparative study of linear and non-linear MIMO feedback configurations. Int J Syst Sci 16(7):789–813
31. Stoorvogel (1992) The H∞ control problem: a state space approach. Prentice-Hall International, New Jersey

Chapter 11
Structural Properties of Model-Following Schemes

The model-following control (MFC) structure has been utilized in this book as a systematic approach to uncertainty reduction. By integrating model-following principles into the robust control of mass–stiffness–damping systems, the introduced scheme actively compensates for model inaccuracies, parameter variations, and external disturbances, thereby significantly improving system performance.

As demonstrated in Sect. 10.4.2, two different model-following configurations were applied to enhance the robustness of model-based control strategies. These configurations effectively reduced the influence of plant uncertainties by forcing the system dynamics to closely follow a desired reference model. While some fundamental structural properties of these configurations were previously introduced, their deeper analytical foundations and performance characteristics were not fully explored.

This chapter aims to further elaborate on the structural properties of these model-following schemes, providing a comprehensive examination of their theoretical underpinnings and practical implications. The focus will be on understanding how the introduced configurations influence system stability, performance robustness, and control flexibility. Additionally, the chapter will explore how these properties can be leveraged for broader applications, extending beyond mass–stiffness–damping systems to a wider class of uncertain dynamic systems.

11.1 Introduction

MFC [1–3] has been widely utilized in control theory for several decades, with extensions into adaptive MFC [4, 5] and variable-structure MFC [6] further enhancing its applicability in dynamic and uncertain environments. A key advantage of MFC lies in its ability to define the desired system behavior through a reference model, rather than relying on explicit performance indices. Since specifying appropriate

H. Zhu, *Model-Based Control of Mass–Stiffness–Damping Systems*, Advances in Industrial Control, https://doi.org/10.1007/978-3-031-97592-9_11

performance indices is often challenging in practical control applications, MFC offers a more intuitive and structured alternative by enforcing a predefined response through a reference model. Given its inherent ability to shape system dynamics in a desired manner, MFC naturally emerges as a preferred approach for control systems operating in uncertain conditions.

In a MFC system, the plant is regulated to closely approximate the behavior of a specified reference model, which serves as an integral part of the control design. The model-following controller must ensure that the error between the plant response and the model remains zero or sufficiently small as time progresses. Reports from various control applications confirm that MFC is an effective control method with intrinsic robustness against plant uncertainties and external disturbances. However, as demonstrated in [7] and through Examples 10.1–10.4 in Chap. 10, the robustness and performance of MFC systems can be significantly influenced by their specific structural configurations.

Building on the insights from Chap. 10, this chapter provides a deeper structural analysis of MFC for robustness enhancement. It was shown in Sect. 10.4.2 that two different model-following configurations were used to enhance the robustness of mass–stiffness–damping systems, demonstrating the potential of MFC as an uncertainty reduction mechanism. However, certain technical aspects of these configurations were not fully explored. This chapter aims to further analyze and generalize the structural properties of these MFC schemes, offering a more rigorous understanding of their impact on system robustness.

Comparative studies will show that the structure of a MFC system can influence its robustness against modeling uncertainty and external disturbance. It will also be shown that the results presented in this chapter are general purpose and can be used in most of the model-based applications. MFC of mass–stiffness–damping systems with revealing results is presented to illustrate the influence of model-following structure on the resulting performance of the model-based control systems.

11.2 Conventional Structure: Outer-Loop MFC

Consider the MFC system depicted in Fig. 11.1a, which represents a conventional configuration of MFC widely adopted in control applications, as previously discussed in Fig. 10.4. To facilitate the analysis, this chapter focuses on linear time-invariant (LTI) single-input, single-output (SISO) systems and assumes that the given structure is internally stable. In Fig. 11.1a, the reference model G_m specifies the ideal (desired) input–output behavior of the original closed-loop system without the MF compensator, such that:

$$\frac{\mathrm{P}}{(1+\mathrm{PF}_B)} \rightarrow \mathrm{G}_m, \tag{11.1}$$

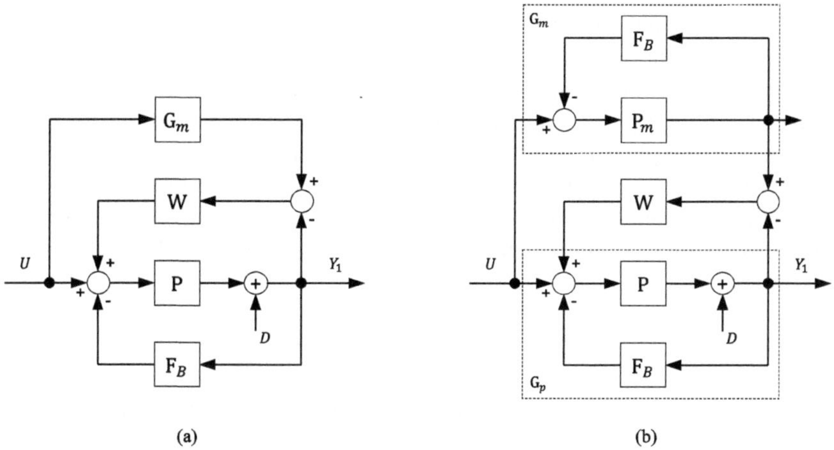

Fig. 11.1 Outer-loop MFC and its equivalent

where P represents the equivalent plant, F_B is the original outer-loop feedback controller, and W is a weight designed for error compensation in the model-following structure.

To gain deeper insight into the structure, suppose that the reference model G_m can be rearranged and rewritten as:

$$G_m = \frac{P_m}{(1 + P_m F_B)}, \tag{11.2}$$

where P_m can be regarded as the nominal dynamic model of the equivalent plant P. By substituting this expression into the model-following framework, the system in Fig. 11.1a can be transformed into an equivalent structure, as shown in Fig. 11.1b. In this transformed configuration, the two subsystems within the dashed circles share an identical form, indicating that most conventional MFC systems can be expressed in this structure. Henceforth, the MFC system shown in Fig. 11.1b will be referred to as the outer-loop model-following control system.

For the outer-loop model-following control system, the corresponding closed-loop transfer function from the command input U to the output Y_1 is obtained as:

$$G_1 = \frac{G_p(1 + G_m W)}{\left(1 + G_p W\right)}, \tag{11.3}$$

where

$$G_p = \frac{P}{(1 + P\,F_B)}, \tag{11.4}$$

represents the original closed-loop transfer function of the system without model-following compensation.

A direct substitution of G_m and G_p into Eq. (11.3) results in a complex transfer function, making analytical evaluation difficult. Instead, by regarding the discrepancy between the actual plant P and its nominal model P_m as modeling uncertainty, the system response can be expressed in a simplified uncertainty framework.

Given that uncertainty models can be transformed mathematically from one representation to another [as illustrated in Eq. (10.6)], this chapter adopts the additive uncertainty model:

$$P = P_m + \Delta, \tag{11.5}$$

where P_m is the modeled (nominal) dynamics of the plant, and Δ represents the additive uncertainty in the system.

Since G_m defines the ideal closed-loop dynamics from U to Y_1, the residual uncertainty Δ_{r1} in the closed-loop transfer function G_1 can be derived as:

$$\Delta_{r1} = G_1 - G_m = \frac{1}{(1 + P_m F_B)} \cdot S_1 \cdot \Delta, \tag{11.6}$$

where

$$S_1 = \frac{1}{1 + P(F_B + W)}, \tag{11.7}$$

denotes the sensitivity operator of the outer-loop model-following control system.

Similarly, for the external disturbance D acting on the system in Fig. 11.1b, the corresponding disturbance response Y_1^D (from D to Y_1) can be determined as:

$$Y_1^D = \frac{1}{1 + P(F_B + W)} D = S_1 \cdot D. \tag{11.8}$$

The results in Eqs. (11.6) and (11.8) reveal an important property of outer-loop model-following control:

- The influence of both the modeling uncertainty Δ and the external disturbance D on the system response is attenuated by a factor of S_1^{-1}.
- This indicates that the sensitivity function S_1 plays a critical role in determining the robustness of the system against uncertainties and disturbances.
- If the system is well-designed such that S_1 is small in critical frequency ranges, then both modeling errors and disturbances are effectively suppressed.

These findings emphasize the robustness-enhancing capability of the model-following structure, providing a quantitative understanding of how uncertainties and disturbances are mitigated in outer-loop model-following control systems. In the following sections, these properties will be further analyzed, along with comparative studies to evaluate different model-following configurations in terms of their robustness characteristics.

11.3 Modified Structure: Enhanced MFC

An alternative MFC configuration is examined here by analyzing the relationship between the plant G_p and the reference model G_m in Fig. 11.1b. A key observation is that their feedback components are structurally identical. By leveraging this equivalence, the feedback controller F_B can be designed to be the same for both the plant and the model. This flexibility arises because the designer has full control over the configuration of G_m and its associated feedback loop, making it possible to align its structure with that of the actual plant system.

Building upon this insight, an alternative model-following scheme can be derived, leading to the modified configuration shown in Fig. 11.2, which resembles the structure previously introduced in Fig. 10.7. This modified configuration is referred to as the inner-loop model-following control system in this chapter. It is assumed that the plant P, its nominal model P_m, the feedback controller F_B, and the compensating weight W in the inner-loop model-following system are identical to those in the outer-loop configuration. The modified scheme is internally stable, but the synthesis of F_B and W is not explicitly addressed in this chapter.

For the system in Fig. 11.2, the transfer function G_2 from the command input U to the output Y_2 is given as:

$$G_2 = \frac{P_c}{(1 + P_c F_B)}, \tag{11.9}$$

where the compensated plant transfer function P_c is:

$$P_c = \frac{P\,(1 + P_m W)}{(1 + PW)}. \tag{11.10}$$

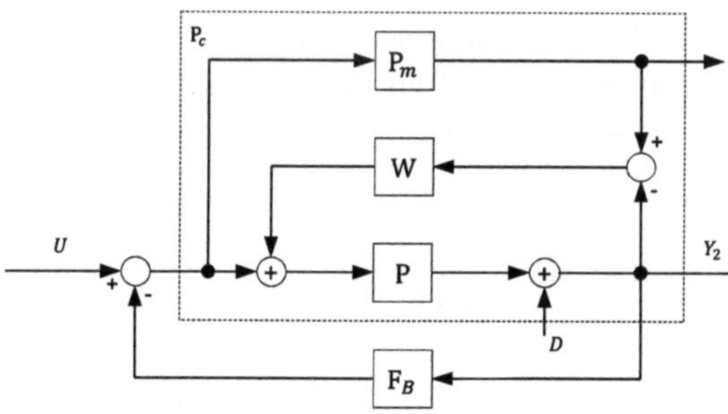

Fig. 11.2 Inner-loop model-following control

Since G_m represents the ideal closed-loop response from U to Y_2, the residual uncertainty Δ_{r2} in G_2 can be expressed as:

$$\Delta_{r2} = G_2 - G_m = \frac{1}{(1 + P_m F_B)} \cdot S_2 \cdot \Delta, \tag{11.11}$$

where the modified sensitivity function S_2 is:

$$S_2 = \frac{1}{1 + P(F_B + W) + PP_m F_B W}. \tag{11.12}$$

This expression reveals that Δ_{r2} follows the same structural form as Δ_{r1} in Eq. (11.6). Similarly, the disturbance response Y_2^D from external disturbance D to output Y_2 is:

$$Y_2^D = \frac{1}{1 + P(F_B + W) + PP_m F_B W} D = S_2 \cdot D. \tag{11.13}$$

This is structurally identical to the disturbance response Y_1^D in Eq. (11.8), indicating that in the inner-loop model-following control system, both modeling uncertainty and external disturbances are attenuated by a factor of S_2^{-1}. The key distinction between the inner-loop and outer-loop configurations lies in the difference between their respective sensitivity functions S_2 and S_1.

By comparing S_1 from Eq. (11.7) and S_2 from Eq. (11.12), it is evident that the denominator of S_2 contains an additional term $PP_m F_B W$. This additional term provides an opportunity for enhanced robustness. Under appropriate design of W, it follows that:

$$|1 + P(F_B + W)| < |1 + P(F_B + W) + PP_m F_B W|, \quad \forall \ \omega \in \Omega, \tag{11.14}$$

where Ω is the frequency range of interest. This inequality suggests that the inner-loop model-following control configuration achieves a higher loop gain in the desired frequency range, leading to:

$$|S_1(j\omega)| > |S_2(j\omega)|, \quad \forall \ \omega \in \Omega, \tag{11.15}$$

which implies that the inner-loop structure yields better sensitivity reduction. As a direct consequence, from Eqs. (11.6) to (11.11), the residual uncertainty in the inner-loop case satisfies:

$$|\Delta_{r1}(j\omega)| > |\Delta_{r2}(j\omega)|, \quad \forall \ \omega \in \Omega. \tag{11.16}$$

Similarly, from Eqs. (11.8) to (11.13), the disturbance response satisfies:

$$\left|Y_1^D(j\omega)\right| > \left|Y_2^D(j\omega)\right|, \quad \forall \ \omega \in \Omega. \tag{11.17}$$

These results confirm that under the same weighting function W, the inner-loop model-following control system achieves greater robustness compared to the outer-loop counterpart. Specifically, the inner-loop approach exhibits stronger attenuation of both modeling uncertainty and external disturbances.

It is important to note that the validity of these inequalities depends on the appropriate design of the weight W, which may vary across different plant systems. While the precise selection of W is beyond the scope of this chapter, the general trend observed in Eqs. (11.16) and (11.17) demonstrates that the inner-loop model-following configuration is fundamentally more robust than the outer-loop alternative.

Numerical results in Chap. 10, particularly Examples 10.3 and 10.4, support these theoretical findings. The simulations illustrate that the inner-loop model-following control system consistently exhibits superior performance, reducing the adverse effects of plant uncertainties and external disturbances more effectively than the outer-loop model-following approach. This observation reinforces the analytical conclusions drawn in this chapter.

Overall, the inner-loop model-following control scheme offers a structural advantage by embedding the compensatory mechanism directly within the primary feedback loop of the plant. This modification enhances loop gain, improves sensitivity characteristics, and strengthens the robustness of the system against both parameter variations and disturbances. By properly designing W, the inner-loop configuration can achieve performance levels unattainable with conventional outer-loop model-following control.

11.4 Equivalence Condition

As shown in the previous analysis, the inner-loop and outer-loop model-following control configurations exhibit different levels of robustness against uncertainties, despite using the same compensating weight W. The inner-loop structure inherently achieves higher loop gain, leading to better attenuation of modeling uncertainties and external disturbances. However, under certain conditions, the input–output performance of both configurations can be identical. The necessary conditions for achieving this equivalence are formally stated in Property 11.1.

Property 11.1 (Equivalence Condition) *The outer-loop and inner-loop model-following control systems achieve input–output equivalence if the weight for the outer-loop MFC is designed as*

$$W_1 = K\frac{P_f}{G_m}, \tag{11.18}$$

and the weight for the inner-loop model-following control is given by

$$W_2 = K\frac{P_f}{P_m}, \tag{11.19}$$

where P_f *serves as a pre-filter to ensure the physical realizability of* W_1 *and* W_2.

Proof: From the input–output relations of the outer-loop model-following control in Eq. (11.3) and the inner-loop model-following control in Eq. (11.9), it follows that when the weights are chosen as

$$W_1 = K\frac{P_f}{G_m},$$

and

$$W_2 = K\frac{P_f}{P_m},$$

the resulting transfer function for both configurations becomes

$$G_1 = G_2 = \frac{(1 + KP_f)P\,P_m}{(P_m + KP_fP) + (1 + KP_f)F_BP\,P_m}, \tag{11.20}$$

which establishes that the input–output behavior of the two MFC systems is identical. □

This finding highlights that the choice between the inner-loop and outer-loop model-following configurations should be guided by practical considerations, such as computational complexity, system constraints, and implementation feasibility. While the inner-loop configuration inherently offers superior robustness against modeling uncertainties and external disturbances, the outer-loop approach may be preferable in scenarios where modifying inner-loop dynamics is impractical or where simplicity in controller design is prioritized. The equivalence established in Property 11.1 ensures that, under appropriate weight selection, both configurations yield identical input–output performance. This result was further corroborated by numerical examples in Examples 10.1 and 10.3 of Chap. 10, as well as in [7].

It is well-understood that any reconfiguration of a control loop intended to enhance performance will inevitably influence the stability characteristics of the system. However, such changes should not be regarded as inherently detrimental. Instead, they offer an opportunity to refine the dynamic behavior of the system in alignment with specific control objectives. The inner-loop model-following scheme introduced in this chapter exemplifies this principle. While it may yield a narrower stability margin compared to the outer-loop configuration, it provides a compelling trade-off by significantly enhancing robustness to system uncertainties.

Although stability robustness and performance robustness are closely linked in control theory, they represent distinct facets of the response of a system to uncertainty or perturbation. Their key differences are summarized in Table 11.1.

Table 11.1 Comparison of stability robustness and performance robustness

Aspect	Stability robustness	Performance robustness
Goal	Maintain stability under uncertainty	Achieve control objectives under uncertainty
Outcome	Binary (stable/unstable)	Graded (degraded or acceptable performance)
Evaluation	Gain/phase margins, nyquist, μ-analysis	$\mathcal{H}^\infty$ norms, sensitivity functions, transient response
Failure mode	Instability	Slower response, higher error, increased oscillations
Acceptable trade-off	Often slightly sacrificed for performance	Sometimes traded off to preserve stability

Stability robustness refers to the ability of the system to remain stable despite modeling inaccuracies, external disturbances, or parameter variations. It addresses a binary outcome—whether the system retains stability under uncertainty—and is typically assessed using gain and phase margins, Bode/Nyquist criteria, or structured singular value (μ) analysis. For instance, a suspension system with shifting dynamics due to tire-pressure variations can remain stable if designed with strong stability robustness.

Performance robustness, on the other hand, concerns the preservation of desired control performance—such as tracking accuracy, disturbance rejection, or transient behavior—despite uncertainty. Even when stability is maintained, performance may degrade due to unmodeled dynamics or system variations. Metrics like sensitivity functions, $\mathcal{H}^\infty$ norms, settling time, and overshoot are used to quantify this aspect. For example, a robotic manipulator may stay stable under a payload change but exhibit slower tracking, indicating reduced performance robustness.

Given the inherent trade-off between these two forms of robustness, optimal control design involves balancing them based on application-specific priorities. No single configuration can maximize both simultaneously; practical implementation requires informed compromises. The reconfiguration strategy introduced in this chapter is not presented as a one-size-fits-all solution but rather as a flexible approach for enhancing system performance. Its suitability should be evaluated relative to the stability and robustness demands of the application at hand.

11.5 Model-Following Control of MKC Systems

The analysis above demonstrates that the dynamic performance of the two model-following schemes differs under the same weight W. To further investigate these differences, an illustrative control problem involving a mass–stiffness–damping system is considered. The open-loop dynamics of a single-degree-of-freedom system, which has been decoupled and linearized through model-based methods [7], can be expressed as:

$$\mathrm{P} = \frac{c}{s^2 + as + b}, \tag{11.21}$$

where s is the Laplace operator, and a, b, and c are real constants. This formulation represents a broad class of engineering systems, including mass–spring–damper systems. Suppose the ideal model of the plant system is given by:

$$\mathrm{P}_m = \frac{1}{s^2}, \tag{11.22}$$

which corresponds to a double integrator. In an ideal setting with perfect modeling, the parameters should satisfy $a = 0$, $b = 0$, and $c = 1$, implying that the inner-loop pre-compensated dynamics of an individual degree-of-freedom system with negligible actuator dynamics should ideally match this form.

To enhance robustness, the compensating weight W and the outer-loop feedback controller F_B are designed as:

$$\mathrm{W} = k_1 s + k_2, \tag{11.23}$$

which is a PD (position-velocity) compensator utilizing the position and velocity errors between the model and the plant, and:

$$\mathrm{F}_B = k_v s + k_p, \tag{11.24}$$

which is a PD feedback controller for closed-loop control. Substituting these into the transfer functions derived earlier for the two model-following schemes, the resulting system dynamics for both configurations take the general form:

$$\mathrm{G} = \frac{n_1 s^2 + n_2 s + n_3}{m_1 s^4 + m_2 s^3 + m_3 s^2 + m_4 s + m_5}, \tag{11.25}$$

where the coefficients differ between the two configurations. Specifically, for the outer-loop model-following system:

$$n_{11} = c,$$

$$n_{12} = c(k_1 + k_v),$$

$$n_{13} = c\left(k_2 + k_p\right),$$

$$m_{11} = 1,$$

$$m_{12} = a + k_v + c(k_1 + k_v),$$

$$m_{13} = b + k_p + ak_v + c\left(k_2 + k_p + k_1k_v + k_v^2\right),$$

$$m_{14} = k_p[a + c(k_1 + k_v)] + k_v\left[b + c\left(k_2 + k_p\right)\right],$$

$$m_{15} = k_p\left[b + c\left(k_2 + k_p\right)\right].$$

For the inner-loop model-following system:

$$n_{21} = c,$$

$$n_{22} = ck_1,$$

$$n_{23} = ck_2,$$

$$m_{21} = 1,$$

$$m_{22} = a + c(k_1 + k_v),$$

$$m_{23} = b + c\left(k_2 + k_p + k_1k_v\right),$$

$$m_{24} = c\left(k_1k_p + k_2k_v\right),$$

$$m_{25} = ck_2k_p.$$

Under perfect modeling conditions ($a = 0$, $b = 0$, $c = 1$), both configurations reduce to:

$$\mathrm{G}_m = \frac{1}{s^2 + k_v s + k_p}, \tag{11.26}$$

indicating identical nominal performance. However, discrepancies arise under modeling uncertainties and external disturbances.

11.5.1 Steady-State Effect of Modeling Uncertainty

To evaluate steady-state accuracy, consider a unit-step input. The final value for the outer-loop model-following system is:

$$y_{1\infty} = \frac{n_{13}}{m_{15}} = \frac{c(k_2 + k_p)}{k_p[b + c(k_2 + k_p)]}, \tag{11.27}$$

while for the inner-loop model-following system:

$$y_{2\infty} = \frac{n_{23}}{m_{25}} = \frac{1}{k_p}, \tag{11.28}$$

which matches the ideal response:

$$y_{m\infty} = \frac{1}{k_p}. \tag{11.29}$$

The result indicates that under modeling inaccuracies ($b \neq 0$, $c \neq 1$), the outer-loop configuration exhibits steady-state error, while the inner-loop configuration does not. This phenomenon was also illustrated in Examples 10.2 and 10.4.

11.5.2 Steady-State Effect of External Disturbance

Examining the disturbance response, we derive that for a unit-step disturbance, the steady-state response of the outer-loop model-following system is:

$$y_{1\infty}^D = \frac{b}{b + c(k_2 + k_p)}, \tag{11.30}$$

implying incomplete compensation for constant disturbances. In contrast, the inner-loop model-following system achieves perfect disturbance rejection:

$$y_{2\infty}^D = 0. \tag{11.31}$$

These results highlight an important structural advantage of the inner-loop configuration. When the ideal model P_m is a double integrator, the outer-loop model-following system exhibits steady-state error in response to modeling mismatches or disturbances because the integrators of the model are outside the feedback loop [8–10]. In contrast, the inner-loop configuration places the integrators within the feedback loop, eliminating steady-state error. This characteristic is also evident in the additional term PP_mF_BW in the sensitivity function S_2. The following numerical example further illustrates these properties.

Example 11.1 Suppose the mass–stiffness–damping system parameters are set as $a = 10$, $b = 100$, and $c = 10$. These values are deliberately chosen to be large to highlight the impact of model uncertainty. The control gains for the closed-loop and model-following controllers are set as $k_v = 20$, $k_p = 100$; $k_1 = 30$, and $k_2 = 100$. Using these parameters, the transfer functions G_m, G_1, and G_2 are derived for comparison.

Since the step response of a dynamic system provides insight into both its transient and steady-state behavior, a unit-step input is applied to both model-following configurations to evaluate their performance. The results are presented in Fig. 11.3 {Solid line: y_m. Dashed line: y_1. Dotted line: y_2}, where the solid line represents the response y_m of the ideal system G_m, the dashed line corresponds to the response y_1 of the outer-loop model-following system G_1, and the dotted line shows the response y_2 of the inner-loop model-following system G_2. The results clearly indicate that under model mismatches, the inner-loop model-following system significantly outperforms the outer-loop counterpart. The outer-loop configuration exhibits a substantial steady-state error, while the inner-loop configuration effectively eliminates it.

To further examine frequency-domain robustness, the frequency responses of the sensitivity operators S_1 (dashed line) and S_2 (dotted line) are shown in Fig. 11.4a {Dashed line: S_1. Dotted line: S_2}. It is evident that:

$$|S_2(j\omega)| \ll |S_1(j\omega)|, \quad \forall\, \omega < 10,$$

where $\omega = 10$ is determined by the closed-loop gain $k_p = 100$ of the feedback controller F_B. This result confirms that the inner-loop model-following control configuration is significantly more robust in terms of both uncertainty reduction and disturbance attenuation across the relevant frequency band.

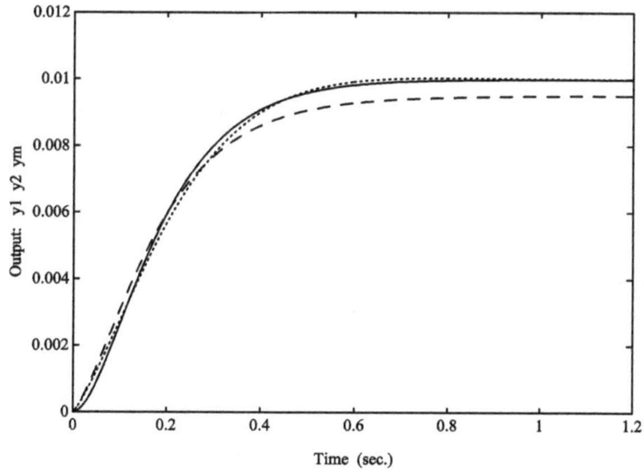

Fig. 11.3 Unit-step response of the model-following systems

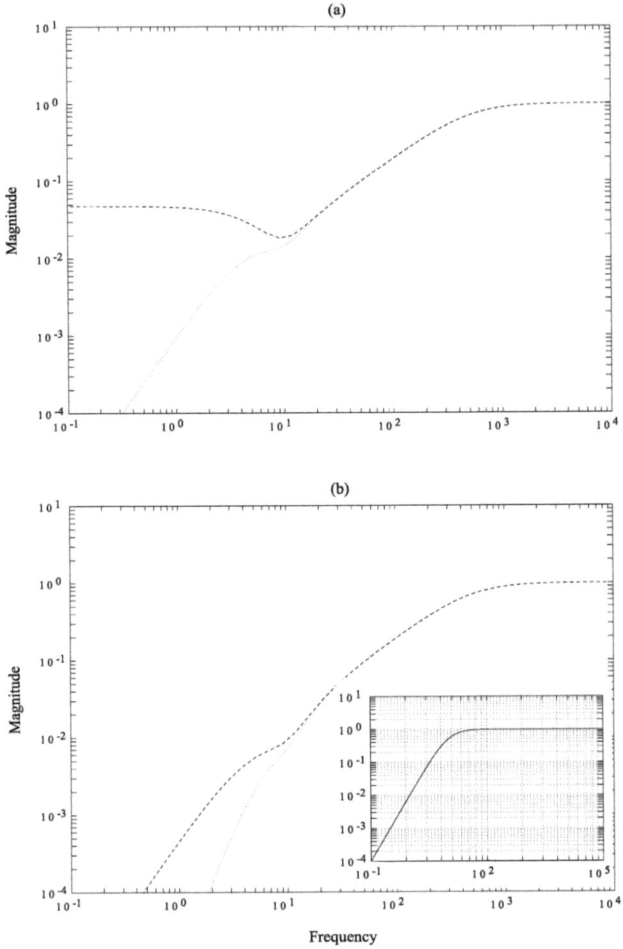

Fig. 11.4 Frequency response of sensitivity function

The residual uncertainty Δ_{ri} (for $i = 1, 2$) in a model-following system is given by:

$$\Delta_{ri} = \mathrm{S}_{ri} \cdot \Delta,$$

where

$$\mathrm{S}_{ri} = \frac{1}{(1 + \mathrm{P}_m \mathrm{F}_B)} \cdot \mathrm{S}_i,$$

represents the sensitivity operator for uncertainty reduction. The frequency responses of the resulting uncertainty reduction sensitivity operators S_{ri} (for $i = 1, 2$) are displayed in the main figure of Fig. 11.4b {Dashed line: S_{r1}. Dotted line: S_{r2}}, while the frequency response of the term $(1 + P_m F_B)^{-1}$ is shown in the inset of Fig. 11.4b. Notably, the low-frequency characteristics differ from those in Fig. 11.4a, further emphasizing the superior robustness of the inner-loop model-following control system.

11.6 Summary

This chapter provides a detailed exploration of the structural properties of MFC schemes, highlighting their influence on robustness against uncertainties and external disturbances. Both analytical derivations and numerical studies confirm that the specific configuration of a model-following system significantly affects its performance in dynamic and steady-state conditions. By applying MFC to a mass–stiffness–damping system, the comparative study offers valuable insights into how different configurations impact stability, steady-state accuracy, and overall system robustness.

Given the widespread use of MFC in various engineering applications, the insights presented in this chapter provide essential guidance for designing structurally simple yet robust control systems. The analysis underscores the importance of selecting an appropriate model-following architecture to achieve improved disturbance rejection and uncertainty attenuation. Notably, many model-following controllers are formulated in the state space and employ state error compensation strategies. The results derived from the mass–stiffness–damping system serve as a generalizable framework for enhancing the steady-state performance of such systems. By understanding the structural implications of different model-following configurations, control system designers can make informed decisions to optimize performance while maintaining implementation feasibility.

References

1. Monopoli RV (1981) Model following control of gas turbine engines. Trans ASME J Dyn Syst Measur Control 103(3):285–289
2. FitzSimons PM (1990) A robust model following controller. In: Proceedings of the 1990 American control conference, vol 3, pp 2626–2631
3. Wei JH, Shih MC, Chiu CP (1991) Model-following PID control using hyperstability theory. JSME Int J Ser 3 Vibr Control Eng Eng Ind 34(3):377–381
4. Dubowsky S, DesForges DT (1979) The application of model-referenced adaptive control to robotic manipulators. Trans ASME J Dyn Syst Measur Control 101(3):193–200
5. Landau YD (1979) Adaptive control: the model reference approach. Marcel Dekker
6. Zinober ASI, Ei-Ghezawi OME, Billings SA (1982) Multivariable variable-structure adaptive model-following control systems. IEE Proceed D Control Theory Appl 129(1):6–12
7. Zhu HA, Teo CL, Hong GS, Poo AN (1992) An enhanced scheme for the model-based control of robot manipulators. Int J Control 56(6):1243–1261

8. Sandell N, Athans M (1973) On "type L" multivariable linear systems. Automatica 9(1):131–136
9. Morari M (1985) Robust stability of systems with integral control. IEEE Trans Autom Control 30(6):574–577
10. Gu G, Yedavalli RK, Robust stability of "type L" feedback systems. In: Proceedings of the 29th IEEE conference on decision and control, vol 1, pp 423–424

Chapter 12
Enhanced Feedforward Control

Feedforward control is a fundamental technique in control engineering, widely used for improving reference tracking and disturbance rejection in dynamic systems. Unlike feedback control, which reacts to system deviations after they occur, feedforward control anticipates the required corrective actions based on a predictive model of the system dynamics. When designed correctly, feedforward control can significantly enhance system performance by preemptively compensating for known disturbances and reference trajectory variations.

However, as highlighted in previous chapters, conventional feedforward control is inherently model-dependent, making it highly sensitive to modeling inaccuracies. Since the feedforward controller directly relies on an inverse model of the plant, any discrepancies between the actual system and its mathematical representation can lead to degraded performance, steady-state errors, or even instability. In real-world applications, factors such as parameter variations, unmodeled dynamics, and external disturbances further exacerbate this issue, limiting the practical effectiveness of conventional feedforward schemes.

To address these challenges, this chapter focuses on robustness enhancement strategies for model-based feedforward control. The discussion builds on the concepts introduced in earlier chapters, leveraging model-following and uncertainty-reduction techniques to improve the reliability of feedforward controllers under uncertain conditions. By integrating robustness-enhancing modifications, the presented approaches aim to mitigate the adverse effects of model inaccuracies, ensuring that the system maintains high-performance tracking and disturbance rejection despite uncertainties. Through analytical studies and practical case examples, this chapter provides a comprehensive framework for designing feedforward control systems that are both effective and resilient to modeling errors.

H. Zhu, *Model-Based Control of Mass–Stiffness–Damping Systems*, Advances in Industrial Control, https://doi.org/10.1007/978-3-031-97592-9_12

12.1 Introduction

Feedforward control (FFC) is a widely employed technique in industrial automation and control, often implemented as an enhancement to conventional closed-loop systems. Its primary function is to improve system performance by enabling precise reference trajectory tracking and mitigating the impact of measurable disturbances before they affect the system output [1–4]. The theoretical foundation of feedforward control suggests that if an accurate mathematical model of the plant is available, the method can, at least in principle, achieve perfect control without relying on feedback corrections [5, 6]. However, practical implementation challenges often prevent this ideal performance from being realized.

Feedforward control is typically applied in two major categories of control problems. The first application is reference tracking [5, 7–11], as discussed in Chap. 7. In this context, a properly designed feedforward controller preemptively determines the control input needed to drive the system along a desired trajectory. Unlike feedback control, which reacts to deviations after they occur, feedforward control directly specifies the response dynamics of the system. However, it does not inherently contribute to system stability, as it operates independently of closed-loop feedback mechanisms [12]. The second, and often more critical, application is disturbance rejection, as explored in Chap. 9. When measurable disturbances frequently affect a system, feedforward control provides a proactive means of counteracting their effects before they propagate through the system [10, 11, 13, 14]. This is particularly advantageous over conventional feedback control, which only begins correcting for disturbances after they have influenced the output. By directly compensating for anticipated disturbances, feedforward control can significantly improve transient response and reduce deviations from the desired operating point.

Despite these advantages, two fundamental challenges limit the effectiveness of conventional feedforward control implementations. The first issue is sensitivity to plant perturbations. The performance of a feedforward controller is highly dependent on an accurate plant model, as its transfer function is typically derived from the dynamics of the plant. However, real-world systems are subject to parameter variations, nonlinearities, and unmodeled dynamics, making perfect modeling infeasible. As an open-loop approach, feedforward control lacks the inherent robustness of feedback mechanisms, meaning that any deviation from the assumed plant model can degrade performance and reduce control accuracy. The second issue is the responding delay introduced by pre-filtering dynamics. In many practical applications, perfect feedforward compensation requires predictive elements that introduce additional dynamics, often in the form of pre-filters. These pre-filters are necessary to ensure realizability and stability, but their residual dynamics can introduce response delays, thereby limiting the effectiveness of the feedforward compensation.

To overcome these limitations, this chapter introduces a modified feedforward control scheme that incorporates model refinement techniques to enhance robustness [15]. By implementing targeted structural modifications, the presented approach effectively mitigates the impact of modeling inaccuracies and increases the resilience of feedforward control against plant perturbations. Both theoretical analysis and experimental validation confirm that the introduced enhancements significantly improve the robustness and effectiveness of feedforward control while minimizing its inherent limitations. The chapter provides a detailed synthesis of the modified feedforward control architecture, emphasizing its straightforward implementation and minimal structural adjustments to existing control frameworks. Moreover, the versatility of the introduced enhancements ensures broad applicability across various feedforward control applications, making it a practical and scalable solution for improving performance in real-world engineering systems.

12.2 Conventional FFC

Feedforward control schemes are widely utilized for two primary control objectives: reference tracking and disturbance rejection. Despite their differing applications, both share the same fundamental principle—compensating for anticipated system behavior before deviations occur. This section provides a concise introduction to feedforward control while assuming that the necessary conditions regarding plant dynamics and feedforward controller assumptions are satisfied.

12.2.1 Reference Tracking

Two conventional feedforward control configurations for reference tracking are illustrated in Fig. 12.1. Beyond regulation problems, reference tracking (or servo control) is a fundamental requirement in many control applications. Feedforward schemes enable the states and/or outputs of a plant to follow a specified trajectory with minimal error.

In the system depicted in Fig. 12.1a, **P** represents the plant, $\mathbf{C}_B$ is the feedback controller responsible for closed-loop stability, and $\mathbf{C}_F$ is the feedforward controller

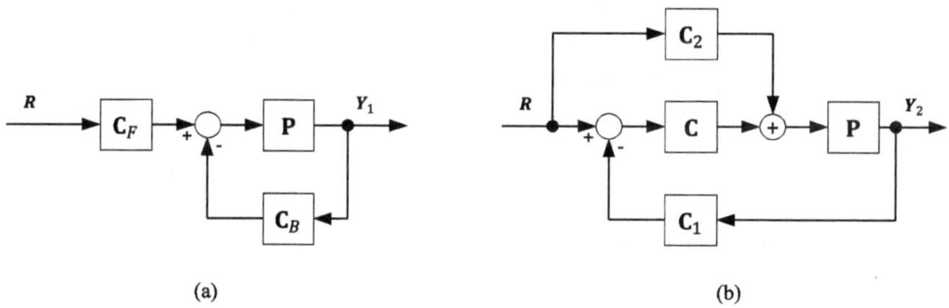

Fig. 12.1 FFC for reference tracking

(pre-filter) for reference tracking. The corresponding input–output relationship from the reference input $\boldsymbol{R}$ to the output $\boldsymbol{Y}_1$ is expressed as:

$$\boldsymbol{Y}_1 = (\mathbf{I} + \mathbf{P}\mathbf{C}_B)^{-1}\mathbf{P}\mathbf{C}_F\boldsymbol{R}. \tag{12.1}$$

If the feedforward controller is designed as:

$$\mathbf{C}_F := \mathbf{P}^{-1} + \mathbf{C}_B, \tag{12.2}$$

then perfect tracking is achieved, i.e., $\boldsymbol{Y}_1 = \boldsymbol{R}$. The trajectory tracking controllers for mass–stiffness–damping systems presented in Sect. 7.2 follow this design approach.

In Fig. 12.1b, $\mathbf{P}$ is the plant, $\mathbf{C}$ and $\mathbf{C}_1$ are feedback controllers, and $\mathbf{C}_2$ serves as the feedforward controller [9]. The transfer function from $\boldsymbol{R}$ to $\boldsymbol{Y}_2$ is given by:

$$\boldsymbol{Y}_2 = (\mathbf{I} + \mathbf{P}\mathbf{C}\mathbf{C}_1)^{-1}\mathbf{P}(\mathbf{C} + \mathbf{C}_2)\boldsymbol{R}. \tag{12.3}$$

For perfect tracking, where $\boldsymbol{Y}_2 = \boldsymbol{R}$, the corresponding feedforward controller should be:

$$\mathbf{C}_2 := \mathbf{P}^{-1} + \mathbf{C}(\mathbf{C}_1 - \mathbf{I}). \tag{12.4}$$

The integral and feedforward compensation controller in Sect. 7.3.2 is structured in this manner.

By comparing Eqs. (12.1) and (12.3), it can be seen that if:

$$\mathbf{C}_B := \mathbf{C}\mathbf{C}_1 \quad \text{and} \quad \mathbf{C}_F := \mathbf{C} + \mathbf{C}_2, \tag{12.5}$$

then the two configurations in Fig. 12.1a, b are equivalent [10].

This type of feedforward control is often associated with model-following control [5, 10, 11] due to its reference-tracking nature. It is also referred to as two-parameter compensation [6, 16] or two-degrees-of-freedom control [11, 17] because the feedforward and feedback controllers can be independently designed for tracking and stability enhancement, respectively. In practical applications, control systems must balance both servo and regulation requirements. The advantage of these feedforward configurations is their ability to decouple these objectives, allowing feedback to be tuned for robustness against plant uncertainties and disturbances while feedforward control ensures precise reference tracking.

12.2.2 Disturbance Rejection

Two conventional feedforward control schemes for disturbance rejection are illustrated in Fig. 12.2. When a measurable or estimable disturbance occurs frequently, feedforward control improves the quality of control by preemptively compensating

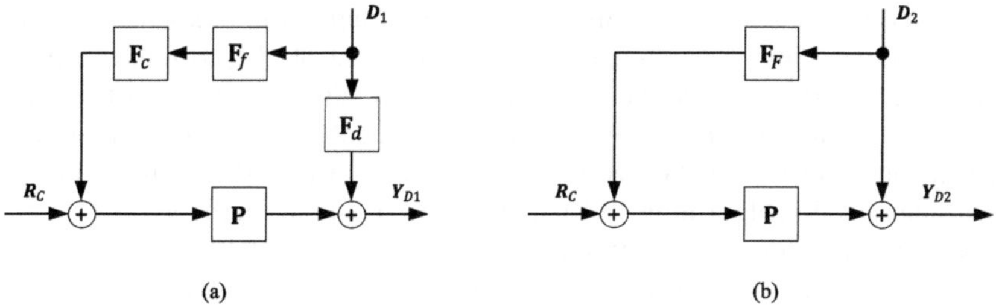

Fig. 12.2 FFC for disturbance rejection

for the disturbance. Unlike feedback control, which reacts to a disturbance only after it affects the system, feedforward control acts proactively, mitigating its impact before it propagates through the system.

In Fig. 12.2a, $\mathbf{P}$ represents the plant, $\mathbf{F}_d$ is the disturbance-related dynamics, $\mathbf{F}_f$ is the disturbance measurement filter, $\mathbf{F}_c$ is the feedforward controller, $\boldsymbol{D}_1$ is the disturbance, and $\boldsymbol{R}_C$ is the closed-loop control. The controlled output is $\boldsymbol{Y}_{D1}$.

If the disturbance $\boldsymbol{D}_1$ is detected by $\mathbf{F}_f$ in real time, the feedforward controller $\mathbf{F}_c$ can preemptively adjust the plant input to nullify its effect on $\boldsymbol{Y}_{D1}$. Assuming $\boldsymbol{R}_C = 0$, the disturbance response is:

$$\boldsymbol{Y}_{D1} = \left(\mathbf{F}_d + \mathbf{P}\mathbf{F}_c\mathbf{F}_f\right)\boldsymbol{D}_1. \tag{12.6}$$

For perfect disturbance rejection, i.e., $\boldsymbol{Y}_{D1} = 0$, the ideal feedforward controller is:

$$\mathbf{F}_c := -\mathbf{P}^{-1}\left(\mathbf{F}_d\mathbf{F}_f^{-1}\right). \tag{12.7}$$

This scheme relies entirely on open-loop compensation, where disturbance measurements feed directly into the controller, generating counteracting control inputs before disturbances impact system performance.

A simplified feedforward disturbance rejection scheme is shown in Fig. 12.2b, where $\mathbf{P}$ is the plant, $\mathbf{F}_F$ is the feedforward controller, and $\boldsymbol{D}_2$ is the equivalent disturbance. Assuming $\boldsymbol{R}_C = 0$, the disturbance response is:

$$\boldsymbol{Y}_{D2} = (\mathbf{I} + \mathbf{P}\mathbf{F}_F)\boldsymbol{D}_2. \tag{12.8}$$

For perfect disturbance rejection, where $\boldsymbol{Y}_{D2} = 0$, the required feedforward controller is:

$$\mathbf{F}_F := -\mathbf{P}^{-1}. \tag{12.9}$$

Since real-world disturbances are often unmeasurable or unpredictable, feedforward control is typically combined with feedback control [14] to improve robustness against uncertainties. Predictive techniques can also be employed to estimate disturbances from measurable signals, allowing feedforward control to mitigate the predictable portion of the disturbance before it affects the system [11]. The disturbance rejection algorithms in Chap. 9 follow a similar methodology.

However, as an open-loop strategy, feedforward control heavily depends on an accurate plant model—an ideal condition rarely met in practical applications. Model mismatches and plant uncertainties can significantly degrade its effectiveness. Furthermore, as evident from Eqs. (12.2), (12.4), (12.7), and (12.9), feedforward control fundamentally requires computing the inverse dynamics of the plant. In real-world systems, these inverses are often non-causal or physically unrealizable, necessitating the use of pre-filters to approximate the inverse dynamics. This introduces a lead–lag compensator [14], which, while improving realizability, also introduces delays and residual dynamics that can compromise compensation performance.

These challenges highlight the need for a more robust feedforward configuration that mitigates model inaccuracies and enhances practical applicability. The next sections introduce a modified feedforward structure that incorporates model refinement techniques to improve robustness, reduce model dependence, and extend the effectiveness of feedforward control across a wide range of engineering systems.

12.3 Enhanced FFC

Since the fundamental principles of feedforward control for reference tracking and disturbance rejection are similar, and the introduced robustness enhancement scheme is applicable to both, the discussion in this chapter focuses on the feedforward structure for disturbance rejection shown in Fig. 12.2b without loss of generality.

Building on the uncertainty reduction scheme introduced in Chap. 10, an enhanced feedforward control structure is presented in Fig. 12.3. The modified scheme extends the original feedforward configuration in Fig. 12.2b by introducing additional components within the dotted box, which serve as an uncertainty reduction mechanism. Here, $\mathbf{P}_M$ represents a reference model associated with the plant $\mathbf{P}$, and $\mathbf{W}$ is a weight for error compensation. It is important to emphasize that $\mathbf{P}_M$ is a flexible design parameter and can differ from the nominal model $\mathbf{P}_m$ of the plant. The system is assumed to be internally stable.

The purpose of this modification is to force the subsystem within the dotted box to behave like $\mathbf{P}_M$, thereby improving plant dynamics and achieving a more desirable system response. By incorporating these additional components, the enhanced feedforward scheme improves robustness against plant perturbations and enhances performance.

If the additional components within the dotted box were absent, corresponding to the conventional feedforward control system in Fig. 12.2b, the transfer function from disturbance $\boldsymbol{D}$ to output $\boldsymbol{Y}_D$ would be:

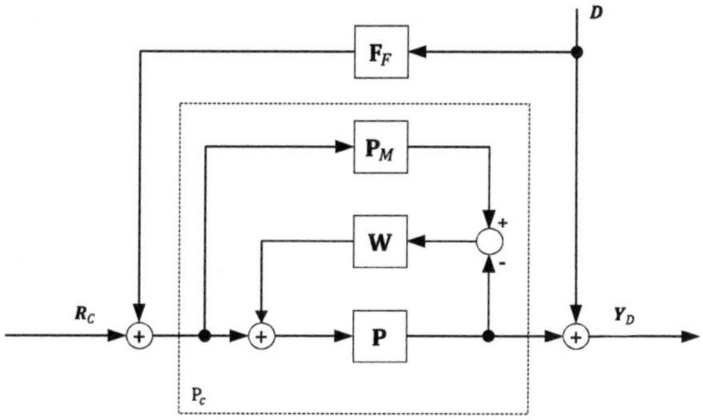

Fig. 12.3 Structure of the enhanced FFC

$$\mathbf{G}_1^D := \mathbf{I} + \mathbf{P}\mathbf{F}_F, \tag{12.10}$$

which is directly dependent on the plant dynamics. For robustness analysis, it is standard practice to model plant perturbations as [16, 17]:

$$\mathbf{P} := [\mathbf{I} + \Delta_m]\mathbf{P}_m + \Delta_a, \tag{12.11}$$

where $\mathbf{P}_m$ represents the nominal plant model, while Δ_m and Δ_a denote multiplicative and additive uncertainties, respectively. From Eq. (12.10), it is clear that these uncertainties significantly affect $\mathbf{G}_1^D$, and as the uncertainties increase, the performance of the conventional feedforward control scheme deteriorates.

In contrast, for the modified feedforward scheme in Fig. 12.3, the corresponding disturbance transfer function is:

$$\mathbf{G}_2^D := \mathbf{I} + \mathbf{P}_c\mathbf{F}_F, \tag{12.12}$$

where the compensated plant dynamics $\mathbf{P}_c$ is given by:

$$\mathbf{P}_c := (\mathbf{I} + \mathbf{P}\mathbf{W})^{-1}\mathbf{P}(\mathbf{I} + \mathbf{W}\mathbf{P}_M). \tag{12.13}$$

If the reference model $\mathbf{P}_M$ is set equal to the nominal plant model $\mathbf{P}_m$, i.e., $\mathbf{P}_M = \mathbf{P}_m$, then substituting into Eq. (12.13) yields:

$$\mathbf{P}_c = \left[\mathbf{I} + (\mathbf{I} + \mathbf{P}\mathbf{W})^{-1}\Delta_m\right]\mathbf{P}_m + (\mathbf{I} + \mathbf{P}\mathbf{W})^{-1}\Delta_a. \tag{12.14}$$

Comparing Eq. (12.14) with Eq. (12.11), it becomes evident that both uncertainties Δ_m and Δ_a are now reduced by a factor of $(\mathbf{I} + \mathbf{P}\mathbf{W})$, meaning that the modified disturbance transfer function $\mathbf{G}_2^D$ is less sensitive to plant uncertainties than $\mathbf{G}_1^D$.

By substituting Eq. (12.13) into Eq. (12.12), the modified disturbance transfer function can also be expressed as:

$$\mathbf{G}_2^D = (\mathbf{I} + \mathbf{P}\mathbf{W})^{-1}(\mathbf{I} + \mathbf{P}\mathbf{F}_F) + (\mathbf{I} + \mathbf{P}\mathbf{W})^{-1}\mathbf{P}\mathbf{W}(\mathbf{I} + \mathbf{P}_M\mathbf{F}_F). \tag{12.15}$$

If the loop gain **PW** is sufficiently large, then an approximation follows:

$$\mathbf{G}_2^D \approx (\mathbf{I}+\mathbf{PW})^{-1}(\mathbf{I}+\mathbf{PF}_F)+(\mathbf{I}+\mathbf{P}_M\mathbf{F}_F)=(\mathbf{I}+\mathbf{PW})^{-1}\mathbf{G}_1^D+(\mathbf{I}+\mathbf{P}_M\mathbf{F}_F). \tag{12.16}$$

As will be discussed further, the second term in Eq. (12.16) will be negligible or vanish if $\mathbf{P}_M\mathbf{F}_F \rightarrow [-\mathbf{I}]$, leading to the approximation:

$$\mathbf{G}_2^D \rightarrow (\mathbf{I}+\mathbf{PW})^{-1}\mathbf{G}_1^D.$$

This result indicates that under the modified feedforward scheme, the disturbance response is approximately reduced by an additional factor of $(\mathbf{I}+\mathbf{PW})$ compared to the conventional feedforward scheme.

The structural modification for uncertainty reduction in Fig. 12.3 is slightly different from that in Chap. 10. Here, the disturbance is intentionally left outside the uncertainty reduction loop to allow a clearer observation of the performance enhancement. While this provides an insightful analysis of uncertainty reduction, the structure could also be adjusted to incorporate the disturbance within the loop, yielding alternative results. However, such variations are beyond the scope of this discussion.

12.4 Synthesis of the Enhanced FFC

The synthesis of the key components $\mathbf{F}_F$, $\mathbf{P}_M$, and $\mathbf{W}$ in the modified feedforward control scheme plays a crucial role in achieving enhanced robustness and performance. Different design choices for these components lead to variations in system behavior. The following sections discuss their selection and implementation.

12.4.1 Feedforward Controller Design

The conventional feedforward controller $\mathbf{F}_F$ is ideally designed as:

$$\mathbf{F}_F := -\mathbf{P}_m^{-1}, \tag{12.17}$$

which is derived from the inverse dynamics of the plant model. However, in practical systems, this expression is often unrealizable. If any element $\mathrm{p}_m^{\mathrm{inv}}$ in $\mathbf{P}_m^{-1}$ has a numerator order exceeding its denominator by n, the resulting controller will require nth order differentiation, which is mathematically valid but physically impractical.

To address this issue, a pre-filter is commonly introduced in the scalar form:

$$p_f = \left(T_{\mathrm{pf}}s + 1\right)^{-n},$$

where T_{pf} is the time constant of the pre-filter. Incorporating this pre-filter into the feedforward controller gives:

$$\mathbf{F}_F := -\mathbf{P}_m^{-1}\mathbf{P}_f, \tag{12.18}$$

where $\mathbf{P}_f$ is the corresponding pre-filter matrix. However, this introduces additional dynamics that delay the disturbance rejection response, a well-known drawback in conventional feedforward schemes. In the case of perfect modeling ($\mathbf{P}_m = \mathbf{P}$), the disturbance transfer function becomes:

$$\mathbf{G}_1^D = \mathbf{I} - \mathbf{P}_f, \tag{12.19}$$

which clearly illustrates the undesirable effect of the pre-filter. The introduced modified scheme mitigates this issue by appropriately designing the reference model $\mathbf{P}_M$.

12.4.2 Reference-Model Selection

The choice of the reference model $\mathbf{P}_M$ significantly influences the robustness and performance of the system. Two possible reference-model configurations are considered.

A natural choice is to set the reference model equal to the nominal plant model:

$$\mathbf{P}_{M1} := \mathbf{P}_m. \tag{12.20}$$

In the presence of model mismatch between $\mathbf{P}_m$ and $\mathbf{P}$, this choice ensures that the compensated dynamics $\mathbf{P}_c$ approximate the nominal model.

In the case of perfect modeling ($\mathbf{P}_m = \mathbf{P}$), the disturbance transfer function simplifies to:

$$\begin{aligned}\mathbf{G}_{21}^D &= (\mathbf{I} + \mathbf{PW})^{-1}(\mathbf{I} + \mathbf{PF}_F) + (\mathbf{I} + \mathbf{PW})^{-1}\mathbf{PW}\big(\mathbf{I} - \mathbf{P}_f\big) \\ &\approx (\mathbf{I} + \mathbf{PW})^{-1}(\mathbf{I} + \mathbf{PF}_F) + \big(\mathbf{I} - \mathbf{P}_f\big).\end{aligned} \tag{12.21}$$

If the residual dynamics of $\mathbf{P}_f$ are sufficiently small, the last term in Eq. (12.21), which arises from pre-filtering dynamics, can be neglected. Consequently, it is standard practice to design the time constants of the pre-filters as small as possible to minimize their impact. However, due to physical constraints and stability considerations, reducing these time constants is not always feasible. Even in the ideal case where the plant model perfectly matches the actual system ($\mathbf{P}_m = \mathbf{P}$), the introduction of $\mathbf{P}_f$ into the feedforward controller inevitably introduces a discrepancy between $\mathbf{F}_F$

and the exact inverse dynamics of $\mathbf{P}_m$. This discrepancy leads to a delayed response in disturbance compensation. As a result, the presence of pre-filtering dynamics $\mathbf{P}_f$ remains an inherent limitation of conventional feedforward control schemes, as well as the modified scheme that uses $\mathbf{P}_M = \mathbf{P}_m$ as its reference model.

To overcome the limitations of $\mathbf{P}_{M1}$, an alternative reference model is introduced:

$$\mathbf{P}_{M2} := \mathbf{P}_f^{-1}\mathbf{P}_m. \tag{12.22}$$

This choice ensures:

$$\mathbf{P}_{M2} = -\mathbf{F}_F^{-1}, \tag{12.23}$$

which integrates the inverse dynamics of $\mathbf{P}_f$ into the reference model. As a result, the compensated plant dynamics $\mathbf{P}_c$ approximate $\mathbf{P}_f^{-1}\mathbf{P}_m$, allowing the feedforward controller and reference model to effectively cancel each other out. Consequently, the disturbance transfer function reduces to:

$$\mathbf{G}_{22}^D = (\mathbf{I} + \mathbf{PW})^{-1}(\mathbf{I} + \mathbf{PF}_F). \tag{12.24}$$

This formulation introduces two significant improvements. First, the disturbance is attenuated by an additional factor of $(\mathbf{I} + \mathbf{PW})$, enhancing robustness against plant uncertainties. Second, the undesirable residual dynamics from the pre-filter in Eq. (12.15) are eliminated, ensuring a faster response compared to the scheme with $\mathbf{P}_{M1}$. The resulting system dynamics are often of lower order and simpler, making $\mathbf{P}_{M2}$ the preferable choice for most applications.

12.4.3 Error Compensator Design

The weight $\mathbf{W}$ is synthesized in a straightforward manner:

$$\mathbf{W} := \mathbf{HK}, \tag{12.25}$$

where $\mathbf{K}$ is a gain parameter, and $\mathbf{H}$ is an error compensator. The gain $\mathbf{K}$ must be sufficiently large to ensure effective compensation, but internal stability constraints must also be considered.

A natural choice for $\mathbf{H}$ is:

$$\mathbf{H} := \mathbf{P}_m^{-1}\mathbf{P}_f, \tag{12.26}$$

which results in:

$$\mathbf{H} = -\mathbf{F}_F. \tag{12.27}$$

With this design and assuming perfect modeling ($\mathbf{P}_m = \mathbf{P}$) and $\mathbf{P}_{M2}$ as the reference model, the disturbance transfer function simplifies to:

$$\mathbf{G}_{22}^{D} = \left(\mathbf{I} + \mathbf{P}_f\mathbf{K}\right)^{-1}\left(\mathbf{I} - \mathbf{P}_f\right). \tag{12.28}$$

Comparing this with Eq. (12.19), it is evident that the modified scheme offers a significantly faster disturbance rejection response.

Other variations of $\mathbf{H}$, such as state-error-based weights derived from plant-model discrepancies, may also yield satisfactory performance, but these alternatives are not discussed in this chapter.

The discussion above demonstrates that the robustness and performance of the conventional feedforward control structure in Fig. 12.2b can be significantly enhanced using the modified configuration in Fig. 12.3. The general methodology applies not only to the specific case considered but can also be extended to other feedforward configurations. The presented modifications mitigate the effects of plant uncertainties, reduce response delays introduced by pre-filters, and improve disturbance rejection capabilities. These enhancements make the feedforward scheme more practical and reliable for a broad range of engineering applications.

12.5 Design Examples

To validate the effectiveness of the introduced enhanced feedforward control scheme, design examples and experiments for disturbance rejection in a mass–stiffness–damping system are presented. The configuration in Fig. 12.3 is employed to compare the performance of the conventional feedforward control scheme and its enhanced counterpart.

In Chap. 9, two feedforward disturbance rejection algorithms for mass–stiffness–damping systems were introduced, where the ideal plant dynamics in the forward path was either a pure integrator (Eq. 9.8) or a unity-gain (Eq. 9.14). Here, a generalized first-order plant model is considered:

$$\mathrm{P} = \frac{k}{Ts + 1}, \tag{12.29}$$

where $k > 0$ represents an equivalent plant gain, and T is an equivalent time constant. This formulation generalizes the system behavior: for large T, the system approximates an integrator, while for small T, it behaves as a pure gain. Variations in k reflect plant uncertainties. The modeled plant dynamics is given as:

$$\mathrm{P}_m = \frac{k_m}{T_m s + 1}, \tag{12.30}$$

where $k_m > 0$ and T_m are estimated parameters of k and T, respectively.

For both the conventional and enhanced feedforward schemes, the feedforward controller F_F is designed as:

$$F_F = -\frac{T_m s + 1}{k_m\left(T_{pf} s + 1\right)}, \tag{12.31}$$

where T_{pf} is the time constant of the pre-filter. Clearly, this controller follows the conventional form $F_F = -P_m^{-1} P_f$.

Under perfect modeling conditions ($P_m = P$), the disturbance transfer function for the conventional feedforward scheme is derived as:

$$\frac{Y_1}{D} = \frac{T_{pf} s}{T_{pf} s + 1}, \tag{12.32}$$

which highlights the delay effect introduced by the pre-filter. The steady-state disturbance response for a unit-step disturbance ($d = 1$, $r_C = 0$) is:

$$y_{1\infty} = 0, \tag{12.33}$$

indicating zero steady-state error. However, under modeling inaccuracies, the disturbance transfer function becomes:

$$\frac{Y_1}{D} = \frac{k_m T_{pf} T s^2 + \left[k_m\left(T_{pf} + T\right) - k T_m\right] s + (k_m - k)}{k_m T_{pf} T s^2 + k_m\left(T_{pf} + T\right) s + k_m}, \tag{12.34}$$

with a steady-state error:

$$y_{1\infty} = \frac{k_m - k}{k_m}, \tag{12.35}$$

showing that significant model mismatches ($k_m - k$) lead to large steady-state errors.

For the enhanced scheme, the error compensator H in the weight W is designed as:

$$H = \frac{T_m s + 1}{k_m\left(T_{pf} s + 1\right)}, \tag{12.36}$$

which follows the same form as F_F. Two reference models P_M are considered.

Using P_{M1}, the reference model is set as:

$$P_{M1} = \frac{k_m}{T_m s + 1}. \tag{12.37}$$

From Eq. (12.13), the compensated plant becomes:

$$\mathrm{P}_{c1} = \frac{k_m k\left[T_{\mathrm{pf}} s + (\mathrm{K}+1)\right]}{k_m T_{\mathrm{pf}} T s^2 + \left[k_m\left(T_{\mathrm{pf}} + T\right) - k\mathrm{K}T_m\right]s + (k_m - k\mathrm{K})}, \tag{12.38}$$

leading to the disturbance transfer function:

$$\frac{Y_2}{D} = \frac{n_1 s^3 + n_2 s^2 + n_3 s + n_4}{m_1 s^3 + m_2 s^2 + m_3 s + m_4}, \tag{12.39}$$

where the coefficients are:

$$\begin{aligned}
n_1 &= k_m T_{\mathrm{pf}}^2 T \\
n_2 &= T_{\mathrm{pf}}\left[k_m\left(T_{\mathrm{pf}} + 2T\right) + k(\mathrm{K}-1)T_m\right] \\
n_3 &= k_m\left(2T_{\mathrm{pf}} + T\right) + k\left[(\mathrm{K}-1)T_{\mathrm{pf}} - T_m\right] \\
n_4 &= k_m - \mathrm{K} \\
m_1 &= k_m T_{\mathrm{pf}}^2 T \\
m_2 &= T_{\mathrm{pf}}\left[k_m\left(T_{\mathrm{pf}} + 2T\right) + k\mathrm{K}T_m\right] \\
m_3 &= k_m\left(2T_{\mathrm{pf}} + T\right) + k\mathrm{K}\left(T_{\mathrm{pf}} + T_m\right) \\
m_4 &= k_m + k\mathrm{K}.
\end{aligned}$$

Under perfect modeling conditions, we obtain:

$$\mathrm{P}_{c1} = \mathrm{P} = \mathrm{P}_m. \tag{12.40}$$

Thus, the disturbance transfer function remains identical to Eq. (12.32). Under model mismatch, the steady-state error for a unit-step disturbance is:

$$y_{2\infty} = \frac{k_m - k}{k_m + k\mathrm{K}}, \tag{12.41}$$

which is smaller than $y_{1\infty}$ when $\mathrm{K} > 0$, showing improved robustness.

For the second enhanced scheme, the reference model is chosen as:

$$\mathrm{P}_{M2} = \frac{k_m\left(T_{\mathrm{pf}} s + 1\right)}{T_m s + 1}. \tag{12.42}$$

Using P_{M2}, the compensated plant is:

$$\mathrm{P}_{c2} = \frac{k_m k(\mathrm{K}+1)\left(T_{\mathrm{pf}} s + 1\right)}{k_m T_{\mathrm{pf}} T s^2 + \left[k_m\left(T_{\mathrm{pf}} + T\right) + k\mathrm{K}T_m\right]s + (k_m + k\mathrm{K})}. \tag{12.43}$$

The disturbance transfer function becomes:

$$\frac{Y_3}{D} = \frac{k_m T_{\mathrm{pf}} T s^2 + \left[k_m \left(T_{\mathrm{pf}} + T\right) - k T_m\right] s + (k_m - k)}{k_m T_{\mathrm{pf}} T s^2 + \left[k_m \left(T_{\mathrm{pf}} + T\right) + k \mathrm{K} T_m\right] s + (k_m + k\mathrm{K})}, \tag{12.44}$$

which is of lower order than Y_2/D, simplifying implementation. The steady-state error remains:

$$y_{3\infty} = y_{2\infty}, \tag{12.45}$$

confirming that both enhanced schemes reduce steady-state error. However, the use of P_{M2} yields superior transient response.

For perfect modeling, Eq. (12.28) gives:

$$\frac{Y_3}{D} = \frac{T_{\mathrm{pf}} s}{T_{\mathrm{pf}} s + (1 + \mathrm{K})} = \frac{\frac{T_{\mathrm{pf}}}{(1+\mathrm{K})} s}{\frac{T_{\mathrm{pf}}}{(1+\mathrm{K})} s + 1}. \tag{12.46}$$

In comparison with Eq. (12.32), both the disturbance response magnitude and settling time are reduced by a factor of $(1 + \mathrm{K})$, indicating a substantial enhancement in compensation performance.

This analysis confirms that the introduced enhanced feedforward control scheme substantially improves disturbance rejection compared to conventional approaches. The use of P_{M2} as the reference model provides additional benefits by mitigating pre-filter delays and improving transient response. The enhanced scheme effectively addresses the robustness limitations of conventional feedforward control, making it more resilient to model uncertainties and disturbances in real-world applications.

Example 12.1 **(*Accurate Modeling*)** In this example, the performance of the enhanced FFC scheme is compared with that of the conventional scheme under the condition of accurate modeling. The mass–stiffness–damping system discussed earlier is used, with system parameters set as follows: the plant and its model parameters are $k_m = k = 1$ and $T_m = T = 0.001$ s. The time constant of the pre-filter is chosen as $T_{\mathrm{pf}} = 0.0001$ s., and the gain K of the weight W is set to $\mathrm{K} = 10$.

For a unit-step disturbance input ($d = 1$, $r_C = 0$), the compensation result y_1 of the conventional feedforward control is depicted in Fig. 12.4 {Solid line: y_1, y_2 (with P_{M1}). Dotted line: y_3 (with P_{M2})} as a solid line. In this case of accurate modeling, the response y_2, corresponding to the enhanced feedforward scheme with the reference model P_{M1}, completely overlaps with y_1 (solid line), as expected. This occurs because under perfect modeling conditions, the compensated plant dynamics are identical to the original system, i.e., $\mathrm{P}_{c1} = \mathrm{P} = \mathrm{P}_m$, leading to the same disturbance transfer function as given in Eq. (12.32).

The response y_3, representing the enhanced feedforward scheme with the reference model P_{M2}, is shown as a dotted line in Fig. 12.4. The results clearly demonstrate

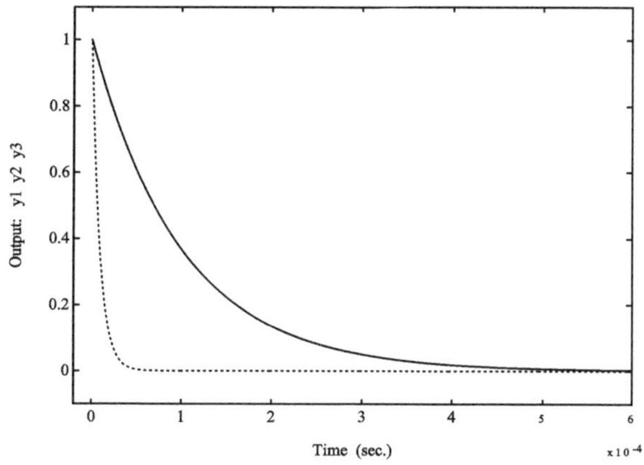

Fig. 12.4 Disturbance rejection: accurate model

the superior performance of the enhanced scheme with P_{M2}, which achieves significantly faster disturbance rejection compared to both the conventional scheme and the enhanced scheme using P_{M1}.

The key takeaway from Fig. 12.4 is that even under perfect modeling conditions, the enhanced feedforward scheme with the reference model P_{M2} provides a much faster response than the conventional scheme or the enhanced scheme using P_{M1}. This confirms that the modified scheme with P_{M2} not only improves robustness against modeling inaccuracies but also enhances transient performance by accelerating disturbance rejection, making it an effective choice for high-performance control applications.

Example 12.2 (***Inaccurate Modeling***) This example examines the impact of model mismatch on the performance of both conventional and enhanced feedforward control schemes. The actual system parameters are set as $k \in [0.5, 2]$ and $T = 0.005$ s., while all other parameters remain the same as those in Example 12.1. A large deviation in the system gain k is deliberately introduced to evaluate the robustness of the control schemes under significant modeling errors.

For a unit-step disturbance input ($d = 1$, $r_C = 0$), the compensation result y_1 of the conventional feedforward control is shown in Fig. 12.5{a: $k = 0.5$), (b: $k = 2$ Solid line: y_1. Dashed line: y_2 (with P_{M1}). Dotted line: y_3 (with P_{M2}} as a solid line. As expected, performance degradation is severe due to model mismatch. The resulting steady-state errors are $+0.5$ for $k = 0.5$ and -1.0 for $k = 2$, which confirms the sensitivity of the conventional scheme to model inaccuracies and its substantial loss in control effectiveness.

The disturbance response y_2 of the enhanced feedforward control with P_{M1} as the reference model is represented by the dashed line in Fig. 12.5. Under model mismatch conditions, its steady-state errors are significantly reduced to $+0.0833$ for $k = 0.5$ and -0.0476 for $k = 2.0$. Compared to the conventional scheme, the enhanced feedforward control effectively minimizes both steady-state and transient errors, demonstrating improved robustness against model uncertainties.

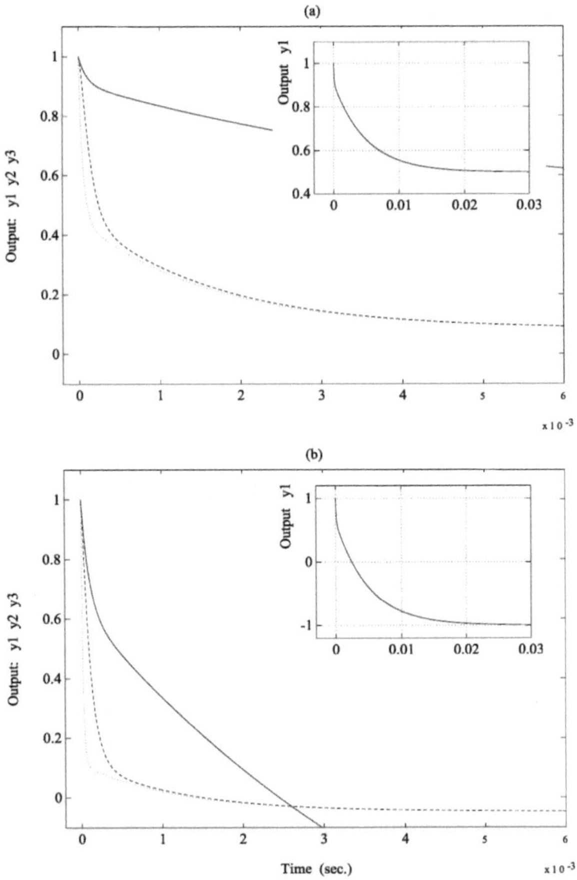

Fig. 12.5 Disturbance rejection: inaccurate model

The disturbance response y_3 of the enhanced feedforward control with P_{M2} as the reference model is shown as the dotted line in Fig. 12.5. Even under substantial model mismatch, y_3 exhibits a faster response than y_2, confirming the superior transient performance of the enhanced scheme with P_{M2}.

It is evident from Fig. 12.5 that the conventional feedforward control scheme is highly sensitive to model mismatch, leading to large steady-state errors and a noticeable degradation in performance. The enhanced feedforward scheme with P_{M1} significantly improves robustness by effectively minimizing steady-state errors even in the presence of plant variations. The enhanced scheme with P_{M2} not only mitigates steady-state errors but also achieves faster disturbance rejection, demonstrating superior dynamic performance. These results illustrate that the introduced enhanced feedforward scheme substantially improves robustness and performance, making it a more reliable solution for real-world applications where model inaccuracies are inevitable.

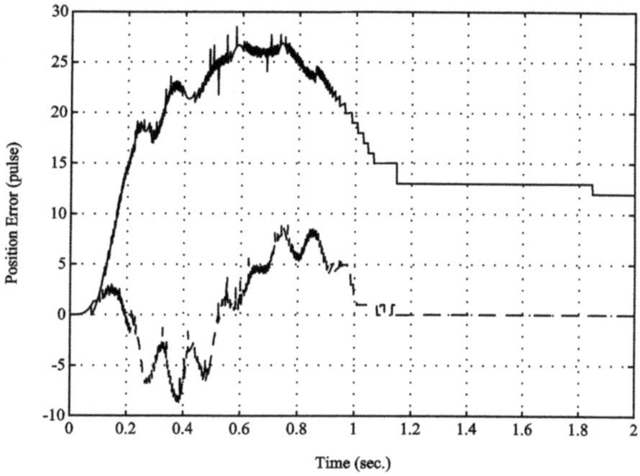

Fig. 12.6 Experimental results for disturbance rejection: Case 1

Experiment 12.1

The effectiveness of the enhanced feedforward control scheme was further validated through experimental testing on the testbed described in Appendix D. The experimental results are presented in Fig. 12.6 {Dashed line: Enhanced. Solid line: Conventional}. In this experiment, the manipulator was required to move from its initial lower vertical position $q_0 = -\pi/2$ to its final horizontal position $q_f = 0$ within the time interval $[0, t_f] = [0, 1.0]$ s, following the trajectory specified in Eq. (C.1).

To emulate realistic operating conditions, joint friction was deliberately left uncompensated, and gravity compensation was intentionally set to be inaccurate. These non-ideal effects introduced significant disturbances, impacting tracking accuracy and steady-state performance. As shown by the solid line in Fig. 12.6, the conventional feedforward control exhibited noticeable tracking errors and residual steady-state deviations due to its sensitivity to modeling inaccuracies and unmodeled disturbances.

By implementing the enhanced feedforward control scheme, the undesirable tracking and steady-state errors were significantly reduced. As indicated by the dashed line in Fig. 12.6, the improved control strategy effectively compensated for the unmodeled disturbances, achieving near-perfect tracking with zero steady-state error. The experimental results confirm that the enhanced feedforward scheme provides superior robustness and performance over the conventional approach, making it a more reliable solution for practical applications involving dynamic uncertainties and plant variations.

Experiment 12.2

This experiment follows the same motion task as Experiment 12.1, with the joint friction deliberately left uncompensated to emulate real-world disturbances. However, unlike Experiment 12.1, the gravity effect is also left uncompensated, introducing an additional source of dynamic uncertainty. The experimental results are presented in Fig. 12.7 {Dashed line: Enhanced. Solid line: Conventional}.

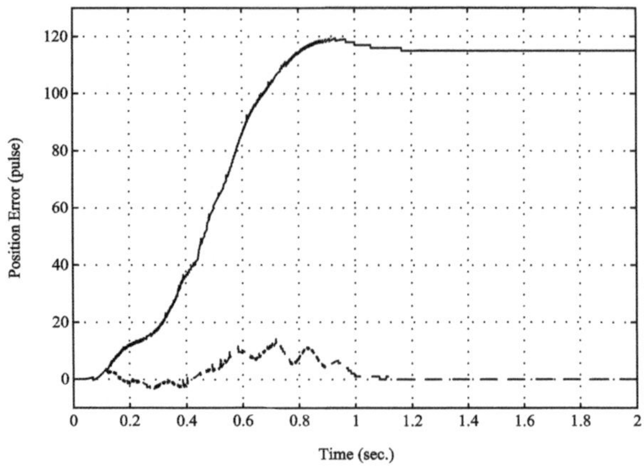

Fig. 12.7 Experimental results for disturbance rejection: case 2

With these non-ideal effects, particularly the uncompensated gravity torque, the conventional feedforward control exhibits significant tracking deviations and large steady-state errors, as shown by the solid line in Fig. 12.7. The inability of the conventional scheme to compensate for gravity-induced disturbances highlights its sensitivity to model inaccuracies and external perturbations.

By applying the enhanced feedforward control scheme, tracking errors were significantly minimized, and steady-state performance was substantially improved. As illustrated by the dashed line in Fig. 12.7, the enhanced control method successfully mitigates the gravity-induced errors, achieving precise trajectory tracking with zero steady-state error. These results further confirm the robustness and effectiveness of the enhanced feedforward scheme in compensating for unmodeled disturbances and improving overall system performance in practical applications.

Experiment 12.3

In this experiment, the manipulator, as described in Appendix D, is tasked with moving from its initial lower vertical position $q_0 = -\pi/2$ to the final horizontal position $q_f = 0$ within the time interval $\left[0, t_f\right] = [0, 1.0]$ s, following the trajectory defined in Eq. (C.1). Joint friction is deliberately left uncompensated to introduce a disturbance, and the gravity compensation is intentionally set to be inaccurate.

At $t = 2.0$ s., an additional payload is attached to the end effector, introducing both payload variation and external disturbance. The experimental results are presented in Fig. 12.8 {Dashed line: Enhanced. Solid line: Conventional}. Under these non-ideal conditions, the conventional feedforward control exhibits substantial tracking errors, as shown by the solid line in Fig. 12.8.

During the initial phase ($t \in [0, 2]$), the tracking errors mirror those observed in Fig. 12.6, indicating similar system behavior under uncompensated friction and imperfect gravity compensation. However, in the interval $t \in [2, 4]$, after the payload is added, the tracking error under conventional feedforward control increases significantly, highlighting its sensitivity to dynamic variations and external disturbances.

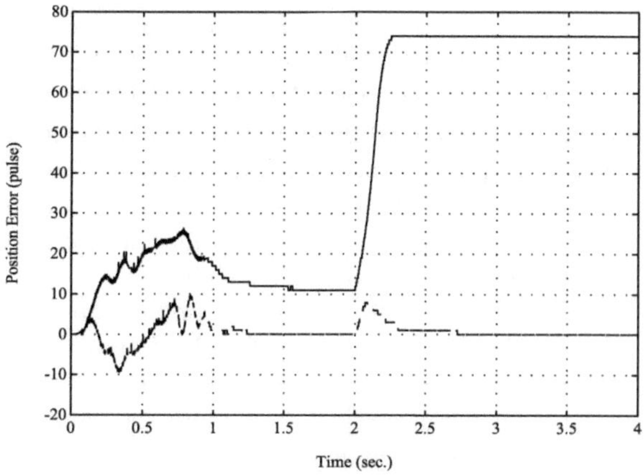

Fig. 12.8 Experimental results for disturbance rejection: Case 3

With the enhanced feedforward scheme, the disturbance effects caused by the additional payload are effectively mitigated. The results, represented by the dashed line in Fig. 12.8, show that the enhanced scheme maintains precise trajectory tracking, successfully counteracting the payload-induced disturbances while achieving zero steady-state error. These findings further validate the robustness of the enhanced feedforward approach in handling dynamic uncertainties and improving overall system performance.

12.6 Summary

The enhanced feedforward control configuration presented in this chapter effectively addresses the key limitations of conventional feedforward schemes, namely sensitivity to plant perturbations and response delays caused by pre-filtering dynamics. By introducing a structurally modified framework, the presented approach significantly improves robustness and performance while maintaining a straightforward implementation process with minimal additional complexity. The general-purpose nature of the modification ensures broad applicability across various control applications without imposing significant computational or design burdens.

The performance improvements of the enhanced scheme can be characterized in two key aspects. First, its robustness against plant perturbations is substantially increased, allowing it to mitigate the adverse effects of model mismatches and uncertainties that would otherwise degrade conventional feedforward control. Second, even under ideal conditions with perfect modeling, the enhanced feedforward scheme exhibits superior response speed and disturbance rejection capabilities, thereby improving overall system performance beyond the limitations of traditional approaches.

The synthesis and effectiveness of the introduced scheme are thoroughly examined through analytical derivations, numerical simulations, and experimental validation using a mass–stiffness–damping system. These studies consistently demonstrate that the structural modifications lead to tangible improvements in both transient and steady-state performance. The experimental results confirm that the enhanced feedforward control can effectively handle real-world uncertainties, such as frictional disturbances, gravity compensation inaccuracies, and payload variations, all while maintaining high precision and stability.

It is worth noting that advanced robust control methodologies, such as $\mathcal{H}^\infty$ optimization [18, 19], can further refine the design and synthesis of the enhanced control system. These techniques could be leveraged to systematically optimize robustness and disturbance rejection in more complex or highly uncertain environments, extending the practical utility of the introduced approach even further.

References

1. Davison EJ (1973) The feedforward control of linear multivariable time-invariant systems. Automatica 9(5):561–573
2. Davison EJ (1976) The steady-state invertibility and feedforward control of linear time-invariant systems. IEEE Trans Autom Control 21(4):529–534
3. Broussard JR, O'Brien MJ (1980) Feedforward control to track the output of a forced model. IEEE Trans Autom Control 25(4):851–853
4. Tomizuka M, Hu JS, Chiu TC, Kamano T (1992) Synchronization of two motion control axes under adaptive feedforward control. Trans ASME J Dyn Syst Meas Control 114(2):196–203
5. Anderson BDO, Moore JB (1989) Optimal control: linear quadratic methods. Prentice-Hall New Jersey
6. Chen CT (1993) Analog and digital control system design: transfer-function, state-space, and algebraic methods. Saunders College Publishing, HBJ International
7. Broussard JR, Berry PW (1980) The relationship between implicit model following and eigenvalue eigenvector placement. IEEE Trans Autom Control 25(3):591–594
8. Tomizuka M (1987) Zero phase error tracking algorithm for digital control. J Dyn Syst Meas Control 109(1):65–68
9. McLean D (1990) Automatic flight control systems. Prentice-Hall New Jersey
10. Ogata K (2010) Modern control engineering. Prentice-Hall New Jersey
11. Åström KJ, Wittenmark B (1997) Computer controlled systems: theory and design. Prentice-Hall, New Jersey
12. Isermann R (1991) Digital control systems. Vol. 2: Stochastic control, multivariable control, adaptive control, applications. Springer, Berlin
13. Van de Vegte J (1993) Feedback control systems. Prentice-Hall, New Jersey
14. LeBlanc SE, Coughanowr DR (2009) Process systems analysis and control. McGraw-Hill, New York
15. Zhu HA, Teo CL, Hong GS, Poo AN (1992) An enhanced scheme for the model-based control of robot manipulators. Int J Control 56(6):1243–1261
16. Vidyasagar M (2011) Control system synthesis: a factorization approach. Morgan & Claypool, California

17. Morari M, Zafiriou E (1989) Robust process control. Prentice-Hall, New Jersey
18. Kwakernaak H (1993) Robust control and H_∞-optimization—tutorial paper. Automatica 29(2):255–273
19. Grimble MJ (2006) Robust industrial control systems: optimal design approach for polynomial systems. Wiley, New Jersey

Chapter 13
Enhanced Model-Reference Adaptive Control

This chapter presents a decentralized stabilization methodology for large-scale mass–stiffness–damping systems, addressing the complexities of their multivariable, nonlinear, and cross-coupled dynamics. Controlling such systems remains a challenge due to structural couplings, external disturbances, and high-dimensional cross-couplings. By leveraging large-scale system theory, this chapter introduces a framework that enables decentralized control, providing a feasible and scalable approach to regulating these intricate dynamic systems.

A key insight is that large-scale mass–stiffness–damping systems can be structured in a way that allows the application of established large-scale system control techniques. By decomposing the system into manageable subsystems, decentralized control strategies can be systematically designed, avoiding the computational burden and practical limitations of centralized approaches.

This decentralized methodology enhances robustness, adaptability, and computational efficiency, making it well-suited for real-world applications such as rotor-bearing assemblies, structural vibration control, and interconnected mechanical networks. A numerical case study of a large rotor-bearing system demonstrates the effectiveness of the introduced framework, showing its ability to maintain stability, reduce vibrations, and mitigate coupling effects, even under challenging operating conditions.

13.1 Introduction

Recent advancements in active control of mechanical and structural systems, particularly large-scale mass–stiffness–damping systems, have intensified research efforts toward developing scalable and effective control methodologies. These systems are inherently nonlinear, multivariable, and dynamically cross-coupled, posing significant challenges in analysis, synthesis, and real-time control. Their motion is typically

H. Zhu, *Model-Based Control of Mass–Stiffness–Damping Systems*, Advances in Industrial Control, https://doi.org/10.1007/978-3-031-97592-9_13

governed by high-order ordinary differential equations (ODEs), where preserving model fidelity often results in systems with $n \gg 100$ degrees-of-freedom. The high dimensionality and complexity of such models present fundamental obstacles to traditional control strategies.

A conventional approach to handling these complexities is model reduction-based centralized control [1]. While this method can simplify the system for control implementation, it suffers from critical limitations, including the spillover problem, where uncontrolled residual dynamics introduce instability or degrade performance. Such challenges highlight the inherent shortcomings of centralized control in large-scale systems and have motivated the development of decentralized control strategies. Over the past few decades, decentralized control theory has matured significantly, providing a robust framework for managing interconnected systems with improved scalability, fault tolerance, and real-time implementation feasibility [2–4].

This chapter introduces a pole-placement-based decentralized adaptive stabilization approach tailored for large-scale mass–stiffness–damping systems, often referred to as matrix second-order systems. The presented method operates in continuous time and offers a structured yet computationally efficient framework for stabilizing high-dimensional dynamic systems. The methodology integrates adaptive pole-placement techniques, which dynamically adjust system parameters to enhance stability margins and increase robustness against uncertainties, ensuring reliable performance under varying operating conditions.

To achieve robust stabilization, the chapter presents a fully decentralized control scheme that integrates two key strategies:

- *Adaptive Pole-Placement:* A systematic method for dynamically adjusting system poles, improving stability margins, mitigating nonlinear and coupling effects, and enhancing overall system responsiveness.
- *Robust Model-Based Control:* A strategy designed to suppress disturbances, compensate for system uncertainties, and counteract unbalanced dynamic responses that commonly arise in complex mass–stiffness–damping systems.

These two methodologies work synergistically to enhance stability robustness, improve transient response, and strengthen disturbance rejection in decentralized controllers. The introduced approach is particularly advantageous for large-scale implementations, where system-wide cross-couplings and uncertainties present significant challenges.

A numerical case study of a practical rotor-bearing system is provided to validate the effectiveness of the introduced decentralized control framework. The results demonstrate that the introduced stabilization strategy successfully mitigates instability, enhances damping characteristics, and effectively suppresses disturbances and unbalanced responses in large-scale mass–stiffness–damping systems. The feasibility and efficiency of the decentralized approach make it a compelling alternative to traditional centralized methods, especially in applications where computational efficiency, modularity, and robustness are critical.

13.2 Decentralized Control

The stabilization of large-scale multivariable systems remains a significant challenge due to their cross-coupled dynamics, high dimensionality, and sensitivity to disturbances. Consider a linear, time-invariant, multivariable, continuous-time dynamic system governed by

$$\dot{\boldsymbol{x}}(t) = \mathbf{A}\boldsymbol{x}(t) + \mathbf{B}\boldsymbol{u}(t), \tag{13.1}$$

where $\boldsymbol{x}(t)$ and $\boldsymbol{u}(t)$ are $n \times 1$ and $l \times 1$ state and control vectors, respectively, and matrices $\mathbf{A}$ and $\mathbf{B}$ are real, constant, and of appropriate dimensions. When such a centralized multivariable system consists of r cross-coupled subsystems controlled by r local control stations, it can be expressed in a decentralized information structure as

$$\dot{\boldsymbol{x}}_i(t) = \mathbf{A}_{ii}\boldsymbol{x}_i(t) + \mathbf{B}_i\boldsymbol{u}_i(t) + \sum_{j=1, j\neq i}^{r} \mathbf{A}_{ij}\boldsymbol{x}_j(t), \quad i = 1, 2, \ldots, r, \tag{13.2}$$

where each subsystem follows

$$\dot{\boldsymbol{x}}_i(t) = \mathbf{A}_{ii}\boldsymbol{x}_i(t) + \mathbf{B}_i\boldsymbol{u}_i(t), \quad i = 1, 2, \ldots, r, \tag{13.3}$$

where $\boldsymbol{x}_i(t)$ and $\boldsymbol{u}_i(t)$ are $n_i \times 1$ and $l_i \times 1$ state and control vectors, respectively, and the system satisfies $n = n_1 + n_2 + \cdots + n_r$ and $l = l_1 + l_2 + \cdots + l_r$. The feedback control law for each subsystem is designed as

$$\boldsymbol{u}_i(t) = \mathbf{K}_i\boldsymbol{x}_i(t), \quad i = 1, 2, \ldots, r, \tag{13.4}$$

where $\mathbf{K}_i$ is of dimension $l_i \times n_i$. Applying (13.4) to (13.2) results in the closed-loop system

$$\dot{\boldsymbol{x}}_i(t) = (\mathbf{A}_{ii} + \mathbf{B}_i\mathbf{K}_i)\boldsymbol{x}_i(t) + \sum_{j=1, j\neq i}^{r} \mathbf{A}_{ij}\boldsymbol{x}_j(t), \quad i = 1, 2, \ldots, r. \tag{13.5}$$

The goal of decentralized stabilization is to select local feedback gains $\mathbf{K}_i$ such that the overall system (13.5) is asymptotically stable, meaning that all poles of the system lie in the open left-half complex plane.

Building on the decentralized information structure introduced above, this section examines the decentralized control of mass–stiffness–damping systems. Large-scale mechanical systems—such as rotating machinery, structural assemblies, and flexible multibody configurations—are inherently multivariable, nonlinear, and dynamically cross-coupled [1]. While their mathematical models can effectively capture the complex dynamics of these systems, the development of practical active control strategies remains limited. This is primarily due to the challenges posed by nonlinear behavior, high dimensionality, and strong inter-subsystem couplings, which complicate both controller synthesis and real-time implementation in real-world applications.

In general, the equations of motion for a mass–stiffness–damping system with n-degrees-of-freedom can be described by a matrix-second-order system

$$\mathbf{M}\ddot{\boldsymbol{x}} + \mathbf{C}\dot{\boldsymbol{x}} + \mathbf{K}\boldsymbol{x} = \mathbf{F}_C\boldsymbol{f}_C + \mathbf{F}_D\boldsymbol{f}_D, \tag{13.6}$$

where $\boldsymbol{x}$ is an $n \times 1$ displacement vector, $\boldsymbol{f}_C$ and $\boldsymbol{f}_D$ are control force and external disturbance force vectors, respectively, and $\mathbf{M}$, $\mathbf{C}$, $\mathbf{K}$, $\mathbf{F}_C$, and $\mathbf{F}_D$ are real matrices of appropriate size.

Through structural analysis of mass–stiffness–damping systems, it can be shown that such systems are decomposable into multiple subsystems comprising both coupled and decoupled dynamic components. As a result, the decentralized control formulation for such systems takes the form

$$\begin{aligned} m_{ii}\ddot{x}_i + c_{ii}\dot{x}_i + k_{ii}x_i + \left(\sum_{j=1, j\neq i}^{n} m_{ij}\ddot{x}_j + \sum_{j=1, j\neq i}^{n} c_{ij}\dot{x}_j + \sum_{j=1, j\neq i}^{n} k_{ij}x_j \right) \\ = f_{Ci} + f_{Di}, \quad i = 1, 2, \ldots, n. \end{aligned} \tag{13.7}$$

If $m_{ii} \neq 0$, then by normalizing the mass term, the equation simplifies to

$$\ddot{x}_i + \frac{c_{ii}}{m_{ii}}\dot{x}_i + \frac{k_{ii}}{m_{ii}}x_i + \frac{1}{m_{ii}}(\bullet) = \frac{1}{m_{ii}}(f_{Ci} + f_{Di}), \quad i = 1, 2, \ldots, n, \tag{13.8}$$

where $(\bullet)$ denotes the cross-coupling terms bracketed in Eq. (13.7). Letting f'_{Di} represent the combined effect of disturbances and cross-couplings, and omitting the subscript i for clarity, the system can be rewritten as a state-space representation

$$\begin{bmatrix} \dot{x} \\ \ddot{x} \end{bmatrix} = \begin{bmatrix} 0 & 0 \\ -\frac{k}{m} & -\frac{c}{m} \end{bmatrix} \begin{bmatrix} x \\ \dot{x} \end{bmatrix} + \begin{bmatrix} 0 \\ \frac{1}{m} \end{bmatrix} (f_C + f'_D). \tag{13.9}$$

For simplicity and without loss of generality, adaptive pole-placement control is considered for the subsystem described by Eq. (13.9), which is subject to external disturbances and dynamic cross-coupling effects from other subsystems. A key consideration in decentralized control design is the presence of fixed modes—modes that cannot be shifted by any decentralized feedback—which can fundamentally limit stabilizability. The theoretical background on fixed modes is discussed in [5], and specific results confirming their absence in systems of this class are presented in

[6]. Furthermore, the controllability and observability properties of mass–stiffness–damping systems, as analyzed in Chap. 5 of this book, provide essential criteria for assessing the feasibility and effectiveness of the introduced control design in this decentralized framework.

13.3 Adaptive Pole-Placement

Since stability is a primary design objective in large-scale mass–stiffness–damping systems—particularly those subject to complex cross-couplings, nonlinearities, and external disturbances—adaptive pole-placement is adopted in this chapter as a robust stabilization strategy. This method enables the dynamic adjustment of feedback gains in real-time to maintain or improve the stability margin of the system, even under time-varying operating conditions or structural uncertainty. A substantial body of literature has addressed adaptive pole-placement control, offering both theoretical foundations and practical algorithms [7–9]. Its core advantage lies in the ability to shape the closed-loop pole locations of the system adaptively, thereby improving damping characteristics, transient response, and robustness.

To support the implementation of this method in large-scale mass–stiffness–damping systems, the unified model identification algorithm introduced in Chap. 3 is particularly well-suited. This algorithm estimates the parameters of linear continuous-time systems using sampled input–output data, making it compatible with practical data acquisition constraints in real-world systems.

In particular, the recursive least-squares (RLS) algorithm [10] can be used to identify the key system parameters in Eq. (13.9): namely, the normalized stiffness coefficient $-k/m$, damping coefficient $-c/m$, and mass inverse $1/m$. These parameters form the basis for real-time model updating in the adaptive scheme.

When the RLS identification is combined with the classical pole-placement design [8], a continuous-time decentralized adaptive pole-placement control law can be formulated for each subsystem in the large-scale system. This hybrid approach allows decentralized controllers to respond to evolving subsystem dynamics while maintaining closed-loop stability of the overall system.

The resulting adaptation algorithm, which updates the control gains in real-time based on online parameter estimation and desired pole configuration, is presented as Algorithm 13.1. This decentralized adaptation law provides a scalable and computationally efficient framework for stabilizing large-scale mass–stiffness–damping systems subject to cross-coupling and external perturbations.

Algorithm 13.1

Step 1: Parameter Estimation

Using the recursive estimation algorithm outlined in Chap. 3, the sampled data of displacement x, velocity $\dot{x}$, and control input u at time kT are used to estimate the normalized parameters $-k/m$, $-c/m$ and $1/m$ in system (13.9). These are then used to compute equivalent estimates $\hat{m}$, $\hat{k}$, and $\hat{c}$ of the physical parameters m, k, and c, respectively:

$$\hat{m} = \frac{1}{\widehat{\left(\frac{1}{m}\right)}}, \tag{13.10a}$$

$$\hat{k} = \hat{m}\widehat{\left(\frac{k}{m}\right)}, \tag{13.10b}$$

$$\hat{c} = \hat{m}\widehat{\left(\frac{c}{m}\right)}. \tag{13.10c}$$

Step 2: Control Design

Following the pole-placement method, we define a desired closed-loop characteristic polynomial $\mathrm{T}(\lambda)$, typically of stable second-order form:

$$\mathrm{T}(\lambda) = \lambda^2 + a_1\lambda + a_0.$$

The control law is constructed as:

$$fc(t) = \left[-\acute{k}\ -\acute{c}\right]\begin{bmatrix} x \\ \dot{x} \end{bmatrix}, \tag{13.11}$$

where $\acute{k}$ and $\acute{c}$ represent the actively assigned stiffness and damping, designed to place the poles of the closed-loop system at desired locations. These active parameters must satisfy:

$$\mathrm{T}(\lambda) = \lambda^2 + \frac{\acute{c}+c}{m}\lambda + \frac{\acute{k}+k}{m}. \tag{13.12}$$

To implement this adaptively, we substitute the estimated parameters $\hat{m}$, $\hat{k}$ and $\hat{c}$ in place of m, k and c in Eq. (13.12). Solving for the updated values $\hat{\acute{k}}$ and $\hat{\acute{c}}$, we arrive at the practical adaptive control law:

$$fc(t) = \left[-\hat{\acute{k}}\ -\hat{\acute{c}}\right]\begin{bmatrix} x \\ \dot{x} \end{bmatrix}, \tag{13.13}$$

where the parameters $\hat{\acute{k}}$ and $\hat{\acute{c}}$ satisfy:

$$\mathrm{T}(\lambda) = \lambda^2 + \frac{\hat{\acute{c}}+\hat{c}}{\hat{m}}\lambda + \frac{\hat{\acute{k}}+\hat{k}}{\hat{m}}. \tag{13.14}$$

Step 3. Adaptive Control Execution

The control law from Eq. (13.13) is applied to the subsystem (13.9). At each time step kT, the parameter estimation (Step 1) is updated using new sampled data, and the control law is recomputed accordingly. This closed-loop adaptive procedure ensures that the response of the system continuously aligns with the desired pole-placement, even in the presence of parameter variations and external disturbances.

The selection of the desired polynomial $\mathrm{T}(\lambda)$ should ensure sufficient damping and stability margins appropriate for the application. The estimation and control update interval T should be chosen based on system dynamics and computational limitations to ensure convergence and real-time feasibility. This algorithm provides a decentralized mechanism that enables local controllers to adapt independently based on local measurements, making it highly suitable for large-scale interconnected mass–stiffness–damping systems.

The decentralized adaptive controller developed based on Algorithm 13.1 achieves complete rejection of external disturbances and cross-coupling effects under ideal conditions. This leads to the following result, stated as Property 13.1.

Property 13.1 *Consider the mass–stiffness–damping system described by Eq.* (13.9). *If the adaptive control algorithm presented in Algorithm* 13.1 *is applied, then the resulting closed-loop system satisfies*

$$\mathrm{T}(\lambda)x(t) = 0, \tag{13.15}$$

where $\mathrm{T}(\lambda)$ *is the target characteristic polynomial specified by the pole-placement strategy, and* $x(t)$ *is the displacement response of the system. Under this control law, for any bounded external disturbance* $f_D'(t)$, *the system exhibits asymptotic rejection, and the following holds:*

$$\lim_{t \to \infty} x(t) = 0. \tag{13.16}$$

This result confirms that the decentralized adaptive controller ensures asymptotic stability and complete disturbance rejection for the controlled subsystem.

Proof We begin by substituting the adaptive control law from Eq. (13.13) into the subsystem dynamics given by Eq. (13.9). This yields the closed-loop differential equation:

$$\ddot{x} + \frac{c + \hat{c}}{m}\dot{x} + \frac{k + \hat{k}}{m}x = \frac{1}{m}f_D'. \tag{13.17}$$

To capture the effect of parameter estimation errors, we express the true system parameters in terms of their estimated counterparts plus perturbation terms:

$$\left(\frac{k}{m}\right) = \widehat{\left(\frac{k}{m}\right)} + \delta_{k/m}, \tag{13.18a}$$

$$\left(\frac{c}{m}\right) = \widehat{\left(\frac{c}{m}\right)} + \delta_{c/m}, \tag{13.18b}$$

$$\left(\frac{1}{m}\right) = \widehat{\left(\frac{1}{m}\right)} + \delta_{1/m}. \tag{13.18c}$$

Here, $\delta_{k/m}$, $\delta_{c/m}$, and $\delta_{1/m}$ denote the estimation errors in the respective coefficients. Defining the parameter error vector:

$$\tilde{\boldsymbol{\theta}} = \boldsymbol{\theta} - \hat{\boldsymbol{\theta}} = \begin{bmatrix} -\delta_{k/m} & -\delta_{c/m} & \delta_{1/m} \end{bmatrix}. \tag{13.19}$$

Substituting Eqs. (13.18a)–(13.18c) into Eq. (13.17), and noting from Eq. (13.14) that $\hat{\hat{k}}$ and $\hat{\hat{c}}$ are computed using $\hat{k}$, $\hat{c}$, and $\hat{m}$, we obtain:

$$\mathrm{T}(\lambda)x + \left(\delta_{k/m}x + \delta_{c/m}\dot{x}\right) + \delta_{1/m}\left(\hat{\hat{k}}x + \hat{\hat{c}}\dot{x}\right) = \frac{1}{m}f_D'. \tag{13.20}$$

Using the control input from Eq. (13.13), i.e., $f_C = -\hat{\hat{k}}x - \hat{\hat{c}}\dot{x}$, we can write:

$$\mathrm{T}(\lambda)x + \left(\delta_{k/m}x + \delta_{c/m}\dot{x}\right) - \delta_{1/m}f_C = \frac{1}{m}f_D'. \tag{13.21}$$

Rewriting the left-hand side, we factor it using the error vector $\tilde{\boldsymbol{\theta}}$:

$$-\left(\delta_{k/m}x + \delta_{c/m}\dot{x}\right) + \delta_{1/m}f_C = \tilde{\boldsymbol{\theta}} \cdot \boldsymbol{\phi}(t), \tag{13.22}$$

where the regressor vector is defined as:

$$\boldsymbol{\phi}(t) = \begin{bmatrix} x(t) & \dot{x}(t) & f_C(t) \end{bmatrix}^{\mathrm{T}}. \tag{13.23}$$

Substituting Eq. (13.22) into Eq. (13.21), the closed-loop system becomes:

$$\mathrm{T}(\lambda)x(t) = \tilde{\boldsymbol{\theta}} \cdot \boldsymbol{\phi}(t) + \frac{1}{m}f_D'(t). \tag{13.24}$$

Now consider the RLS estimation of $\boldsymbol{\theta}$ over a finite window of N data points. From [10], the estimation error $\tilde{\boldsymbol{\theta}}$ is expressed as:

$$\tilde{\boldsymbol{\theta}} = \boldsymbol{f}'_{DN}\boldsymbol{\Phi}_N^{\mathrm{T}}\left(\boldsymbol{\Phi}_N\boldsymbol{\Phi}_N^{\mathrm{T}}\right)^{-1}, \tag{13.25}$$

where the disturbance data vector and regressor matrix are defined as:

$$\boldsymbol{f}'_{DN} = \begin{bmatrix} \frac{1}{m}f_D'(t+\Delta t) & \frac{1}{m}f_D'(t+2\Delta t) & \cdots & \frac{1}{m}f_D'(t+N\Delta t) \end{bmatrix} \tag{13.26a}$$

$$\mathbf{\Phi}_N = \left[\boldsymbol{\phi}(t+\Delta t) \;\; \boldsymbol{\phi}(t+2\Delta t) \;\; \cdots \;\; \boldsymbol{\phi}(t+N\Delta t) \right]. \tag{13.26b}$$

From Eq. (13.25), we have the identity:

$$\left(\tilde{\boldsymbol{\theta}}\mathbf{\Phi}_N + \boldsymbol{f}'_{DN}\right)\mathbf{\Phi}_N^{\mathrm{T}} = 0. \tag{13.27}$$

Assuming $N \geq n + l + 1$ (here, $n = 2$, $l = 1$), and that $\mathbf{\Phi}_N$ is of full row rank, it follows that:

$$\tilde{\boldsymbol{\theta}} \cdot \boldsymbol{\phi}(t) + \frac{1}{m} f'_D(t) = 0. \tag{13.28}$$

Substituting Eq. (13.28) into Eq. (13.24), we arrive at the closed-loop homogeneous equation:

$$\mathrm{T}(\lambda)x(t) = 0.$$

Since $\mathrm{T}(\lambda)$ is a stable polynomial by design, all roots of the characteristic equation lie in the open left-half complex plane. Therefore, the system is asymptotically stable, and the displacement response satisfies:

$$\lim_{t\to\infty} x(t) = 0.$$

This concludes the proof. □

Property 13.1 highlights a key strength of the introduced decentralized adaptive pole-placement scheme: its ability to achieve ideal closed-loop performance even in the presence of plant uncertainties, external disturbances, and dynamic cross-coupling. Specifically, the resulting adaptive system maintains the same structural behavior—as defined by its characteristic polynomial and asymptotic dynamics—as a nominal system with perfectly known parameters and no external perturbations or cross-couplings.

This performance is achieved through the integration of RLS estimation and adaptive pole-placement. The RLS algorithm continuously identifies normalized stiffness, damping, and inertial parameters from sampled data, while the pole-placement mechanism dynamically adjusts the control law to match a predefined, stable characteristic polynomial. Crucially, the estimation errors themselves generate corrective actions that, in steady state, cancel the impact of unknown disturbances and coupling effects. As a result, the system output asymptotically satisfies the desired homogeneous differential equation, thereby ensuring disturbance and cross-coupling rejection.

From a systems-theoretic viewpoint, this behavior can be interpreted through the internal model principle of robust control: the adaptive controller effectively embeds a model of the disturbance dynamics within the identification and control loop. Unlike fixed-gain decentralized controllers—which may degrade under modeling

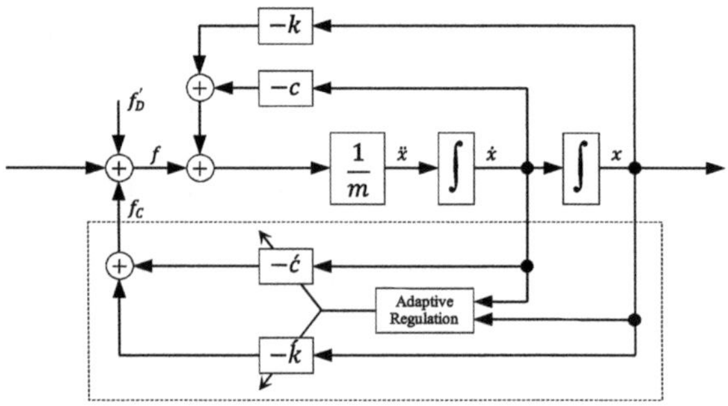

Fig. 13.1 Adaptive pole-placement for decentralized control

errors or dynamic coupling—the adaptive strategy enables real-time compensation, preserving both stability and performance under uncertain operating conditions.

Thus, the introduced approach ensures convergence of the controlled system to a nominal model not only under ideal conditions, but also when subject to persistent disturbances and structural cross-couplings. It offers a scalable, computationally efficient alternative to centralized methods, well-suited for high-dimensional mass–stiffness–damping systems such as rotor-bearing assemblies, large-scale flexible structures, and interconnected mechanical networks.

Based on the preceding algorithm, the configuration of the decentralized adaptive pole-placement control system is illustrated in Fig. 13.1. The active damping $\hat{c}$ and active stiffness $\hat{k}$ are continuously updated through an adaptive mechanism that integrates RLS estimation with online parameter assignment via pole-placement.

13.4 Enhanced MRAC

As evident from the structural diagram in Fig. 13.1, if the adaptation mechanism—responsible for online parameter estimation and control law adjustment—is disabled or disconnected, the resulting control law naturally reverts to a conventional fixed-gain state feedback configuration. This is consistent with standard practice in adaptive control, where the adaptation loop is typically deactivated under conditions where persistent excitation (PE) of the input signal is not guaranteed. In such cases, continued parameter updating can lead to unbounded parameter drift or degraded performance, violating the stability and robustness guarantees of the adaptive scheme.

To address this, a complete adaptive control framework must incorporate a mechanism for switching between adaptive and non-adaptive modes, often governed by a monitoring logic based on excitation conditions or signal quality metrics. However, when adaptation is temporarily suspended, maintaining acceptable system performance—particularly in the presence of uncertainties or faults—requires an auxiliary controller to take over.

In this context, a model-following-based robust and fault-tolerant control (FTC) scheme—such as the one presented in [11]—serves as an appropriate fallback

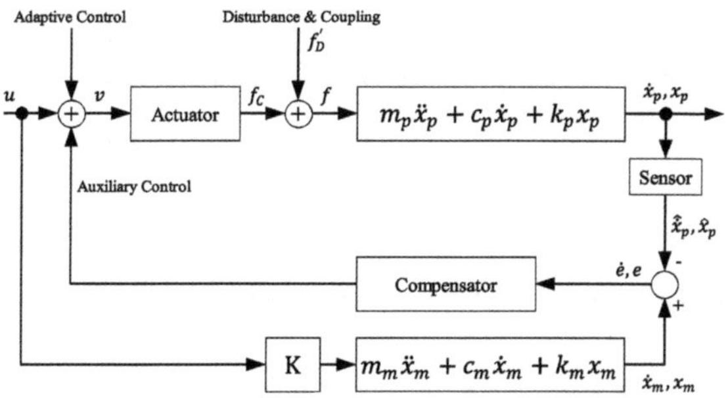

Fig. 13.2 Enhanced model-reference adaptive control

strategy. To this end, an enhanced model-reference adaptive control (MRAC) structure is shown in Fig. 13.2. As illustrated, the auxiliary control loop ensures continuity of robust stabilization and tracking performance by enforcing model-reference behavior, even when adaptive updates are temporarily disabled. The model-following architecture inherently improves robustness to plant variations and external disturbances by regulating the system output toward a predefined reference model. In addition, it can be configured to accommodate a bounded class of actuator or sensor faults, supporting reliable performance under degraded conditions.

The integration of model-following control into the adaptive framework not only enhances overall reliability but also ensures fault tolerance and performance retention during adaptation-off phases. This hybrid configuration aligns with modern robust adaptive control practices and is particularly beneficial for large-scale or safety–critical systems where uninterrupted performance is essential.

13.5 Performance of Decentralized Adaptive Control

To demonstrate the practical utility and performance of the decentralized adaptive pole-placement control scheme, a numerical case study is presented using a realistic large-scale rotor-bearing system.

The modeled rotor system consists of 10 discrete degrees-of-freedom and is supported by two tilting-pad journal bearings at each end. The complete state-space model of this multivariable system is of 26th order, capturing the relevant dynamics including shaft flexibility and bearing cross-couplings. The simplified physical layout is shown in Fig. 13.3.

Control is applied at both rotor ends through four local actuators positioned along the translational directions: x_1, y_1, x_5, y_5. This setup reflects standard industrial practice where direct access to the rotor is limited and control must be implemented through the bearing supports.

Since rotor damping is typically low, the system is inherently susceptible to instability. In fact, the rotor-bearing system used in this example becomes unstable at approximately 12,600 rpm (corresponding to a critical frequency $w \approx 1319$ rad/s).

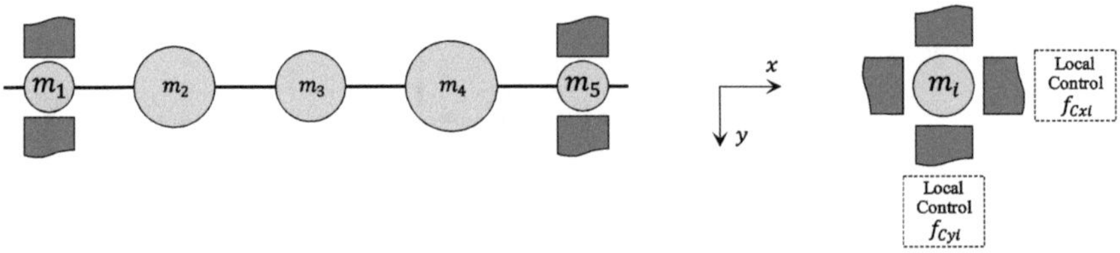

Fig. 13.3 Controlled rotor-bearing system

Hence, stabilization is sought by adaptively adjusting the active damping coefficients $\acute{c}_i$ at the actuator locations.

Figure 13.4 illustrates the closed-loop response of the controlled system under decentralized adaptive damping when subjected to step changes in displacement reference. All initial conditions are set to zero. The plots show displacement versus time, allowing clear visualization of stability improvement through active damping control.

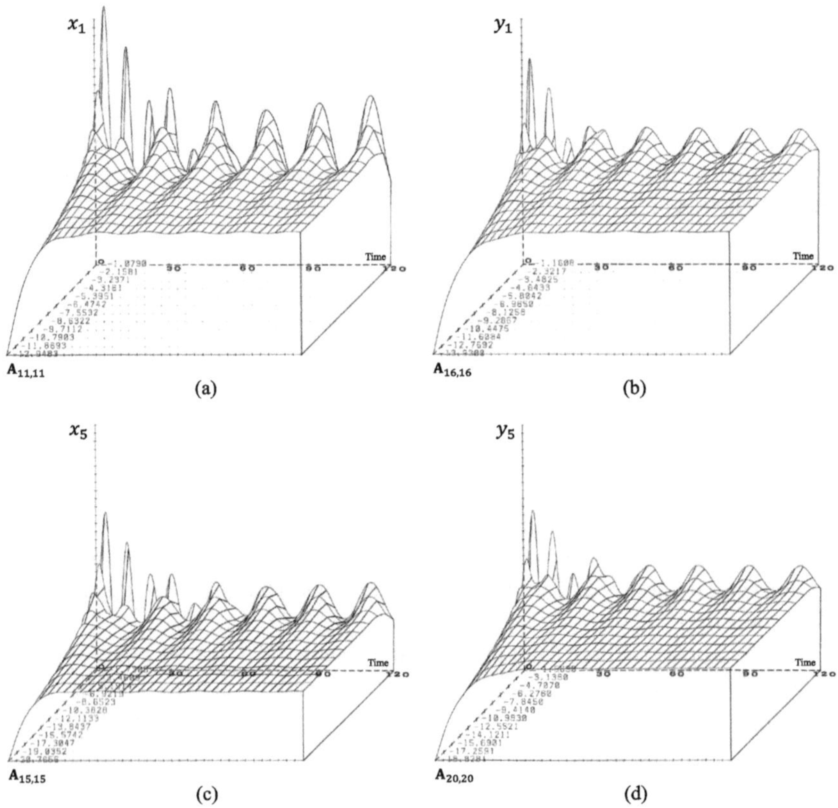

Fig. 13.4 Displacement-damping-time graph of the closed-loop system

As Fig. 13.4 demonstrates, increasing active damping results in enhanced system stability and improved settling behavior. The adaptive controllers succeed in stabilizing the system and regulating displacement against external disturbances and cross-couplings.

To provide deeper insight, Table 13.1 lists the dominant eigenvalues of the closed-loop system for varying levels of active damping.

Figure 13.5 presents the locations of the first 20 eigenvalues as damping is adjusted progressively.

Figure 13.6 further visualizes the evolution of active damping parameters alongside the key real eigenvalues of the system, offering a dynamic perspective on the stabilizing effect.

This case study demonstrates that the decentralized adaptive pole-placement scheme introduced in this chapter can successfully stabilize an inherently unstable large-scale mass–stiffness–damping system. The results confirm that the control objectives—improved damping, disturbance rejection, and overall system stabilization—are effectively achieved in accordance with the design intent.

13.6 Summary

This chapter presents a systematic framework for decentralized adaptive stabilization of mass–stiffness–damping systems, which are representative of a broad class of large-scale, interconnected mechanical and structural systems. Recognizing the intrinsic challenges posed by high dimensionality, structural coupling, and uncertainty, the chapter introduces a decomposition-based approach that reformulates the global dynamics into a set of tractable, second-order local subsystems.

By applying adaptive pole-placement techniques to a specific local subsystem, the control strategy dynamically adjusts the control gains in real time, based on parameter estimates obtained through RLS identification. This allows each controller to compensate for plant uncertainties and external disturbances while maintaining system stability and desired performance.

Compared to conventional centralized approaches, the decentralized adaptive strategy introduced here offers several key advantages:

- *Scalability:* The method is inherently suited for high-order systems with many degrees-of-freedom, making it practical for real-world applications involving complex machinery or infrastructure.
- *Robustness:* Through continuous adaptation, the control law remains effective even in the presence of parameter variations, dynamic coupling, and unmodeled disturbances.
- *Computational Tractability:* The control is implemented locally, avoiding the complexity and communication overhead associated with centralized high-order controllers.

Table 13.1 Eigenvalues of the closed-loop system

	$\mathbf{A}_{11,11}$: -1.0790 $\mathbf{A}_{15,15}$: -1.7305	$\mathbf{A}_{16,16}$: -1.1608 $\mathbf{A}_{20,20}$: -1.5690	$\mathbf{A}_{11,11}$: -2.1581 $\mathbf{A}_{15,15}$: -3.4609	$\mathbf{A}_{16,16}$: -2.3217 $\mathbf{A}_{20,20}$: -3.1380	$\mathbf{A}_{11,11}$: -3.2371 $\mathbf{A}_{15,15}$: -5.1914	$\mathbf{A}_{16,16}$: -3.4825 $\mathbf{A}_{20,20}$: -4.7070	$\mathbf{A}_{11,11}$: -4.3161 $\mathbf{A}_{15,15}$: -6.9219	$\mathbf{A}_{16,16}$: -4.6433 $\mathbf{A}_{20,20}$: -6.2760	$\mathbf{A}_{11,11}$: -5.3951 $\mathbf{A}_{15,15}$: -8.6523	$\mathbf{A}_{16,16}$: -5.8042 $\mathbf{A}_{20,20}$: -7.8450	$\mathbf{A}_{11,11}$: -6.4742 $\mathbf{A}_{15,15}$: -10.3828	$\mathbf{A}_{16,16}$: -6.9650 $\mathbf{A}_{20,20}$: -9.4140
1	*0.0124*	*0.3053*	*0.0011*	*0.3021*	-0.0110	0.3035	-0.0216	0.3102	-0.0242	0.3274	-0.0228	0.3336
2	*0.0124*	*-0.3053*	*0.0011*	*-0.3021*	-0.0110	-0.3035	-0.0216	-0.3102	-0.0242	-0.3274	-0.0228	-0.3336
3	*0.0041*	*0.3050*	-0.0071	0.3055	-0.0169	0.3106	-0.0229	0.3190	-0.0271	0.3210	-0.0265	0.3304
4	*0.0041*	*-0.3050*	-0.0071	-0.3055	-0.0169	-0.3106	-0.0229	-0.3190	-0.0271	-0.3210	-0.0265	-0.3304
5	*0.0218*	*0.6358*	-0.1231	0.5205	-0.1806	0.4168	-0.2055	0.3449	-0.2139	0.3051	-0.2096	0.2578
6	*0.0218*	*-0.6358*	-0.1231	-0.5205	-0.1806	-0.4168	-0.2055	-0.3449	-0.2139	-0.3051	-0.2096	-0.2578
7	-0.0215	3.9739	-0.0320	3.9582	-0.0341	3.9461	-0.0324	3.9365	-0.0296	3.9301	-0.0267	3.9258
8	-0.0215	-3.9739	-0.0320	-3.9582	-0.0341	-3.9461	-0.0324	-3.9365	-0.0296	-3.9301	-0.0267	-3.9258
9	-0.0204	3.9749	-0.0319	3.9597	-0.0336	3.9449	-0.0317	3.9358	-0.0289	3.9297	-0.0261	3.9256
10	-0.0204	-3.9749	-0.0319	-3.9597	-0.0336	-3.9449	-0.0317	-3.9358	-0.0289	-3.9297	-0.0261	-3.9256
11	-0.1344	1.6710	-0.1516	1.5650	-0.1139	1.5085	-0.0864	1.4889	-0.0688	1.4805	-0.0570	1.4762
12	-0.1344	-1.6710	-0.1516	-1.5650	-0.1139	-1.5085	-0.0864	-1.4889	-0.0688	-1.4805	-0.0570	-1.4762
13	-0.1279	1.7003	-0.1369	1.5504	-0.1031	1.5051	-0.0794	1.4881	-0.0639	1.4804	-0.0533	1.4763
14	-0.1279	-1.7003	-0.1369	-1.5504	-0.1031	-1.5051	-0.0794	-1.4881	-0.0639	-1.4804	-0.0533	-1.4763
15	-0.0282	-0.7193	-0.2048	-0.6108	-0.2294	-0.4793	-0.2292	-0.3835	-0.2314	-0.2935	-0.2448	-0.2361
16	-0.0282	0.7193	-0.2048	0.6108	-0.2294	0.4793	-0.2292	0.3835	-0.2314	0.2935	-0.2448	0.2361
17	-0.0761	-0.8021	-0.2301	-0.5954	-0.3208	-0.4429	-0.3475	-0.3231	-0.3423	-0.2282	-0.3259	-0.1486
18	-0.0761	0.8021	-0.2301	0.5954	-0.3208	0.4429	-0.3475	0.3231	-0.3423	0.2282	-0.3259	0.1486
19	-0.1117	-0.9216	-0.3233	-0.7069	-0.3947	-0.5051	-0.4095	-0.3338	-0.3987	-0.1923	-0.3776	-0.0654
20	-0.1117	0.9216	-0.3233	0.7069	-0.3947	0.5051	-0.4095	0.3338	-0.3987	0.1923	-0.3776	0.0654
21	-0.4859	0.0000	-0.4941	0.0000	-0.5069	0.0000	-0.5282	0.0000	-0.5630	0.0000	-0.6074	0.0000
22	-0.5767	0.0000	-0.5862	0.0000	-0.6030	0.0000	-0.6373	0.0000	-0.7019	0.0000	-0.7732	0.0000
23	-1.5214	0.0000	-2.2637	0.0000	-3.2376	0.0000	-4.2857	0.0000	-5.3557	0.0000	-6.4330	0.0000
24	-1.8888	0.0000	-2.6439	0.0000	-3.5988	0.0000	-4.6833	0.0000	-5.8133	0.0000	-6.9602	0.0000
25	-1.9812	0.0000	-3.1722	0.0000	-4.6444	0.0000	-6.1976	0.0000	-7.7683	0.0000	-9.3429	0.0000
26	-2.0421	0.0000	-3.3604	0.0000	-5.0718	0.0000	-6.8137	0.0000	-8.5576	0.0000	-10.2997	0.0000

(continued)

Table 13.1 (continued)

	$A_{11,11}$: -7.5532 $A_{15,15}$: -12.1133	$A_{16,16}$: -8.1258 $A_{20,20}$: -10.9830	$A_{11,11}$: -8.6322 $A_{15,15}$: -13.8437	$A_{16,16}$: -9.2867 $A_{20,20}$: -12.5521	$A_{11,11}$: -9.7112 $A_{15,15}$: -15.5742	$A_{16,16}$: -10.4475 $A_{20,20}$: -14.1211	$A_{11,11}$: -10.7903 $A_{15,15}$: -17.3047	$A_{16,16}$: -11.6084 $A_{20,20}$: -15.6901	$A_{11,11}$: -11.8693 $A_{15,15}$: -19.0352	$A_{16,16}$: -12.7692 $A_{20,20}$: -17.2591	$A_{11,11}$: -12.9483 $A_{15,15}$: -20.7656	$A_{16,16}$: -13.9300 $A_{20,20}$: -18.8281
1	-0.0206	0.3377	-0.0185	0.3404	-0.0166	0.3421	-0.0150	0.3434	-0.0137	0.3443	-0.0125	0.3450
2	-0.0206	-0.3377	-0.0185	-0.3404	-0.0166	-0.3421	-0.0150	-0.3434	-0.0137	-0.3443	-0.0125	-0.3450
3	-0.0236	0.3363	-0.0207	0.3398	-0.0183	0.3419	-0.0163	0.3434	-0.0147	0.3443	-0.0133	0.3451
4	-0.0236	-0.3363	-0.0207	-0.3398	-0.0183	-0.3419	-0.0163	-0.3434	-0.0147	-0.3443	-0.0133	-0.3451
5	-0.2110	0.2183	-0.2130	0.1849	-0.2147	0.1550	-0.2161	0.1268	-0.2172	0.0985	-0.2182	0.0666
6	-0.2110	-0.2183	-0.2130	-0.1849	-0.2147	-0.1550	-0.2161	-0.1268	-0.2172	-0.0985	-0.2182	-0.0666
7	-0.0240	3.9229	-0.0217	3.9208	-0.0198	3.9193	-0.0181	3.9182	-0.0167	3.9174	-0.0155	3.9167
8	-0.0240	-3.9229	-0.0217	-3.9208	-0.0198	-3.9193	-0.0181	-3.9182	-0.0167	-3.9174	-0.0155	-3.9167
9	-0.0235	3.9228	-0.0213	3.9208	-0.0195	3.9193	-0.0178	3.9182	-0.0164	3.9174	-0.0152	3.9167
10	-0.0235	-3.9228	-0.0213	-3.9208	-0.0195	-3.9193	-0.0178	-3.9182	-0.0164	-3.9174	-0.0152	-3.9167
11	-0.0486	1.4737	-0.0423	1.4722	-0.0374	1.4712	-0.0336	1.4704	-0.0304	1.4699	-0.0278	1.4695
12	-0.0486	-1.4737	-0.0423	-1.4722	-0.0374	-1.4712	-0.0336	-1.4704	-0.0304	-1.4699	-0.0278	-1.4695
13	-0.0456	1.4738	-0.0399	1.4723	-0.0354	1.4713	-0.0318	1.4705	-0.0289	1.4700	-0.0264	1.4696
14	-0.0456	-1.4738	-0.0399	-1.4723	-0.0354	-1.4713	-0.0318	-1.4705	-0.0289	-1.4700	-0.0264	-1.4696
15	-0.2513	-0.1883	-0.2554	-0.1439	-0.2582	-0.0974	-0.2603	-0.0253	-0.1262	0.0000	-0.1133	0.0000
16	-0.2513	0.1883	-0.2554	0.1439	-0.2582	0.0974	-0.2603	0.0253	-0.1327	0.0000	-0.1178	0.0000
17	-0.3100	-0.0740	-0.1967	0.0000	-0.1650	0.0000	-0.1428	0.0000	-0.1812	0.0000	-0.1517	0.0000
18	-0.3100	0.0740	-0.2293	0.0000	-0.1809	0.0000	-0.1525	0.0000	-0.3425	0.0000	-0.3744	0.0000
19	-0.2489	0.0000	-0.3658	0.0000	-0.3954	0.0000	-0.4092	0.0000	-0.4175	0.0000	-0.4231	0.0000
20	-0.4711	0.0000	-0.4974	0.0000	-0.5097	0.0000	-0.5170	0.0000	-0.5218	0.0000	-0.5252	0.0000
21	-0.6480	0.0000	-0.6797	0.0000	-0.7039	0.0000	-0.7226	0.0000	-0.7374	0.0000	-0.7495	0.0000
22	-0.8266	0.0000	-0.8646	0.0000	-0.8925	0.0000	-0.9137	0.0000	-0.9303	0.0000	-0.9437	0.0000
23	-7.5130	0.0000	-8.5939	0.0000	-9.6750	0.0000	-10.7563	0.0000	-11.8373	0.0000	-12.9182	0.0000
24	-8.1144	0.0000	-9.2722	0.0000	-10.4315	0.0000	-11.5919	0.0000	-12.7527	0.0000	-13.9137	0.0000
25	-10.9179	0.0000	-12.4926	0.0000	-14.0665	0.0000	-15.6398	0.0000	-17.2125	0.0000	-18.7848	0.0000
26	-12.0396	0.0000	-13.7777	0.0000	-15.5146	0.0000	-17.2503	0.0000	-18.9853	0.0000	-20.7195	0.0000

Eigenvalues originally associated with instability are indicated in italics, whereas those originally real and dominant are highlighted with dark shading

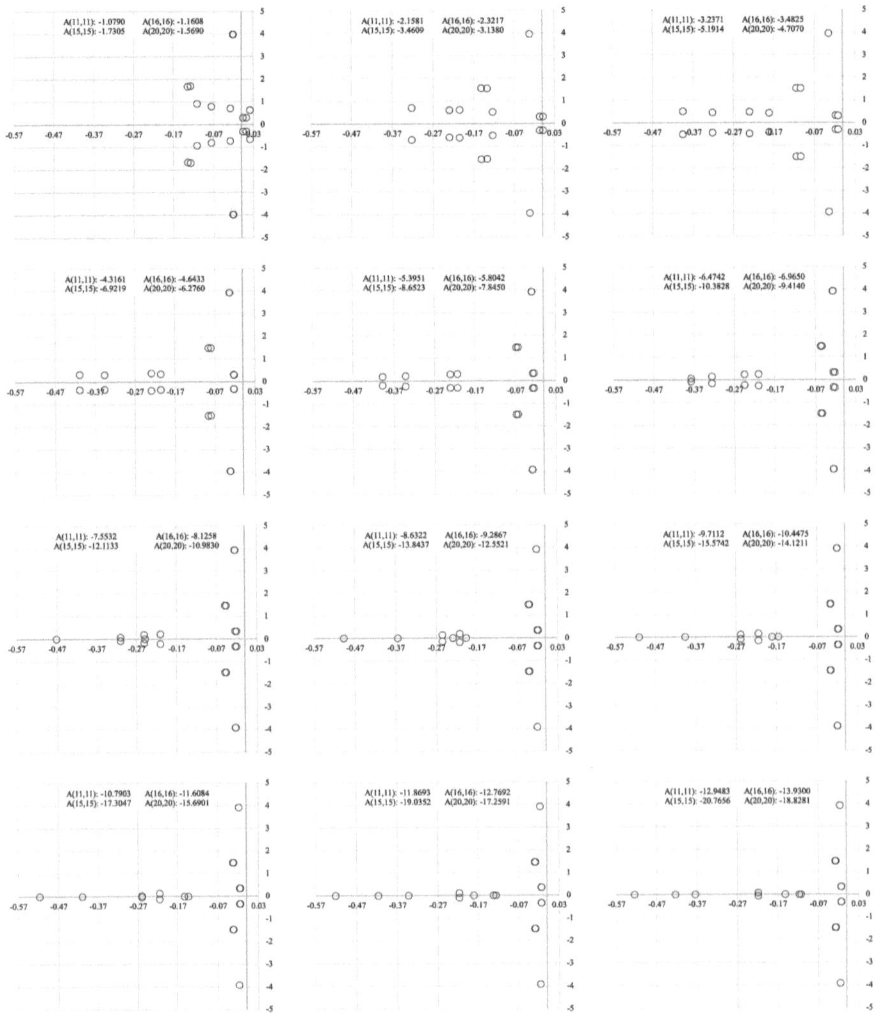

Fig. 13.5 Eigenvalue plot of the closed-loop system

A detailed case study of a 26th-order rotor-bearing system demonstrates the efficacy of the introduced method. The results show that the decentralized controllers effectively stabilize an otherwise unstable system, suppress vibration, and attenuate coupling effects through real-time adjustment of active damping and stiffness parameters.

The approach outlined in this chapter not only achieves reliable stabilization but also aligns with fundamental principles of large-scale system control and robust adaptive design. Moreover, the techniques presented are broadly applicable to other engineering domains, such as flexible multibody systems, robotic platforms, aerospace

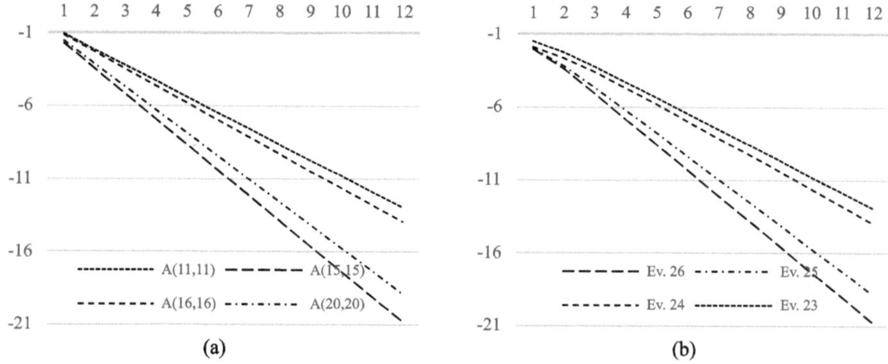

Fig. 13.6 Damping (**a**) and real eigenvalue (**b**) of the closed-loop system

structures, and smart infrastructure networks—where decentralized control and adaptability are critical to performance and reliability.

References

1. Jamshidi M (1997) Large-scale systems—modeling, control and fuzzy logic. Prentice Hall, New Jersey
2. Singh MG, Titli A, Malinowski K (1983) Decentralized control design: an overview. IFAC Proc Volumes 16(12):1–15
3. Levis AH (1987) Challenges to control: a collective view. IEEE Trans Autom Control 32(4):275–285
4. Davison EJ, Aghdam AG, Miller DE (2020) Decentralized control of large-scale systems. Springer, Berlin
5. Wang SH, Davison EJ (1973) On the stabilization of decentralized control systems. IEEE Trans Autom Control 18(5):473–478
6. Davison EJ, Özgüner Ü (1983) Characterizations of decentralized fixed modes for interconnected systems. Automatica 19(2):169–182
7. Åström KJ, Borisson U, Ljung L, Wittenmark B (1977) Theory and applications of self-tuning regulators. Automatica 13(5):457–476
8. Westerberg B, Wittenmark B, Åström KJ (1978) Self-tuning controllers based on pole placement design. Lund Institute of Technology, Research reports TFRT-3148
9. Åström KJ (1983) Theory and applications of adaptive control—a survey. Automatica 19(5):471–486
10. Zhu HA, Hung JC (1990) Identification of dynamic systems: a systematic approach. In: Proceedings of the 16th IEEE International conference on industrial electronics (IECON'90), pp 131–136
11. Zhu HA (1988) Multivariable decentralized adaptive control of rotor-bearing systems. In: Proceedings of the 1988 IEEE International conference on systems, man, and cybernetics, vol 1, pp 215–218

Chapter 14
Enhanced Internal Model Control

A comprehensive treatment of model-based modeling and control would be incomplete without addressing one of the most fundamental strategies in the field: Internal model control (IMC). To that end, this chapter concludes the broader discussions of the book by presenting an enhanced internal model control (enhanced IMC) framework.

Building on the classical IMC architecture, this chapter introduces a structurally modified IMC scheme aimed at improving robustness. The presented modification is both conceptually straightforward and practically effective, requiring only minimal changes to the original configuration. Importantly, the synthesis of the additional components is decoupled from the design of the primary IMC controller, thereby simplifying implementation and preserving the original design objectives of the controller.

Analytical analysis and supporting numerical studies are included to highlight the impact of the presented structure on robustness enhancement. Together, they demonstrate that with minimal structural adjustments, the IMC strategy can be significantly strengthened to meet the demands of uncertain or variable operating environments.

14.1 Introduction

Robustness has emerged as a central theme in the research and development of modern control systems, especially in light of increasing demands for reliability and performance under uncertainty. Over the past few decades, significant progress has been made in both theoretical advancements and practical applications. Among the family of robust control strategies, internal model control (IMC) has undergone extensive development, becoming a mature and versatile methodology. Originally formulated for process control, IMC has since found broad application across various

H. Zhu, *Model-Based Control of Mass–Stiffness–Damping Systems*, Advances in Industrial Control, https://doi.org/10.1007/978-3-031-97592-9_14

domains, including mechanical, electrical, and structural systems [1]. Its compatibility with mass–stiffness–damping systems further supports its status as a viable control strategy in high-performance applications.

The foundational concepts of IMC were first introduced for single-input, single-output (SISO) systems by Garcia and Morari [2], and later extended to multi-input, multi-output (MIMO) systems [3, 4]. These developments were subsequently formalized and expanded into a systematic design methodology through the work of Morari and collaborators [5–8]. Today, IMC stands as both a theoretical framework and a practical tool in a wide range of control engineering applications [9–11].

One of the notable strengths of IMC lies in its ability to provide a unified framework for performance analysis and robustness evaluation [12]. The model-based design of IMC provides greater transparency and simplicity compared to many traditional techniques—even when the target is a conventional PID controller [5]. It also serves as a conceptual bridge between feedback and feedforward control [10], allowing for generalization and reinterpretation of other advanced control techniques within its structure. For instance, Bergh and MacGregor demonstrated that the constrained minimum variance (CMV) controller can be cast into IMC form and proposed an alternative synthesis procedure for the inverse model controller and the filter [13]. Similarly, Grimble and co-workers showed that Linear Quadratic Gaussian (LQG) control can be implemented within the IMC architecture, thereby offering new insights into both LQG and $\mathcal{H}^\infty$ design methods [14].

These developments have contributed to the perception of IMC not only as a specific control structure but also as a general methodological foundation for robust control design. Numerous researchers have drawn connections between IMC and other modern control techniques, suggesting that IMC serves as a basis for extending or unifying disparate robust control approaches. The literature reflects a growing consensus that IMC, with its inherent transparency and intuitive structure, can serve as both a design tool and a platform for theoretical exploration.

The simplicity of IMC makes it especially attractive in practice. It offers both theoretical guarantees [12] and heuristic design guidelines [15] that can be readily applied by engineers. However, as Fisher noted [10], in more realistic scenarios involving plant–model mismatch or system uncertainty, the design and implementation of an IMC controller become substantially more complex. In such cases, even if all individual loop components are stable, the closed-loop IMC system may still exhibit instability. This underscores the importance of addressing uncertainty explicitly during IMC design to ensure both performance and robustness.

To address this need, an enhanced IMC scheme is introduced in this chapter. Building on the conventional IMC framework, the modified scheme features a structural enhancement designed to mitigate plant uncertainty and improve overall system robustness [16]. Importantly, this modification is both simple and modular—the synthesis of the added components is independent of the original IMC controller design. The presented enhancement preserves the favorable characteristics of the standard IMC structure [17], while improving its ability to tolerate disturbances, unmodeled dynamics, and parameter variations.

Both analytical and numerical investigations are presented to validate the effectiveness of the enhanced scheme. While the discussion is initially confined to systems with simplified dynamic models and exogenous inputs, the underlying methodology is general in scope. With appropriate adjustments, the enhanced IMC structure can be extended to a broad range of control applications, including high-order MIMO systems, nonlinear dynamics, and systems with time delays or structural uncertainties.

14.2 Preliminaries of IMC

A brief overview of the key properties and limitations of conventional IMC systems—along with the synthesis procedure for the corresponding IMC controller—is provided below to lay the foundation for subsequent discussions on the enhanced IMC architecture.

14.2.1 Properties of IMC Systems

The fundamental structure of a conventional IMC system is depicted in Fig. 14.1, where:

- $\mathbf{P}$ denotes the actual plant to be controlled.
- $\mathbf{P}_m$ is the nominal (model-based) representation of the plant.
- $\mathbf{C}_{I1}$ is the IMC controller.
- $\boldsymbol{R}$ is the reference input.
- $\boldsymbol{U}$ is the control input applied to the plant.
- $\boldsymbol{Y}$ is the system output.
- $\boldsymbol{D}$ represents the equivalent exogenous disturbance.

For analytical clarity, all signals and system variables are assumed to be of dimension n, and all subsystems (including $\mathbf{P}$, $\mathbf{P}_m$, and $\mathbf{C}_{I1}$) are assumed to be linear and time-invariant (LTI). It is required that both the model $\mathbf{P}_m$ and the IMC controller

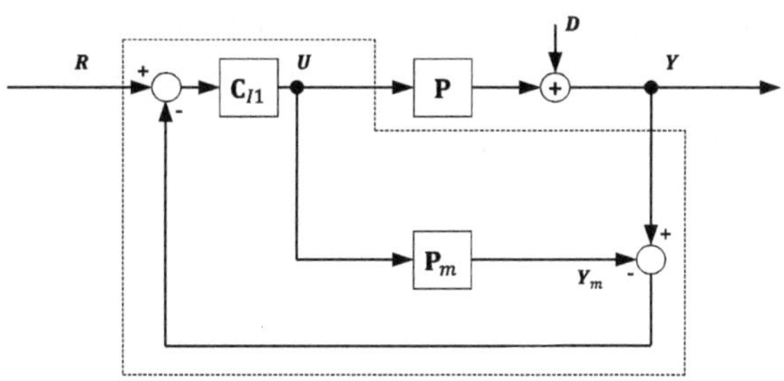

Fig. 14.1 Structure of the conventional IMC

$\mathbf{C}_{I1}$ be stable. The plant $\mathbf{P}$ is also assumed to be stable in this case. For open-loop unstable plants, pre-stabilization techniques [2, 12] can be adopted to enable IMC-based control.

The architecture in Fig. 14.1 captures a defining characteristic of IMC: it leverages an explicit internal model of the plant for control synthesis. This internal model is used to predict plant behavior and cancel known dynamics, thus isolating the impact of uncertainty and disturbances in a structured way.

From the block diagram in Fig. 14.1, the closed-loop transfer function from reference input $\boldsymbol{R}$ to system output $\boldsymbol{Y}$, denoted by $\mathbf{G}_1^R$, is given by:

$$\mathbf{G}_1^R = \left[\mathbf{I} + \mathbf{P}(\mathbf{I} - \mathbf{C}_{I1}\mathbf{P}_m)^{-1}\mathbf{C}_{I1}\right]^{-1}\mathbf{P}(\mathbf{I} - \mathbf{C}_{I1}\mathbf{P}_m)^{-1}\mathbf{C}_{I1}. \tag{14.1}$$

Similarly, the closed-loop transfer function from disturbance input $\boldsymbol{D}$ to output $\boldsymbol{Y}$, denoted by $\mathbf{G}_1^D$, is:

$$\mathbf{G}_1^D = \left[\mathbf{I} + \mathbf{P}(\mathbf{I} - \mathbf{C}_{I1}\mathbf{P}_m)^{-1}\mathbf{C}_{I1}\right]^{-1}. \tag{14.2}$$

By algebraic manipulation and assuming invertibility of the involved terms, these expressions can be simplified into more compact forms. In particular, using the identity:

$$(\mathbf{I} - \mathbf{C}_{I1}\mathbf{P}_m)^{-1}\mathbf{C}_{I1} = \left(\mathbf{C}_{I1}^{-1} - \mathbf{P}_m\right)^{-1},$$

one obtains the equivalent representations:

$$\mathbf{G}_1^R = \left[\mathbf{I} + \left(\mathbf{C}_{I1}^{-1} - \mathbf{P}_m\right)\mathbf{P}^{-1}\right]^{-1}, \tag{14.3}$$

and

$$\mathbf{G}_1^D = \left[\mathbf{I} + \mathbf{P}\left(\mathbf{C}_{I1}^{-1} - \mathbf{P}_m\right)^{-1}\right]^{-1}. \tag{14.4}$$

Several key properties of the IMC system depicted in Fig. 14.1 are examined in the following discussion, with reference to foundational works in the field [2, 12].

Property 14.1 (Dual Stability) *Assume that the model is exact, i.e.,* $\mathbf{P}_m = \mathbf{P}$. *In this case, the stability of both the plant* $\mathbf{P}$ *and the IMC controller* $\mathbf{C}_{I1}$ *is sufficient to ensure the internal stability of the overall IMC system depicted in* Fig. 14.1.

Proof When the model is perfect, i.e., $\mathbf{P}_m = \mathbf{P}$, the control input $\boldsymbol{U}$ and system output $\boldsymbol{Y}$ can be directly derived from the IMC structure in Fig. 14.1 as:

$$\boldsymbol{U} = \mathbf{C}_{I1}\boldsymbol{R} - \mathbf{C}_{I1}\boldsymbol{D}$$

$$\boldsymbol{Y} = \mathbf{P}\mathbf{C}_{I1}\boldsymbol{R} + (\mathbf{I} - \mathbf{P}\mathbf{C}_{I1})\boldsymbol{D}.$$

These expressions clearly indicate that the internal stability of the control input $\boldsymbol{U}$ and the system output $\boldsymbol{Y}$ is governed by the stability of the plant $\mathbf{P}$ and the IMC controller $\mathbf{C}_{I1}$. Therefore, under the condition $\mathbf{P}_m = \mathbf{P}$, the overall IMC system in Fig. 14.1 is guaranteed to be stable provided that both $\mathbf{P}$ and $\mathbf{C}_{I1}$ are stable. □

Property 14.1 is valid strictly under the idealized condition that the plant model $\mathbf{P}_m$ exactly matches the actual plant $\mathbf{P}$, i.e., $\mathbf{P}_m = \mathbf{P}$. Under this condition, the IMC structure ensures that the stability of the overall closed-loop system is guaranteed if both the plant $\mathbf{P}$ and the IMC controller $\mathbf{C}_{I1}$ are stable. However, this desirable property does not generally hold when modeling inaccuracies exist.

If there is a mismatch between $\mathbf{P}_m$ and $\mathbf{P}$, even when all components—$\mathbf{P}$, $\mathbf{P}_m$, and $\mathbf{C}_{I1}$—are individually stable, the overall IMC system can exhibit instability. This occurs because the IMC structure inherently relies on model-based cancelation of plant dynamics, and any deviation from the nominal model disrupts this cancelation, potentially introducing destabilizing feedback paths.

One of the conceptual features of IMC lies in its structural decoupling of performance and stability considerations under perfect modeling. Specifically, when $\mathbf{P}_m = \mathbf{P}$, the stability of the loop is determined entirely by the stability of $\mathbf{P}$ and $\mathbf{C}_{I1}$, allowing the designer to focus exclusively on shaping performance (e.g., tracking, disturbance rejection, robustness) through the design of the controller and associated filters [2].

Furthermore, a critical feature of the IMC architecture is its resilience to input constraints. Because of its feedforward-like structure, IMC remains stable in the presence of bounded (possibly nonlinear) input constraints—such as actuator saturation—provided the constrained control input is also applied to the internal model $\mathbf{P}_m$. This ensures consistency between the model-based prediction and the actual control action, thereby preserving internal loop stability. This feature is particularly useful in practical applications where input saturation or rate limits are unavoidable.

From a control-theoretic perspective, this robustness to constraints aligns with the principle of internal consistency in model-based control and further underscores the importance of accurate plant modeling or appropriate structural augmentation—such as robustness-enhancing filters or compensation schemes—when deploying IMC in systems subject to uncertainty.

Property 14.2 (Perfect Controller) *Assume that the IMC system depicted in* Fig. 14.1 *is closed-loop stable, and that the IMC controller is implemented as the exact inverse of the model, i.e.,*

$$\mathbf{C}_{I1} = \mathbf{P}_m^{-1}.$$

Under these conditions, perfect reference tracking can be achieved, meaning the system output satisfies

$$\boldsymbol{y}(t) = \boldsymbol{r}(t), \forall t > 0,$$

regardless of the presence of disturbances $\mathbf{d}(t)$ *or modeling inaccuracies (i.e., even when* $\mathbf{P}_m \neq \mathbf{P}$*).*

Proof When the IMC controller is chosen as the exact inverse of the model, i.e.,

$$\mathbf{C}_{I1} = \mathbf{P}_m^{-1},$$

then from Eqs. (14.3) and (14.4), the resulting closed-loop transfer functions simplify to:

$$\mathbf{G}_1^R = \mathbf{I}, \quad \mathbf{G}_1^D = 0.$$

This indicates that the system achieves perfect reference tracking and complete rejection of external disturbances. In other words, the output $\boldsymbol{Y}$ exactly follows the reference input $\boldsymbol{R}$, and the influence of any additive disturbance $\boldsymbol{D}$ is entirely eliminated from the system output.

This ideal behavior underpins the theoretical appeal of IMC. The ability to achieve such exact performance outcomes—under the assumptions of perfect model matching $\mathbf{P} = \mathbf{P}_m$, controller stability, and system stability—served as the primary motivation for the development of the IMC methodology. However, practical realizations of this ideal are constrained by plant–model mismatch and non-minimum-phase behavior, which necessitate robustness-enhancing modifications to the basic structure. □

Property 14.2 reflects a fundamental characteristic of the IMC structure: its inherent feedforward formulation. When the IMC controller is chosen as the exact inverse of the plant model, $\mathbf{C}_{I1} = \mathbf{P}_m^{-1}$, the product $\mathbf{P}_m\mathbf{C}_{I1} = \mathbf{I}$ ideally cancels the modeled dynamics, resulting in a control action that enforces $\boldsymbol{Y} = \boldsymbol{R}$, irrespective of the actual plant $\mathbf{P}$ or disturbance $\boldsymbol{D}$, provided that the overall system remains internally stable.

However, this property relies critically on two ideal conditions: (i) the model is perfect ($\mathbf{P}_m = \mathbf{P}$), and (ii) the inverse controller $\mathbf{P}_m^{-1}$ is realizable and stable. In real-world applications, modeling errors and unmodeled dynamics are inevitable, meaning the cancelation $\mathbf{P}_m\mathbf{C}_{I1} \approx \mathbf{I}$ is typically approximate at best. As a result, the closed-loop system may become unstable even when the plant, model, and controller are individually stable—especially in the presence of pole–zero mismatches, dynamic cross-couplings, or non-minimum-phase behavior.

Additionally, exact model inversion is rarely practical. When the plant includes time delays, non-minimum-phase zeros, or has high relative degree, $\mathbf{P}_m^{-1}$ may be non-causal or non-proper, making it physically unrealizable. Direct implementation of such an inverse can result in an ill-posed controller or severe sensitivity to measurement noise and disturbances.

Despite these limitations, $\mathbf{C}_{I1} = \mathbf{P}_m^{-1}$ remains a valuable theoretical construct. It offers a benchmark for ideal performance and serves as a conceptual foundation for designing robust IMC controllers. In practice, realizable approximations of the inverse—regularized via stable low-pass pre-filters or other robustness-enhancing techniques—are used to achieve a balance between performance and stability.

These approximations preserve the central features of IMC: Transparency, design simplicity, and a clear separation between performance objectives and robustness considerations. Ultimately, while the exact inverse controller is seldom implemented directly, it continues to guide IMC design philosophy and provides insight into performance limits, robustness trade-offs, and achievable closed-loop behavior.

14.2.2 Synthesis of IMC Controller

The design of the IMC controller $\mathbf{C}_{I1}$ is the cornerstone of an IMC system. The robustness of the IMC scheme against modeling errors, plant uncertainties, and external disturbances is fundamentally determined by how $\mathbf{C}_{I1}$ is constructed. Since the focus of this chapter is on robustness enhancement, we consider, for simplicity, that the model $\mathbf{P}_m$ of the plant $\mathbf{P}$ is minimum-phase. Extensions to non-minimum-phase systems—those with right-half-plane (RHP) zeros or pure time delays—are possible and can be handled using established methodologies discussed in [12].

Theoretically, the ideal IMC controller is defined as the exact inverse of the plant model, i.e.,

$$\mathbf{C}_{I1} = \mathbf{P}_m^{-1}.$$

However, this ideal structure is rarely implementable in practice. If the plant model $\mathbf{P}_m$ includes elements whose denominators exceed their numerators in order by N, then the corresponding inverse $\mathbf{P}_m^{-1}$ will involve pure differentiation of order N, rendering the controller physically unrealizable in practice. To address this, the common solution is to introduce a low-pass pre-filter p_f to make the controller causal and proper. A typical scalar form of such a pre-filter is:

$$\mathrm{p}_f := \left(T_f s + 1\right)^{-N}, \tag{14.5}$$

where $T_f > 0$ is the time constant and N is the relative degree of the transfer function being inverted.

The implementable form of the IMC controller is then given by:

$$\mathbf{C}_{I1} = \mathbf{P}_m^{-1}\mathbf{P}_{f1}, \tag{14.6}$$

where $\mathbf{P}_{f1}$ is a diagonal matrix of low-pass pre-filters applied to each control channel. This formulation guarantees that $\mathbf{C}_{I1}$ remains proper and physically realizable.

Substituting this form into the closed-loop transfer function derived previously [see Eq. (14.3)], we obtain:

$$\mathbf{G}_1^R = \left[\mathbf{I} + \left(\mathbf{P}_{f1}^{-1} - \mathbf{I}\right)\mathbf{P}_m\mathbf{P}^{-1}\right]^{-1}, \tag{14.7}$$

and for the disturbance-to-output transfer function [cf. Eq. (14.4)]:

$$\mathbf{G}_1^D = \left[\mathbf{I} + \mathbf{P}\mathbf{P}_m^{-1}\left(\mathbf{P}_{f1}^{-1} - \mathbf{I}\right)^{-1}\right]^{-1}. \tag{14.8}$$

In the ideal case where the plant model is exact ($\mathbf{P}_m = \mathbf{P}$), these simplify to:

$$\mathbf{G}_1^R = \mathbf{P}_{f1}, \tag{14.9}$$

$$\mathbf{G}_1^D = \mathbf{I} - \mathbf{P}_{f1}. \tag{14.10}$$

These results underscore a critical insight: when the model is perfect, the closed-loop performance is governed solely by the pre-filter $\mathbf{P}_{f1}$. If the time constant T_f is sufficiently small, such that $\mathbf{P}_{f1} \to \mathbf{I}$ as $t \to \infty$, the system achieves near-perfect reference tracking and disturbance rejection. Hence, the design of $\mathbf{P}_{f1}$ plays a central role in determining the performance and robustness characteristics of the IMC system [10].

While higher-order or non-minimum-phase systems may require more sophisticated pre-filters, the trade-off between design complexity and performance improvement must be carefully evaluated by the control engineer.

The conventional IMC structure (Fig. 14.1) can be reformulated into a standard unity-feedback configuration, enabling a more familiar analysis framework. The closed-loop transfer function $\mathbf{G}_1^R$ can be written as:

$$\mathbf{G}_1^R = (\mathbf{I} + \mathbf{G}_{o1})^{-1}\mathbf{G}_{o1}, \tag{14.11}$$

where the open-loop transfer function $\mathbf{G}_{o1}$ is:

$$\mathbf{G}_{o1} := \mathbf{P}(\mathbf{I} - \mathbf{C}_{I1}\mathbf{P}_m)^{-1}\mathbf{C}_{I1}. \tag{14.12}$$

Similarly, the disturbance transfer function becomes:

$$\mathbf{G}_1^D = (\mathbf{I} + \mathbf{G}_{o1})^{-1}. \tag{14.13}$$

This form reveals that the IMC system is dynamically equivalent to a unity-feedback system (Fig. 14.2), where the equivalent controller $\mathbf{C}$ is given by:

$$\mathbf{C} := (\mathbf{I} - \mathbf{C}_{I1}\mathbf{P}_m)^{-1}\mathbf{C}_{I1} = \left(\mathbf{C}_{I1}^{-1} - \mathbf{P}_m\right)^{-1}. \tag{14.14}$$

Thus, we have:

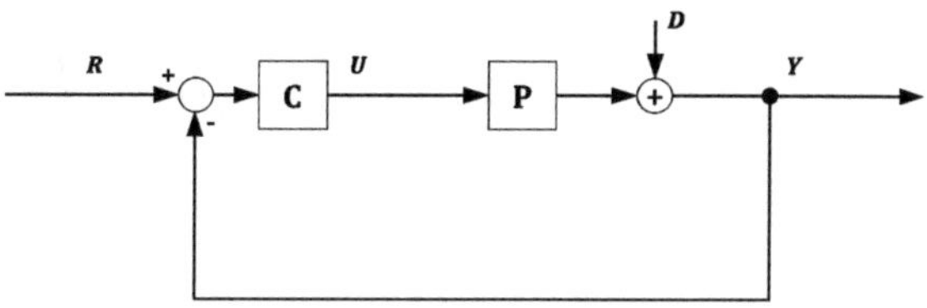

Fig. 14.2 Equivalent control of the IMC in the form of unity-feedback

$$\mathbf{G}_{o1} = \mathbf{PC}.$$

This equivalence provides a deeper understanding of the IMC architecture and enables the use of classical analysis techniques. In many practical applications, the equivalent controller **C** takes the form of a PID controller. Consequently, IMC offers a systematic and transparent method for designing and tuning PID controllers, especially through the single tuning parameter T_f [5]. This stands in contrast to heuristic tuning rules such as Ziegler–Nichols or Cohen–Coon [15], offering improved transparency and robustness.

Assume that $\mathbf{P}_m = \mathbf{P}$, and both **P** and $\mathbf{C}_{I1}$ are stable. Then the equivalent unity-feedback system with controller **C** defined in Eq. (14.14) is internally stable [12]. This result has profound implications. It suggests that instead of searching for stabilizing controllers **C** directly—a process constrained by internal stability requirements—one can instead design any stable $\mathbf{C}_{I1}$ and obtain a stabilizing **C** via Eq. (14.14). This significantly simplifies controller synthesis.

Despite the attractive features of conventional IMC, it remains highly sensitive to modeling inaccuracies. In practice, even modest discrepancies between the true plant and its model can result in significant deviations between the intended and actual control action, undermining performance and potentially destabilizing the system.

While appropriate filter design and careful controller synthesis can improve robustness to some extent [12], the IMC structure still lacks sufficient resilience under large model uncertainties or dynamic disturbances. Therefore, a structural enhancement to the IMC configuration is warranted.

In the next section, an enhanced IMC architecture is introduced. This framework retains the benefits of the conventional IMC design while incorporating mechanisms to explicitly compensate for modeling errors and external disturbances, yielding a significantly more robust control solution.

14.3 IMC with Enhanced Robustness

An enhanced IMC scheme is depicted in Fig. 14.3. Structurally, it extends the conventional IMC framework by introducing an additional parallel pathway that supplements the original control action. Specifically, a complementary model-following (MF) compensator $\mathbf{C}_{I2}$ is introduced, along with a pure gain weight matrix **K** [16,

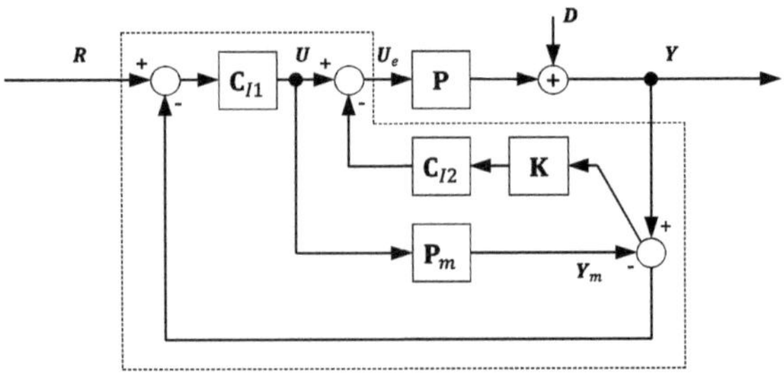

Fig. 14.3 Structure of the enhanced IMC

17]. This added structure enhances robustness by actively compensating for plant–model mismatch through a feedback mechanism that injects corrective action derived from the plant–model output error.

Conceptually, the conventional IMC relies primarily on a feedforward strategy based on the inverse model $\mathbf{P}_m^{-1}$, filtered for realizability and robustness. While effective under nominal conditions, the classical IMC structure becomes increasingly sensitive to model inaccuracies and disturbances. The enhanced IMC scheme in Fig. 14.3 addresses this limitation by embedding a dynamic correction mechanism within the plant–model error path, allowing the system to recover from modeling errors and external perturbations more effectively.

In this configuration, the compensator $\mathbf{C}_{I2}$ operates on the discrepancy between the real plant output and the model output, which inherently captures unmodeled dynamics and disturbances. The resulting signal is then scaled by the gain matrix **K**, and routed back into the control input through the main IMC loop. This feedback-like augmentation essentially blends model-based prediction with real-time model-error correction, combining the benefits of IMC and feedback regulation.

From the enhanced block diagram in Fig. 14.3, the reference-to-output transfer function $\mathbf{G}_2^R$ is derived as:

$$\mathbf{G}_2^R = \left[\mathbf{I} + \left(\mathbf{C}_{I1}^{-1} - \mathbf{P}_m\right)(\mathbf{I} + \mathbf{C}_{I2}\mathbf{K}\mathbf{P}_m)^{-1}(\mathbf{I} + \mathbf{C}_{I2}\mathbf{K}\mathbf{P})\mathbf{P}^{-1}\right]^{-1}. \tag{14.15}$$

This expression shows how the internal dynamics of the controller and the model mismatch influence the closed-loop response to reference inputs.

The disturbance-to-output transfer function $\mathbf{G}_2^D$, which characterizes the disturbance rejection capability of the system, is given by:

$$\mathbf{G}_2^D = \left[\mathbf{I} + \mathbf{P}\left(\mathbf{I} + \mathbf{C}_{I2}\mathbf{K}\mathbf{C}_{I1}^{-1}\right)\left(\mathbf{C}_{I1}^{-1} - \mathbf{P}_m\right)^{-1}\right]^{-1}. \tag{14.16}$$

These expressions reflect how the additional correction path improves the robustness of the system to uncertainties in the plant model and external disturbances. Specifically, the disturbance response now depends not only on the pre-filter and inverse model $\mathbf{C}_{I1}$, but also on the dynamic compensator $\mathbf{C}_{I2}$, which provides additional degrees-of-freedom for shaping the system response.

The term $\left(\mathbf{C}_{I1}^{-1} - \mathbf{P}_m\right)$ represents the mismatch between the actual model inversion and the true plant behavior. In the classical IMC formulation, this mismatch directly affects system stability and performance. In the enhanced scheme, this mismatch is dynamically modified through interaction with $\mathbf{C}_{I2}$ and $\mathbf{K}$, effectively forming a robustifying inner loop that reduces sensitivity to modeling errors.

This enhanced formulation can be interpreted as a disturbance-observer-like mechanism embedded in the IMC structure. The presence of $\mathbf{C}_{I2}$ ensures that the system responds not only to modeled dynamics but also to residual unmodeled effects, thereby providing improved robustness.

The augmentation is simple yet powerful: it requires no redesign of the original IMC controller $\mathbf{C}_{I1}$. This modular design makes the enhanced IMC structure highly attractive for applications where controller certification, reuse, or computational simplicity is essential.

The enhanced IMC system in Fig. 14.3 also satisfies internal stability conditions under certain assumptions, as will be described in Property 14.3. The key requirement remains the stability of the components $\mathbf{P}$, $\mathbf{P}_m$, $\mathbf{C}_{I1}$, and $\mathbf{C}_{I2}$. Additional stability margin is gained through appropriate selection of the gain matrix $\mathbf{K}$, which acts as a tuning parameter to balance robustness and transient performance.

Property 14.3 (Internal Stability) *Assume that the plant model is perfect, i.e.,* $\mathbf{P}_m = \mathbf{P}$. *Then the enhanced IMC system in* Fig. 14.3 is *internally stable if and only if the plant* $\mathbf{P}$, *the original IMC controller* $\mathbf{C}_{I1}$, *the model-following compensator* $\mathbf{C}_{I2}$, *and the term* $(\mathbf{I} + \mathbf{P}\mathbf{C}_{I2}\mathbf{K})^{-1}$ *are all stable.*

Proof Assume that the model is perfect, i.e., $\mathbf{P}_m = \mathbf{P}$. From Fig. 14.3, the internal signals of the enhanced IMC system can be expressed as:

$$\boldsymbol{U} = \mathbf{C}_{I1}\boldsymbol{R} - \mathbf{C}_{I1}(\mathbf{I} + \mathbf{P}\mathbf{C}_{I2}\mathbf{K})^{-1}\boldsymbol{D},$$

$$\boldsymbol{U}_e = \mathbf{C}_{I1}\boldsymbol{R} - (\mathbf{C}_{I1} + \mathbf{C}_{I2}\mathbf{K})(\mathbf{I} + \mathbf{P}\mathbf{C}_{I2}\mathbf{K})^{-1}\boldsymbol{D},$$

$$\boldsymbol{Y} = \mathbf{P}\mathbf{C}_{I1}\boldsymbol{R} + (\mathbf{I} - \mathbf{P}\mathbf{C}_{I1})(\mathbf{I} + \mathbf{P}\mathbf{C}_{I2}\mathbf{K})^{-1}\boldsymbol{D}.$$

These expressions show that the dynamics of $\boldsymbol{U}$, $\boldsymbol{U}_e$, and $\boldsymbol{Y}$ are governed by the stability of $\mathbf{P}$, $\mathbf{C}_{I1}$, $\mathbf{C}_{I2}$, and $(\mathbf{I} + \mathbf{P}\mathbf{C}_{I2}\mathbf{K})^{-1}$. Therefore, the enhanced IMC system is internally stable if and only if these components are stable. □

Similarly to Property 14.1, the internal stability result for the enhanced IMC system reflects a fundamental property of model-based control: in the ideal case where the plant model is exact (i.e., $\mathbf{P}_m = \mathbf{P}$), stability of the overall system becomes a trivial matter—provided that all constituent subsystems (i.e., $\mathbf{P}$, $\mathbf{C}_{I1}$, $\mathbf{C}_{I2}$, and $(\mathbf{I} + \mathbf{P}\mathbf{C}_{I2}\mathbf{K})^{-1}$ are themselves stable. This condition is sufficient to guarantee internal stability of the closed-loop system, regardless of external disturbances or reference inputs.

However, in practical applications, model–plant mismatch is almost inevitable. When $\mathbf{P}_m \neq \mathbf{P}$, the internal dynamics of the closed-loop system become more complex, and the guarantee of stability is no longer automatic. In fact, even if the nominal model and controller are individually stable, mismatches can lead to undesirable pole–zero interactions, non-minimum-phase behavior, or even instability in the overall system. Thus, under model uncertainty, the role of the structural augmentation introduced in the enhanced IMC scheme becomes critical for ensuring both robustness and performance.

The robustness contribution of the added compensatory path—consisting of the model-following compensator $\mathbf{C}_{I2}$ and weight matrix $\mathbf{K}$—is now explored in further detail. At first glance, the closed-loop transfer function of the enhanced IMC system given in Eq. (14.15) appears structurally distinct from that of the conventional IMC system in Eq. (14.3), making a direct comparison of performance difficult.

To facilitate a meaningful comparison, Eq. (14.15) is rewritten as:

$$\mathbf{G}_2^R = \left\{\mathbf{I} + \left(\mathbf{C}_{I1}^{-1} - \mathbf{P}_m\right)\left[\mathbf{P}(\mathbf{I} + \mathbf{C}_{I2}\mathbf{K}\mathbf{P})^{-1}(\mathbf{I} + \mathbf{C}_{I2}\mathbf{K}\mathbf{P}_m)\right]^{-1}\right\}^{-1}. \tag{14.17}$$

By comparing this with the conventional IMC transfer function:

$$\mathbf{G}_1^R = \left[\mathbf{I} + \left(\mathbf{C}_{I1}^{-1} - \mathbf{P}_m\right)\mathbf{P}^{-1}\right]^{-1},$$

it becomes clear that the role of the inverse plant $\mathbf{P}^{-1}$ in the conventional structure has been effectively replaced in the enhanced IMC by a compensated dynamic transformation:

$$\mathbf{P}_{cp} := P(\mathbf{I} + \mathbf{C}_{I2}\mathbf{K}\mathbf{P})^{-1}(\mathbf{I} + \mathbf{C}_{I2}\mathbf{K}\mathbf{P}_m). \tag{14.18}$$

This term can be interpreted as the compensated plant, which replaces the original plant $\mathbf{P}$ in an equivalent IMC representation. As such, the enhanced IMC structure in Fig. 14.3 can be reformulated into the standard IMC framework as shown in Fig. 14.4—where the plant is now $\mathbf{P}_{cp}$, and the controller remains the same.

Crucially, when the plant model is accurate ($\mathbf{P} = \mathbf{P}_m$), we recover $\mathbf{P}_{cp} = \mathbf{P}$, and the enhanced IMC reduces to the conventional IMC. However, when $\mathbf{P} \neq \mathbf{P}_m$, the modified structure provides an internal compensation mechanism that reshapes the

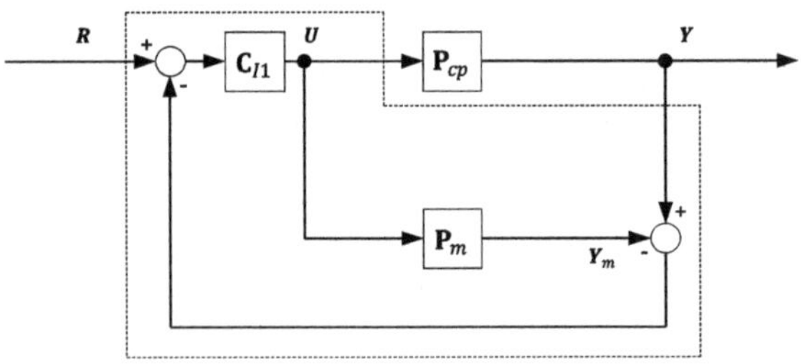

Fig. 14.4 Enhanced IMC system with compensated plant

plant dynamics to improve robustness and reduce the sensitivity of the system to model mismatch.

Thus, the performance comparison between the conventional and enhanced IMC systems is reduced to analyzing the difference between the original plant $\mathbf{P}$ and its compensated counterpart $\mathbf{P}_{cp}$. In this transformed viewpoint, the augmented structure serves not only to regularize the IMC controller (e.g., by avoiding instability due to model mismatch), but also to improve disturbance rejection and uncertainty attenuation in the presence of real-world nonidealities.

In the sections that follow, we analyze the robustness of the enhanced IMC system separately with respect to plant uncertainties and external disturbances. This separation allows for a clearer characterization of how the MF compensator and associated gain $\mathbf{K}$ contribute to the overall robustness properties of the system—framed within the broader context of linear system theory and structured uncertainty analysis.

14.3.1 Robustness to Model Uncertainties

In the robustness analysis of control systems, it is a well-established practice to characterize the perturbed plant dynamics in terms of its nominal model and the associated uncertainties. This modeling framework facilitates a structured evaluation of the sensitivity of the system to imperfections in plant modeling. Specifically, a perturbed plant $\mathbf{P}$ is commonly expressed using both multiplicative and additive uncertainty representations, which capture distinct types of modeling errors. The plant dynamics can thus be written as:

$$\mathbf{P} := [\mathbf{I} + \mathbf{\Delta}_m]\mathbf{P}_m + \mathbf{\Delta}_a. \tag{14.19}$$

Here:

- $\mathbf{P}_m$ denotes the nominal plant model, typically derived from analytical modeling or system identification.
- $\mathbf{\Delta}_m$ represents the multiplicative uncertainty, which models parametric deviations such as gain errors or frequency-dependent deviations relative to the nominal dynamics.
- $\mathbf{\Delta}_a$ denotes the additive uncertainty, which typically captures unmodeled dynamics, such as high-frequency components or neglected nonlinearities.

This decomposition aligns with standard robust control theory [12], where multiplicative uncertainties are particularly effective in capturing relative deviations, while additive uncertainties are suitable for modeling structural discrepancies.

Adopting this representation, we can now proceed to analyze the robustness properties of the enhanced IMC structure. The robustness characteristics of the enhanced IMC configuration can be systematically evaluated in the presence of the uncertainty structure defined in Eq. (14.19).

Accordingly, for the enhanced IMC system architecture, we now formulate the following robustness property:

Property 14.4 Uncertainty Attenuation *For the enhanced IMC system shown in* Fig. 14.3, *both the multiplicative uncertainty* $\mathbf{\Delta}_m$ *and the additive uncertainty* $\mathbf{\Delta}_a$ *associated with the plant* $\mathbf{P}$ *are attenuated in the closed-loop behavior by a factor of* $(\mathbf{I}+\mathbf{PC}_{I2}\mathbf{K})$, *relative to the conventional IMC structure depicted in* Fig. 14.1.

Proof By applying the well-known matrix identity for invertible matrices:

$$\mathbf{A}(\mathbf{I}+\mathbf{BA})^{-1} = (\mathbf{I}+\mathbf{AB})^{-1}\mathbf{A}, \tag{14.20}$$

the compensated plant $\mathbf{P}_{cp}$ in Eq. (14.18) can be equivalently rewritten as:

$$\mathbf{P}_{cp} = (\mathbf{I}+\mathbf{PC}_{I2}\mathbf{K})^{-1}\mathbf{P}(\mathbf{I}+\mathbf{C}_{I2}\mathbf{KP}_m). \tag{14.21}$$

Expanding Eq. (14.21):

$$\begin{aligned}\mathbf{P}_{cp} &= (\mathbf{I}+\mathbf{PC}_{I2}\mathbf{K})^{-1}(\mathbf{P}+\mathbf{PC}_{I2}\mathbf{KP}_m)\\ &= (\mathbf{I}+\mathbf{PC}_{I2}\mathbf{K})^{-1}[(\mathbf{I}+\mathbf{PC}_{I2}\mathbf{K})\mathbf{P}_m+\mathbf{P}-\mathbf{P}_m].\end{aligned}$$

Substituting the uncertainty structure of the original plant $\mathbf{P}$ from Eq. (14.19):

$$\mathbf{P} = [\mathbf{I}+\mathbf{\Delta}_m]\mathbf{P}_m+\mathbf{\Delta}_a,$$

gives:

$$\begin{aligned}\mathbf{P}_{cp} &= (\mathbf{I}+\mathbf{PC}_{I2}\mathbf{K})^{-1}\{(\mathbf{I}+\mathbf{PC}_{I2}\mathbf{K})\mathbf{P}_m+[(\mathbf{I}+\mathbf{\Delta}_m)\mathbf{P}_m+\mathbf{\Delta}_a]-\mathbf{P}_m\}\\ &= (\mathbf{I}+\mathbf{PC}_{I2}\mathbf{K})^{-1}[(\mathbf{I}+\mathbf{PC}_{I2}\mathbf{K})\mathbf{P}_m+\mathbf{\Delta}_m\mathbf{P}_m+\mathbf{\Delta}_a]\\ &= (\mathbf{I}+\mathbf{PC}_{I2}\mathbf{K})^{-1}\{[(\mathbf{I}+\mathbf{PC}_{I2}\mathbf{K})+\mathbf{\Delta}_m]\mathbf{P}_m+\mathbf{\Delta}_a\},\end{aligned}$$

which leads to the compact expression:

$$\mathbf{P}_{cp} = \left[\mathbf{I}+(\mathbf{I}+\mathbf{PC}_{I2}\mathbf{K})^{-1}\mathbf{\Delta}_m\right]\mathbf{P}_m+(\mathbf{I}+\mathbf{PC}_{I2}\mathbf{K})^{-1}\mathbf{\Delta}_a. \tag{14.22}$$

This result demonstrates that both the multiplicative uncertainty $\mathbf{\Delta}_m$ and the additive uncertainty $\mathbf{\Delta}_a$ are attenuated in the compensated plant dynamics by the factor $(\mathbf{I}+\mathbf{PC}_{I2}\mathbf{K})$.

Consequently, compared to the original plant expression in Eq. (14.19), the effective uncertainties in the compensated plant $\mathbf{P}_{cp}$ are significantly reduced. This directly improves the robustness of the closed-loop system against plant–model mismatch and unmodeled dynamics. By appropriate design of the MF compensator $\mathbf{C}_{I2}$ and the gain matrix $\mathbf{K}$, the enhanced IMC structure achieves a higher degree of insensitivity to structural uncertainties, thereby outperforming the conventional IMC in robustness and reliability. □

This result implies that the effective influence of plant–model mismatch and unmodeled dynamics on the overall system performance is substantially attenuated due to the incorporation of the model-following compensator $\mathbf{C}_{I2}$ and the weighting gain matrix $\mathbf{K}$. Specifically, the additional compensatory feedback path serves to reshape the input–output behavior of the uncertain plant, yielding a compensated plant $\mathbf{P}_{cp}$ whose dynamics are closer to the nominal model $\mathbf{P}_m$. This structure not only enhances nominal performance but also contributes significantly to robustness by reducing sensitivity to structured uncertainties, including both multiplicative perturbations and additive unmodeled dynamics.

From a control-theoretic standpoint, the term $(\mathbf{I}+\mathbf{PC}_{I2}\mathbf{K})^{-1}$ in the expression for $\mathbf{P}_{cp}$ acts as a robustifying filter, dynamically modifying the response of the plant to mitigate the impact of uncertainty. As the loop gain $\mathbf{PC}_{I2}\mathbf{K}$ increases, the filtering effect strengthens, further diminishing the contribution of perturbations.

In the limiting case, assuming that $\mathbf{C}_{I2}\mathbf{K}$ is designed with sufficiently large gain and that the feedback loop remains internally stable, we obtain the asymptotic relation:

$$\mathbf{P}_{cp} \to \mathbf{P}(\mathbf{C}_{I2}\mathbf{KP})^{-1}(\mathbf{C}_{I2}\mathbf{KP}_m) = \mathbf{P}_m.$$

This suggests that the compensated plant behaves nearly identically to the nominal model, effectively nullifying model discrepancies in the closed-loop behavior. In practical terms, this enables the IMC controller to operate as if it were regulating the idealized model $\mathbf{P}_m$, thereby preserving desired performance specifications such as reference tracking and disturbance rejection.

Importantly, this approximation holds under the assumption that the compensator and plant combination remain stable and that $\mathbf{C}_{I2}\mathbf{K}$ is properly tuned. Excessively high gain could lead to actuator saturation, noise amplification, or destabilization if not managed appropriately. Thus, the selection of $\mathbf{C}_{I2}$ and $\mathbf{K}$ must balance robustness enhancement with physical realizability and implementation constraints.

Overall, this mechanism illustrates the core advantage of the enhanced IMC structure: By embedding a corrective action derived from model-following principles, it ensures that the plant uncertainty is actively counteracted, improving both nominal performance and robustness without requiring redesign of the original IMC controller.

14.3.2 Robustness to External Disturbances

Now, let us turn to the analysis of robustness to external disturbances in both the conventional and enhanced IMC systems, with particular attention to the contribution of the additional compensatory path in the enhanced structure. As previously discussed, in the context of linear time-invariant (LTI) systems, the effects of parametric uncertainties and exogenous disturbances can often be treated independently due to the superposition principle. Therefore, to isolate and evaluate the impact of external disturbances on system behavior, we consider the idealized scenario in which the plant model is perfectly matched, i.e., $\mathbf{P}_m = \mathbf{P}$.

This simplification is not merely theoretical: it reflects a common practice in robust control analysis, where one investigates the disturbance rejection capabilities under nominal conditions before introducing modeling imperfections. Under this assumption, we are able to clearly distinguish the role of the additional model-following path—comprising the MF compensator $\mathbf{C}_{I2}$ and the weight matrix $\mathbf{K}$—in improving disturbance attenuation.

In this setting, the enhanced IMC architecture is expected to outperform its conventional counterpart by actively filtering and counteracting the effects of the disturbance through the error-driven feedback loop formed by the plant–model mismatch signal. The injected compensation adjusts the system response in real time based on observed deviations, thereby enhancing the closed-loop disturbance rejection capability.

This leads to the following result regarding the improved disturbance robustness of the enhanced IMC structure:

Property 14.5 Disturbance Rejection *Assume that the plant model is perfect, i.e.,* $\mathbf{P}_m = \mathbf{P}$. *Then, for the enhanced IMC system shown in* Fig. 14.3, *the impact of external disturbances on the system output is further attenuated by the factor* $(\mathbf{I} + \mathbf{P}\mathbf{C}_{I2}\mathbf{K})$, *as compared to the conventional IMC structure.*

Proof From Eq. (14.4), the disturbance transfer function $\mathbf{G}_1^D$ of the conventional IMC system shown in Fig. 14.1 can be expressed as:

$$\mathbf{G}_1^D = (\mathbf{I} - \mathbf{P}_m\mathbf{C}_{I1})[\mathbf{I} + (\mathbf{P} - \mathbf{P}_m)\mathbf{C}_{I1}]^{-1}.$$

In the ideal case where the plant model is perfect, i.e., $\mathbf{P}_m = \mathbf{P}$, this simplifies to:

$$\mathbf{G}_1^D = \mathbf{I} - \mathbf{P}\mathbf{C}_{I1}. \tag{14.23}$$

Now consider the enhanced IMC system in Fig. 14.3. From Eq. (14.16), the corresponding disturbance transfer function $\mathbf{G}_2^D$ is given by:

$$\mathbf{G}_2^D = (\mathbf{I} - \mathbf{P}_m\mathbf{C}_{I1})[\mathbf{I} + (\mathbf{P} - \mathbf{P}_m)\mathbf{C}_{I1} + \mathbf{P}\mathbf{C}_{I2}\mathbf{K}]^{-1}.$$

Again, under the assumption of perfect modeling $\mathbf{P}_m = \mathbf{P}$, this becomes:

$$\mathbf{G}_2^D = (\mathbf{I} - \mathbf{P}\mathbf{C}_{I1})(\mathbf{I} + \mathbf{P}\mathbf{C}_{I2}\mathbf{K})^{-1}. \tag{14.24}$$

By comparing Eqs. (14.23) and (14.24), it is evident that in the enhanced IMC configuration, the disturbance sensitivity is attenuated by an additional multiplicative factor of $(\mathbf{I} + \mathbf{P}\mathbf{C}_{I2}\mathbf{K})$. This indicates that, even under perfect modeling conditions, the enhanced structure introduces an active compensatory mechanism that further reduces the sensitivity of the system to external disturbances.

In effect, the additional compensatory path, consisting of the model-following compensator $\mathbf{C}_{I2}$ and gain matrix $\mathbf{K}$, leverages the model-based error feedback to improve disturbance rejection. By dynamically counteracting the disturbance-induced deviations at the plant output, this path enhances robustness without requiring changes to the original IMC controller. As a result, the enhanced IMC architecture exhibits superior disturbance attenuation compared to the conventional IMC design. □

Although the analysis leading to this property is derived under linear system assumptions, the disturbance attenuation effect introduced by the enhanced IMC structure extends, at least qualitatively, to nonlinear systems as well. In particular, the structure of the disturbance rejection pathway—specifically the multiplicative attenuation by the factor $(\mathbf{I} + \mathbf{P}\mathbf{C}_{I2}\mathbf{K})$—suggests that even in the presence of moderate nonlinearities, the additional compensatory loop can actively suppress the influence of disturbances on system output. This inference is supported by the logic underpinning the proofs of Property 14.1 and Property 14.3, where it is explicitly shown that the effect of the disturbance input $\boldsymbol{D}$ is further reduced in the enhanced IMC configuration.

The analytical derivations above confirm that the model-following compensator, characterized by the term $\mathbf{C}_{I2}\mathbf{K}$, introduces an additional dynamic feedback path that acts as a robustifying filter. This mechanism effectively reshapes the plant response in such a way that both modeling uncertainties and external disturbances are attenuated before they propagate to the system output. As a result, compared to the conventional IMC structure, the enhanced scheme exhibits improved disturbance rejection and greater insensitivity to plant–model mismatch.

This result highlights a fundamental benefit of the enhanced IMC architecture: the complementary feedback path not only adds robustness but also decouples disturbance rejection performance from sole reliance on the pre-filter $\mathbf{P}_{f1}$. In conventional IMC systems, the pre-filter dynamics entirely determine the trade-off between disturbance attenuation and system responsiveness. However, in the enhanced scheme, the error-driven compensatory signal—derived from the mismatch between the plant and model outputs—contributes an additional correction mechanism. This design is consistent with principles from disturbance-observer-based control (DOBC) and internal model-based robust control, in which plant–model error signals are used to synthesize corrective actions in real time.

In practical terms, this enhancement enables more aggressive tuning of the pre-filter for system responsiveness without sacrificing robustness to unmodeled dynamics or external inputs—an important feature in high-performance applications where both tracking and disturbance rejection are critical.

14.4 Possible Forms of the Model-Following Compensator

As demonstrated in the preceding analysis, the primary role of the additional model-following compensator $\mathbf{C}_{I2}$ is to shape the dynamics of the compensated plant in such a way that the overall IMC system becomes less sensitive to model uncertainties and more robust against external disturbances. By dynamically correcting the mismatch between the actual plant and its model, $\mathbf{C}_{I2}$ serves as an internal robustness enhancer and disturbance attenuator.

Importantly, $\mathbf{C}_{I2}$ is structurally decoupled from the original IMC controller $\mathbf{C}_{I1}$; hence, it can be synthesized independently, allowing for modular control design. This separation of roles enables flexibility in controller architecture: while $\mathbf{C}_{I1}$ targets nominal performance, $\mathbf{C}_{I2}$ can be tailored specifically to compensate for uncertainty and disturbance effects.

Various realizations of $\mathbf{C}_{I2}$ are possible. Each configuration imparts distinct characteristics to the resulting compensated plant, influencing its transient response, stability margin, and disturbance rejection capabilities. The following sections will explore several design options and analyze their impact on the robustness and dynamic behavior of the enhanced IMC system.

14.4.1 MF Compensator in the Same Form as the IMC Controller

A natural and analytically convenient choice for the additional model-following compensator $\mathbf{C}_{I2}$ is to adopt the same structural form as the IMC controller $\mathbf{C}_{I1}$. Specifically, we define:

$$\mathbf{C}_{I2} = \mathbf{P}_m^{-1}\mathbf{P}_{f2}, \tag{14.25}$$

where $\mathbf{P}_{f2}$ is a pre-filter matrix introduced to ensure the properness and physical realizability of $\mathbf{C}_{I2}$. Substituting this into the compensated plant expression in Eq. (14.18) and performing algebraic simplification yields:

$$\mathbf{P}_{cp1} = \left(\mathbf{P}_m\mathbf{P}^{-1} + \mathbf{P}_{f2}\mathbf{K}\right)^{-1}\left(\mathbf{I} + \mathbf{P}_{f2}\mathbf{K}\right)\mathbf{P}_m. \tag{14.26}$$

This expression illustrates how the dynamics of the compensated plant are actively reshaped through the interaction between the model-based inverse $\mathbf{P}_m^{-1}$, the pre-filter $\mathbf{P}_{f2}$, and the gain matrix $\mathbf{K}$. Under the ideal condition of perfect modeling (i.e., $\mathbf{P} = \mathbf{P}_m$) in a linear time-invariant (LTI) system, the disturbance transfer function of the enhanced IMC scheme becomes:

$$\mathbf{G}_{21}^{D} = \left(\mathbf{I} - \mathbf{P}_{f1}\right)\left(\mathbf{I} + \mathbf{P}_{f2}\mathbf{K}\right)^{-1}. \tag{14.27}$$

This expression, derived from Eqs. (14.24) and (14.25), reveals that the effect of external disturbances is further attenuated by the additional factor $\left(\mathbf{I} + \mathbf{P}_{f2}\mathbf{K}\right)$, which depends on both the structure of $\mathbf{P}_{f2}$ and the magnitude of the gain matrix $\mathbf{K}$. In particular, when $\mathbf{K}$ is sufficiently large and the pre-filter $\mathbf{P}_{f2}$ has a fast dynamic response (i.e., small time constants), the approximation

$$\mathbf{P}_{cp1} \to \mathbf{P}_m \quad \text{and} \quad \mathbf{G}_{21}^{D} \to \big(\mathbf{I} - \mathbf{P}_{f1}\big)(\mathbf{I} + \mathbf{K})^{-1},$$

can be established. This result underscores the improved disturbance rejection capability introduced by the complementary MF compensator, with attenuation enhanced by a factor of $(\mathbf{I} + \mathbf{K})$.

It is important to emphasize that the pre-filter $\mathbf{P}_{f2}$ is not introduced for performance shaping but solely to ensure that $\mathbf{C}_{I2}$ remains a realizable, proper transfer function. While it may be convenient to set $\mathbf{P}_{f2} = \mathbf{P}_{f1}$, doing so introduces additional dynamic lag into the compensatory path, potentially degrading the responsiveness of disturbance rejection and increasing the complexity of the compensated plant dynamics. These residual dynamics, although stabilizing from a realizability standpoint, may limit the fast-acting corrective behavior expected from $\mathbf{C}_{I2}$.

Therefore, from both a design and robustness standpoint, the time constants in $\mathbf{P}_{f2}$ should be minimized as much as practically feasible, subject to constraints imposed by system bandwidth, high-frequency noise sensitivity, and actuator limitations. Minimizing the contribution of $\mathbf{P}_{f2}$ ensures that the compensator remains responsive while preserving the desired robustness and disturbance attenuation properties.

14.4.2 Simplified MF Compensator

The discussion above emphasizes that the use of a model-following compensator $\mathbf{C}_{I2}$ with additional pre-filters $\mathbf{P}_{f2}$, while necessary for physical realizability, may introduce undesirable dynamic characteristics into the compensated plant $\mathbf{P}_{\mathrm{cp}}$. These additional dynamics can increase system complexity—especially in the high-frequency domain—potentially degrading performance or requiring further tuning to maintain stability and robustness. Consequently, under certain practical conditions, it may be advantageous to employ simplified forms of $\mathbf{C}_{I2}$ that lead to a more tractable and better-conditioned overall control system.

A particularly simple and computationally efficient choice is to set

$$\mathbf{C}_{I2} = I, \tag{14.28}$$

especially for lower-order plant systems where aggressive dynamic compensation is not required. Substituting this into Eq. (14.18), the resulting compensated plant becomes:

$$\mathbf{P}_{cp2} = \big(\mathbf{P}^{-1} + \mathbf{K}\big)^{-1}\big(\mathbf{P}_m^{-1} + \mathbf{K}\big)\mathbf{P}_m. \tag{14.29}$$

Compared with the earlier result in Eq. (14.26), it is evident that $\mathbf{P}_{cp2}$ is structurally simpler than $\mathbf{P}_{cp1}$, which reduces the complexity of the resulting open-loop and closed-loop transfer functions. This simplification can lead to easier controller implementation and reduced computational burden, which is particularly beneficial for embedded or resource-constrained applications.

The corresponding disturbance transfer function for this simplified case is:

$$\mathbf{G}_{22}^{D} = \big(\mathbf{I} - \mathbf{P}_{f1}\big)(\mathbf{I} + \mathbf{P}_m\mathbf{K})^{-1}. \tag{14.30}$$

This expression shows that, despite the simplification, the enhanced IMC scheme still provides improved disturbance rejection capabilities relative to the conventional IMC system, with attenuation scaled by $(\mathbf{I} + \mathbf{P}_m\mathbf{K})$.

However, caution is warranted when applying $\mathbf{C}_{I2} = \mathbf{I}$ to higher-order or more dynamically complex systems. In such cases, the compensatory effect may be insufficient to counteract plant–model mismatch or to stabilize the system, particularly in the presence of large uncertainties or lightly damped modes. Furthermore, the simplified compensator may not provide the rapid disturbance rejection achievable with the more sophisticated form $\mathbf{C}_{I2} = \mathbf{P}_m^{-1}\mathbf{P}_{f2}$, especially when the pre-filter $\mathbf{P}_{f2}$ is tuned with fast time constants.

Alternative realizations of $\mathbf{C}_{I2}\mathbf{K}$ are also viable and potentially beneficial. For example, when full or partial state measurements are available, a compensator based on the state error between the plant and its model—i.e., $(\boldsymbol{x} - \boldsymbol{x}_m)$—can be formed. This approach aligns with state-feedback principles and enables dynamic compensation based on measurable deviations, thereby improving model matching and disturbance rejection in real time.

Ultimately, the selection of $\mathbf{C}_{I2}$ reflects a trade-off between complexity and performance. The simplified design offers ease of implementation and reduced computational demand, while the more sophisticated structure provides superior dynamic compensation and faster response. The appropriate choice should be guided by the order of the plant, the desired performance specifications, and practical implementation constraints.

In the next section, a representative example is provided to illustrate the synthesis, implementation, and performance characteristics of the enhanced IMC scheme discussed herein.

14.5 An Illustrative Example

To demonstrate the effectiveness of the enhanced IMC architecture introduced in this chapter, a representative example is presented. The setup, illustrated in Fig. 14.5, consists of a simplified, pre-compensated degree-of-freedom of a mass–stiffness–damping system, including the dynamics of an actuator. The plant dynamics are described by the following expression:

$$Y = \frac{1}{\left(T_p s + 1\right)^2}\left[\frac{k}{\tau s + 1} U + D\right],$$

where:

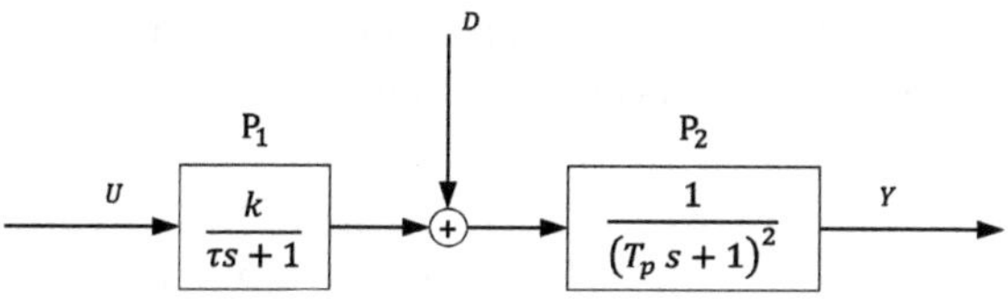

Fig. 14.5 Pre-compensated mass–stiffness–damping system

- Y is the plant output.
- U is the control input generated by either the conventional or enhanced IMC scheme.
- D is an external disturbance.
- k is the actuator gain.
- τ is the time constant of the actuator dynamics.
- T_p is the dominant time constant of the plant.

The transfer function of the overall plant is therefore:

$$\mathrm{P} = \frac{k}{\tau s + 1} \cdot \frac{1}{\left(T_p s + 1\right)^2}.$$

In this case, the nominal model P_m used for IMC design is chosen to approximate the dominant plant dynamics only, excluding the fast actuator mode:

$$\mathrm{P}_m = \frac{1}{(T_m s + 1)^2}.$$

This modeling choice reflects a typical practical scenario, where high-frequency dynamics are neglected due to their small contribution at lower operating frequencies.

From Eq. (14.6), the IMC controller C_{I1} is implemented as:

$$\mathrm{C}_{I1} = \frac{(T_m s + 1)^2}{\left(T_{f1} s + 1\right)^2}.$$

Likewise, based on Eq. (14.25), the complementary model-following compensator is defined as:

$$\mathrm{C}_{I2} = \frac{(T_m s + 1)^2}{\left(T_{f2} s + 1\right)^2}.$$

To simplify the analysis, the following notational substitutions are used:

- $a = (T_m s + 1)^2$.
- $b = \left(T_{f1} s + 1\right)^2$.
- $c = \left(T_{f2} s + 1\right)^2$.

Letting K denote the compensation gain in the MF path, the closed-loop command tracking and disturbance rejection transfer functions can be expressed as:

$$\mathrm{G}^R = \frac{ka[c + \mathrm{K}]}{ka[c + \mathrm{K}b] + c[b - 1](\tau s + 1)\left(T_p s + 1\right)^2}, \tag{14.31}$$

and

$$\mathrm{G}^D = \frac{c[b - 1](\tau s + 1)}{ka[c + \mathrm{K}b] + c[b - 1](\tau s + 1)\left(T_p s + 1\right)^2}. \tag{14.32}$$

Note that Eq. (14.32) is derived using the disturbance response formulation in Eq. (14.33), which accommodates disturbances entering within the plant (rather than strictly at the output):

$$\mathrm{G}_2^D = \left[\mathrm{I} + \mathrm{P}_2\mathrm{P}_1\left(\mathrm{I} + \mathrm{C}_{I2}\mathrm{K}\mathrm{C}_{I1}^{-1}\right)\left(\mathrm{C}_{I1}^{-1} - \mathrm{P}_m\right)^{-1}\right]^{-1}\mathrm{P}_2. \tag{14.33}$$

The simulation parameters are chosen as:

- Plant parameters: $k = 0.8$, $\tau = 0.05$, $T_p = 2$.
- Model parameters: $T_m = 2$.
- Pre-filter time constants: $T_{f1} = 0.1$, $T_{f2} = 0.001$.
- Compensation gains: $\mathrm{K} = 0, 1, 3, 5$.

The case of $\mathrm{K} = 0$ corresponds to the conventional IMC scheme.
Figure 14.6 illustrates the closed-loop response to a step reference input:

- The "x"-line shows the ideal reference tracking performance of a conventional IMC system with perfect modeling (i.e., $\mathrm{P} = \mathrm{P}_m$).

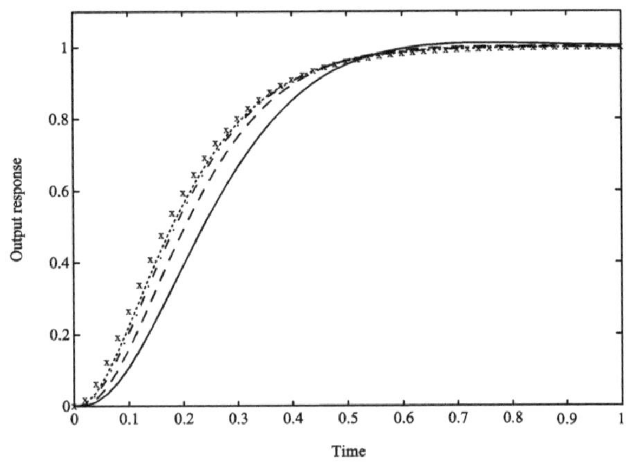

Fig. 14.6 Step input response

- The solid line reflects the degraded performance of the conventional IMC system when modeling mismatch exists ($P \neq P_m$).
- The dashed, dash-dot, and dotted lines correspond to the enhanced IMC system responses with compensation gains K = 1, 3, and 5, respectively.

It is evident that increasing the gain K progressively improves tracking accuracy, reducing the deviation from the ideal response. Figure 14.7 further quantifies this effect, showing the tracking error magnitude. The error reduction approximates the theoretical factor of $(1 + K)$, as predicted in Sect. 14.4. This result confirms that the designer can directly influence robustness to modeling error by adjusting the gain K.

In Fig. 14.8, disturbance rejection performance is evaluated by applying a unit step disturbance with zero reference input:

- The "x"-line shows the disturbance response of the conventional IMC under ideal conditions.

Fig. 14.7 Error of step input response

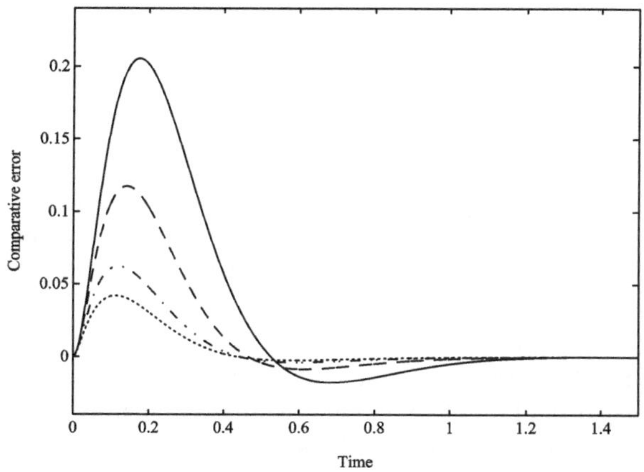

Fig. 14.8 Disturbance response

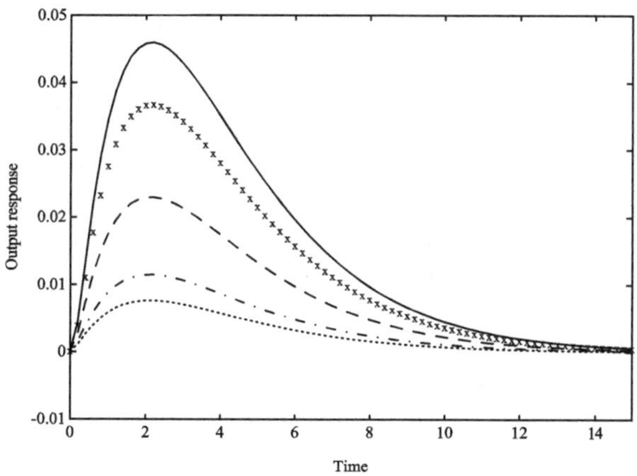

- The solid line reflects the degraded disturbance rejection due to model mismatch.
- The dashed, dash-dot, and dotted lines represent enhanced IMC responses with $K = 1, 3$, and 5, respectively.

Again, increasing K significantly attenuates the impact of disturbances, supporting the analytical findings from Eq. (14.32).

This example demonstrates the clear performance advantage of the enhanced IMC architecture under both model uncertainty and external disturbance conditions. While the plant–model mismatch degrades performance in the conventional IMC approach, the inclusion of the MF compensator in the enhanced scheme allows systematic compensation. Importantly, the degree of robustness enhancement is tunable via the gain parameter K, providing a powerful and flexible design tool.

Despite the added robustness, the structure and synthesis of the enhanced IMC controller remain simple and computationally efficient, making the approach highly practical for real-world implementation in mass–stiffness–damping systems and beyond.

14.6 Summary

This chapter introduces an enhanced IMC scheme aimed at improving the robustness of conventional IMC systems. The presented enhancement is achieved through a structurally simple but conceptually powerful modification of the standard IMC configuration—specifically, by incorporating a complementary model-following compensator and an adjustable gain path. This augmented structure preserves the key advantages of the conventional IMC framework—such as transparency, modularity, and intuitive tuning—while significantly improving resilience to model-plant mismatches and external disturbances.

Analytical derivations and numerical simulations have been conducted to rigorously validate the robustness-improving effects of the presented scheme. These results confirm that the enhanced IMC system achieves improved reference tracking, disturbance attenuation, and reduced sensitivity to structural uncertainties, particularly through the action of the compensatory path. The robustness enhancement is quantifiable and tunable through a gain parameter, providing a flexible means to adjust the level of disturbance rejection and model uncertainty compensation.

Recognizing that advanced mathematical abstraction can be a barrier to practical adoption, especially for application-oriented engineers, the theoretical results in this chapter are framed as well-defined properties. This structured presentation bridges the gap between rigorous analysis and engineering implementation, fostering greater accessibility and facilitating real-world deployment of the enhanced IMC scheme.

While the discussion has been streamlined under certain simplifying assumptions—such as linear time-invariant systems, minimum-phase plant models, and stable dynamics—the underlying methodology is inherently extensible. The enhanced IMC structure is compatible with broader classes of control problems, including non-minimum-phase systems, unstable plants, nonlinear dynamics, and time-delay systems. Such extensions align with current research trends in robust and adaptive control, where IMC is frequently viewed as a unifying framework for integrating model-based compensation with real-time adaptation.

It should be emphasized that although some practical aspects—such as actuator constraints, unmeasured disturbances, and implementation under sensor noise—have not been explored in detail, they represent fertile ground for future refinement of the enhanced IMC architecture. Overall, this chapter lays a solid theoretical and practical foundation for the use of structurally augmented IMC schemes in high-performance, disturbance-sensitive applications across a wide range of engineering systems.

References

1. Isidori A, Marconi L, Serrani A (2003) Robust autonomous guidance: an internal model approach. Springer, Berlin
2. Garcia CE, Morari M (1982) Internal model control. 1. A unifying review and some new results. Ind Eng Chem Process Des Dev 21(2):308–323
3. Garcia CE, Morari M (1985) Internal model control. 2. Design procedure for multivariable systems. Ind Eng Chem Process Des Dev 24(2):472–484
4. Garcia CE, Morari M (1985) Internal model control. 3. Multivariable control law computation and tuning guidelines. Ind Eng Chem Process Des Dev 24(2):484–494
5. Rivera DE, Morari M, Skogestad S (1986) Internal model control. 4. PID controller design. Ind Eng Chem Process Des Dev 25(1):252–265
6. Economou CG, Morari M, Palsson BO (1986) Internal model control. 5. Extension to nonlinear systems. Ind Eng Chem Process Des Dev 25(2):403–411
7. Economou CG, Morari M (1986) Internal model control. 6. Multiloop design. Ind Eng Chem Process Des Dev 25(2):411–419
8. Arkun Y, Hollett J, Canney WM, Morari M (1986) Experimental study of internal model control. Ind Eng Chem Process Des Dev 25(1):102–108
9. Zafiriou E (1987) Recent advances in the use of the internal model control structure for the synthesis of robust multivariable controllers. In: Prett DM, Morari M (eds) The shell process control workshop. Butterworth Publishers, London, pp 251–278
10. Fisher DG (1991) Process control: an overview and personal perspective. Can J Chem Eng 69(1):5–26
11. Henson MA, Seborg DE (1991) An internal model control strategy for nonlinear systems. AIChE J 37(7):1065–1081
12. Morari M, Zafiriou E (1989) Robust process control. Prentice-Hall, New Jersey, USA
13. Bergh LG, MacGregor JF (1987) Constrained minimum variance controllers: internal model structure and robustness properties. Ind Eng Chem Res 26(8):1558–1564
14. Grimble MJ, DeLa Salle S, Ho D (1989) Relationship between internal model control and LQG controller structures. Automatica 25(1):41–53
15. LeBlanc SE, Coughanowr DR (2009) Process systems analysis and control. McGraw-Hill, Columbus, Ohio, USA
16. Zhu HA, Zheng NN (1989) An engineering-oriented robust control scheme for motor drives. In: Proceedings of the 28th IEEE conference on decision and control, USA, pp 2188–2193
17. Zhu HA, Hong GS, Teo CL, Poo AN (1995) Internal model control with enhanced robustness. Int J Syst Sci 26(2):277–293

Chapter 15
Machine Learning in Modeling and Control

Machine learning (ML), as a data-driven approach for modeling and controlling complex and uncertain dynamic systems, has gained significant attention in recent years. As its applications expand across industries, many control practitioners are eager to integrate these techniques into real-world control systems. Given the vast scope of machine learning, which requires multiple books to fully cover its breadth, this chapter will focus on the connection between machine learning, the least-square modeling method, and the adaptive control techniques discussed earlier. Specifically, we will explore the relationship between the recursive least-squares (RLS) method and modern machine learning, examining their theoretical foundations, historical contexts, and how they complement and influence each other in contemporary applications.

15.1 Introduction

The recursive least-squares (RLS) algorithm is a classical method widely used in adaptive control, signal processing, and time-series analysis. Unlike standard batch least-squares estimation, which operates on a fixed dataset, RLS allows for continuous or "online" updating of model parameters. While modern machine learning (ML) is often associated with big data, deep learning, and advanced computational tools, RLS shares several fundamental principles with ML, particularly in its ability to learn patterns from data and update models incrementally.

At first glance, RLS may seem like an "old-school" technique primarily used for adaptive filtering or control. However, its core principles—iterative parameter updating, online learning, and the goal of minimizing error—align closely with modern ML methodologies. In fact, RLS can be considered one of the earliest forms of incremental learning, a subfield of ML that remains vital in contemporary research.

H. Zhu, *Model-Based Control of Mass–Stiffness–Damping Systems*, Advances in Industrial Control, https://doi.org/10.1007/978-3-031-97592-9_15

This chapter explores the relationship between RLS and ML, with a focus on the following:

- A brief background on the RLS method.
- A comparison of the essence of RLS and machine learning.
- The alignment and differences between RLS and modern ML methods.
- Real-world applications and future directions.

Although RLS predates modern ML, its principles of online learning and parameter adaptation form the foundation for many current ML approaches. This chapter delves deeper into:

- The theoretical foundations of both RLS and ML.
- The similarities and key differences between them.
- Areas where RLS remains highly relevant in ML-driven control systems.
- Future opportunities for integrating RLS within emerging ML frameworks.

15.2 The Least-Squares Methods

The least-squares criterion is a fundamental optimization method and learning principle in parameter estimation and system identification. In least-squares estimation, the goal is to find a set of parameters $\boldsymbol{\theta}$ that minimize the squared error between the predictions of a model and the observed data. For a linear model:

$$y(k) = \boldsymbol{\theta}\boldsymbol{x}(k) + e(k), \tag{15.1}$$

where:

- $y(k) \in \mathbb{R}$ is the scalar output (or measurement) of the system at time step k.
- $\boldsymbol{x}(k) \in \mathbb{R}^{n\times 1}$ is a column vector of n inputs (or features) at time step k.
- $\boldsymbol{\theta} \in \mathbb{R}^{1\times n}$ is a row vector of n parameters to be estimated.
- $e(k) \in \mathbb{R}$ represents the residual error (or measurement noise) in the model.

15.2.1 Batch Least-Squares (BLS) Method

Given a dataset of N points (input–output pairs) $\{[\boldsymbol{x}(1), y(1)], \ldots, [\boldsymbol{x}(N), y(N)]\}$, define:

- $\mathbf{X} \in \mathbb{R}^{N\times n}$ as an $N \times n$ design matrix whose kth row is $\boldsymbol{x}^{\mathrm{T}}(k)$, $k = 1, \ldots, N$.
- $\boldsymbol{y} \in \mathbb{R}^{N\times 1}$ as an $N \times 1$ output vector whose kth entry is $y(k)$, $k = 1, \ldots, N$.

$$\mathbf{X} = \begin{bmatrix} \boldsymbol{x}^{\mathrm{T}}(1) \\ \boldsymbol{x}^{\mathrm{T}}(2) \\ \vdots \\ \boldsymbol{x}^{\mathrm{T}}(N) \end{bmatrix}, \quad \boldsymbol{y} = \begin{bmatrix} y(1) \\ y(2) \\ \vdots \\ y(N) \end{bmatrix} \tag{15.2}$$

The batch least-squares solution estimates $\boldsymbol{\theta}$ using all available data. It finds $\boldsymbol{\theta}$ by minimizing the sum of squared errors,

$$\min_{\boldsymbol{\theta}} \sum_{k=1}^{N} \left[y(k) - \boldsymbol{\theta}\boldsymbol{x}(k)\right]^2. \tag{15.3}$$

This is the standard batch least-squares (BLS) formulation—all data points are weighted equally.

The normal equation in matrix form typically produces a column vector solution for $\boldsymbol{\theta}^{\mathrm{T}}$. For $\boldsymbol{\theta}$ as a row vector, we can write:

$$\boldsymbol{\theta}^{\mathrm{T}} = \left(\mathbf{X}^{\mathrm{T}}\mathbf{X}\right)^{-1}\mathbf{X}^{\mathrm{T}}\boldsymbol{y} \quad \Rightarrow \quad \boldsymbol{\theta} = \left[\left(\mathbf{X}^{\mathrm{T}}\mathbf{X}\right)^{-1}\mathbf{X}^{\mathrm{T}}\boldsymbol{y}\right]^{\mathrm{T}}, \tag{15.4}$$

provided the matrix $\left(\mathbf{X}^{\mathrm{T}}\mathbf{X}\right)^{-1}$ is invertible (the inverse exists). Either version is valid, depending on whether we treat $\boldsymbol{\theta}$ as a row or column vector.

The BLS method for parameter identification as a well-known technique is often utilized for system modeling when input (or feature) and output data of the target system are acquired in batches.

15.2.2 Recursive Least-Squares (RLS) Method

Unlike batch methods, RLS is an online estimation and learning algorithm, capable of sequentially updating model parameters as new data arrive. RLS addresses situations where data arrive sequentially (e.g., in real-time processes). Rather than recomputing the parameter estimates from scratch each time a new data point arrives, the RLS algorithm updates $\boldsymbol{\theta}$ (and an associated covariance matrix $\mathbf{P}$) incrementally and recursively as new data arrive. Assume:

- $\mathbf{P}(k)$ is an $n \times n$ covariance matrix at time k.
- f is the forgetting factor ($0 < f \leq 1$).

The standard RLS update equations can be written as follows, ensuring $\boldsymbol{\theta}$ remains a row vector and $\boldsymbol{x}(k)$ a column vector. A typical RLS formulation is:

1. Initialize the estimate $\boldsymbol{\theta}(0)$ and an associated covariance matrix $\mathbf{P}(0)$.
2. For each new data point $[\boldsymbol{x}(k), y(k)]$, the update equations for RLS are:

   (a). Compute a "gain" vector $\boldsymbol{g}(k)$ based on $\mathbf{P}(k-1)$ and $\boldsymbol{x}(k)$:

$$\boldsymbol{g}(k) = \left[f + \boldsymbol{x}^{\mathrm{T}}(k)\mathbf{P}(k-1)\boldsymbol{x}(k)\right]^{-1}\mathbf{P}(k-1)\boldsymbol{x}(k) \tag{15.5}$$

$\boldsymbol{g}(k)$ is an $n \times 1$ column vector.

(b). Update the estimate parameter row vector $\boldsymbol{\theta}(k)$ using the new information:

$$\boldsymbol{\theta}(k) = \boldsymbol{\theta}(k-1) + \left[y(k) - \boldsymbol{\theta}(k-1)\boldsymbol{x}(k)\right]\boldsymbol{g}^{\mathrm{T}}(k) \tag{15.6}$$

- $\boldsymbol{\theta}(k-1)$ is an $1 \times n$ row vector,
- $\boldsymbol{g}^{\mathrm{T}}(k)$ is an $1 \times n$ row vector,
- $\left[y(k) - \boldsymbol{\theta}(k-1)\boldsymbol{x}(k)\right]$ is a scalar.

(c). Update the covariance matrix $\mathbf{P}(k)$:

$$\mathbf{P}(k) = \left[\mathbf{P}(k-1) - \boldsymbol{g}(k)\boldsymbol{x}^{\mathrm{T}}(k)\mathbf{P}(k-1)\right]f^{-1}. \tag{15.7}$$

These equations illustrate how $\boldsymbol{\theta}$ is adapted at each time step to fit the most recent data while progressively "forgetting" older measurements (regulated by f). Here, the forgetting factor f is often chosen slightly below 1 to give more weight to recent data. Since the forgetting factor f prioritizes recent data, it makes RLS suitable for non-stationary environments. RLS thus performs an ongoing, adaptive refinement of parameters, which is particularly useful in non-stationary or time-varying environments. RLS is particularly useful in real-time or streaming scenarios where the system might be time-varying or subject to non-stationary disturbances. RLS has a relevant role in advanced modeling and control because it enables real-time modeling and adaptive control by continuously refining model estimates. It naturally aligns with online learning principles used in ML as well.

In summary, BLS solves for $\boldsymbol{\theta}$ using the entire batch of dataset at once, leading to a closed-form solution via the normal equations. RLS updates $\boldsymbol{\theta}$ incrementally each time a new data pair $\left[\boldsymbol{x}(k), y(k)\right]$ arrives, making it well-suited for online or real-time modeling applications and adaptive control.

15.3 Machine Learning and Online Learning Paradigms

ML has been evolving continuously and consists of a broad class of algorithms that extract patterns from data and make predictions. It encompasses algorithms that enable systems to learn patterns or behaviors from data without explicit programming. It includes a broad set of approaches:

- *Supervised Learning:* Training on and learning from labeled data (e.g., linear regression, logistic regression, neural networks).
- *Unsupervised Learning:* Identifying hidden patterns in data without the use of labels (e.g., clustering, dimensionality reduction such as Principal Component Analysis).

- *Reinforcement Learning:* Learning optimal actions through interaction with an environment, that is, via trial and error in an environment (e.g., Q-learning, policy gradients).

15.3.1 Online Versus Batch Learning

A key distinction in ML is online learning versus batch learning. While many popular ML algorithms are trained in batch mode (collecting a dataset upfront and then training), that is, the model is trained on a complete dataset before deployment, online learning processes data incrementally, that is, the model continuously updates as new data arrive. Each new sample can update the model in real time.

Stochastic gradient descent (SGD) is one of the most renowned online/batch hybrid methods in modern ML, iteratively updating parameters after each new data point (or mini-batch).

RLS is an early example of online learning, much like modern SGD, which updates model parameters incrementally.

15.3.2 Model Complexity and Expressiveness

Machine learning methods nowadays often deal with high-dimensional data and complex models (e.g., deep neural networks), which can approximate nonlinear relationships. Classical methods, including linear regression and its variants (like RLS), remain fundamental for lower-dimensional problems or for interpretability and computational efficiency. Generally,

- RLS handles linear models efficiently, making it useful in real-time applications where computational cost is a concern.
- Deep learning models provide greater expressiveness but often require high-dimensional data and extensive computation.

15.4 The Relationship Between RLS and Modern Machine Learning

The RLS method can be viewed as a legitimate online (incremental) learning algorithm within the broader ML framework. It iteratively updates model parameters in real time to minimize prediction errors—a process that aligns with fundamental ML principles of continual adaptation and data-driven model refinement.

15.4.1 Similarities and Differences

RLS, an early form of online learning, predates many modern algorithms but aligns conceptually with online supervised learning in key aspects:

- *Incremental Parameter Updates:* RLS updates the parameter estimate and covariance matrix with each data point, similar to how SGD updates parameters with new mini-batches.
- *Adaptation to Non-stationary Data:* RLS, especially with a forgetting factor, adapts to changing statistical properties, paralleling modern online learning techniques that re-estimate parameters when distributions shift, such as concept drift in machine learning.

The shared core principle of RLS and ML is: Learning from data. A key requirement for an algorithm to be considered "machine learning" is that it learns from data to improve its predictive or decision-making capabilities. RLS does precisely this:

- *Least-Squares Criterion:* Like standard linear regression in machine learning, RLS minimizes a sum of squared errors between model predictions and observed outputs.
- *Adaptive Updates:* Unlike batch methods that require the entire dataset at once, RLS updates parameters incrementally whenever a new data point arrives. This continuous adaptation is a hallmark of online learning in ML.

From the online learning perspective, online learning in ML involves updating a model as new data streams in, rather than waiting to collect a large batch of data. Common modern examples include SGD and its variants. RLS versus SGD:

- RLS typically derives its parameter update rules from a closed-form matrix-based approach.
- SGD uses gradient-based steps to move parameters in the direction that reduces error.
- Both are incremental and adapt to incoming data.

From a conceptual standpoint, RLS fulfills the same goal as online ML: it refines parameters in real time based on new observations, thereby "learning" continuously.

RLS and ML also share the common mathematical and algorithmic foundations.

- *Cost Function Minimization*

  In machine learning, many algorithms revolve around minimizing a cost function, such as the Mean Squared Error (MSE) or a cross-entropy loss. For RLS, the weighted or exponentially forgetting least-squares:

$$\min_{\boldsymbol{\theta}} \sum_{k=1}^{N} f^{N-k}\left[y(k) - \boldsymbol{\theta}\boldsymbol{x}(k)\right]^2 \tag{15.8}$$

  is used, where $\boldsymbol{\theta}$ is the parameter vector, f is a forgetting factor ($0 < f \leq 1$), and $\boldsymbol{x}(k)$ represents the input features at time k. This objective is entirely in line with typical supervised machine learning setups that aim to reduce a prediction error term. Compared to the unweighted form in Eq. (15.3), this version incorporates a forgetting factor $f \in (0, 1]$, which emphasizes recent data by assigning greater weight to newer observations—thus enabling the model to adapt more effectively to time-varying dynamics.

- *Parameter Updates*

  RLS performs iterative parameter updates (similar to incremental ML methods). On receiving a new data point $[\boldsymbol{x}(k), y(k)]$:

  - Compute the Kalman-gain-like vector $\mathbf{g}(k)$.
  - Update $\boldsymbol{\theta}(k)$ based on the prediction error:

  $$\boldsymbol{\theta}(k) = \boldsymbol{\theta}(k-1) + \left[y(k) - \boldsymbol{\theta}(k-1)\boldsymbol{x}(k)\right]\boldsymbol{g}^{\mathrm{T}}(k) \tag{15.9}$$

  - Adjust the covariance matrix $\mathbf{P}(k)$ to "remember" new data more than old data (depending on f).

  This step-by-step process directly parallels other online ML algorithms that sequentially adjust parameters to reduce error.

RLS and ML are closely linked in various real-world applications:

- *Adaptive Control:* Real-time adjustments in control laws in RLS closely mirror the goals of online reinforcement learning.
- *Signal Processing:* RLS serves as an adaptive filter, learning to filter signals in non-stationary environments, akin to the ML challenge of adapting to concept drift.
- *Time-Series Prediction:* RLS is widely used in forecasting for finance, weather, and communications, which are key areas of machine learning.
- *Resource-Constrained Settings:* RLS is computationally efficient and interpretable, making it ideal for embedded systems, where many high-dimensional ML algorithms are impractical.

The main aspects distinguishing RLS from modern "deep" ML are as follows:

- *Model Complexity:* RLS traditionally uses a linear model. Modern machine learning frequently involves nonlinear models (e.g., deep neural networks).
- *Data Dimensionality:* RLS scales well but can become computationally heavy when the feature dimension is very large (due to matrix inversions). Deep learning, while it can handle high-dimensional data, often relies on gradient-based updates and GPU acceleration.
- *Interpretability:* RLS yields interpretable weights in a linear model, often an advantage over black-box neural networks.

Despite these differences in scope and complexity, the underlying principle of adapting parameters in response to data is fundamentally shared.

A high-level overview on the similarities and differences between RLS and modern machine learning approaches is provided in Table 15.1.

Table 15.1 Comparison of RLS and modern machine learning

Aspect	RLS	Modern ML (online learning)
Model type	Typically, linear parametric models	Linear or nonlinear (e.g., deep neural networks)
Update method	Closed-form matrix updates	Iterative gradient-based updates (SGD)
Complexity and efficiency	Efficient for moderate dimensions	Can handle high-dimensional data but may be computationally complex and intensive
Adaptability	Excels in real-time, streaming data	Designed for large-scale streaming data but often approximates updates
Interpretability	Parameters directly correspond to regression weights	Interpretation depends on model choice; neural networks can be less interpretable

15.4.2 RLS in Contemporary Machine Learning Applications

Despite being sometimes overshadowed by more complex neural networks, RLS remains relevant in:

- *Adaptive Control:* Real-time parameter estimation and control adjustment are crucial.
- *Signal Processing:* Useful for streaming real-time data in communications or sensor networks.
- *Online Regression:* A key tool for situations dominated by linear relationships, where interpretability and computational efficiency are priorities.
- *Resource-Constrained AI:* More manageable than large-scale gradient-based methods in real-time or embedded systems where deep learning is not feasible.

Some recent research has also integrated RLS concepts with advanced machine learning frameworks to form hybrid learning systems:

- *Adaptive Neural Networks:* Using an RLS-like approach to rapidly fine-tune final layers in neural networks under streaming conditions [1–3].
- *Sequential Bayesian Learning:* Viewing RLS through a Bayesian lens to bridge classical control and probabilistic ML methods, RLS can be interpreted through a Bayesian method with Gaussian priors, bridging the gap to probabilistic ML methods like Kalman filters, Gaussian processes, or variational inference [4–7].

15.5 Related Research and Literature

For references, a curated overview of books, review articles, and research papers that discuss or illuminate the connections between the RLS method and machine learning, particularly in the context of online (incremental) learning, adaptive filtering, and real-time parameter estimation is provided here. These references explain how RLS fits within the broader scope of machine learning, how it relates to classical and modern online algorithms, and why it remains relevant in various ML and signal processing tasks.

As foundational texts on RLS and adaptive filters, Haykin [8] is a seminal text covering RLS in depth. This classic textbook introduces adaptive filters, including the RLS algorithm, and links them to concepts like least-squares estimation and learning from streaming data. Sayed [9, 10] present a comprehensive treatment of adaptive filters including RLS variants, discussing their derivation, theoretical properties, and practical applications. Includes insight into how RLS can be interpreted in a Bayesian or incremental learning framework.

On connections to machine learning and online learning, Theodoridis [11] covers a broad range of ML methods, including Bayesian interpretations and optimization perspectives. The sections on linear and Bayesian modeling shed light on how RLS can be viewed as an online Bayesian update or a special case of sequential parameter estimation. Bottou [12] discusses SGD and its relevance to large-scale machine learning. Although this article focuses on SGD, it provides context for online/iterative learning. Contrasting SGD with RLS helps clarify their similarities (both update parameters incrementally) and differences (matrix-based vs. gradient-based).

As specific articles bridging RLS and ML concepts, Brandt and Santa-Clara [13] demonstrates how RLS-like updates can be used in a dynamic machine learning context (portfolio optimization), showing real-time adaptation of model parameters and bridging financial ML methods with classical adaptive filtering. Though focused on kernel-based methods, Kivinen et al. [14] situates online learning (and thereby classical algorithms like RLS) in a broader ML setting, showing the theoretical underpinnings of incremental learning and how simpler adaptive methods can extend to nonlinear feature spaces. Sayed and Kailath [15] discusses RLS in detail and includes broader insights into learning theory within signal processing contexts. Helpful to see how RLS is taught alongside fundamental ML concepts (estimation, convergence, and optimization).

As supplemental works for further context, while the main text is more focused on batch methods (linear models, regularization, trees, etc.), Hastie et al. [16] provides the statistical foundation that underlies many ML algorithms. The appendices and sections on linear models can help draw parallels to how RLS updates parameters in a streaming scenario. A seminal ML textbook that, while not focusing heavily on RLS, Mitchell [17] discusses incremental learning paradigms. Chapters on version space, incremental algorithms, and the bias-variance trade-off help illustrate why methods like RLS fit within an ML framework. RLS as a special case of the Kalman filters, Kalman [18] provides a Bayesian view of sequential estimation. These references offer further insight into the theory and applications of RLS, as well as its connections to modern machine learning approaches, particularly in the realm of online and adaptive learning.

These references collectively highlight how recursive least-squares fits within the broader landscape of machine learning—particularly under the umbrella of online (incremental) learning and adaptive filtering. While RLS has long been a staple in signal processing and control theory, it remains relevant in modern ML for applications involving real-time or streaming data. By reviewing these resources, one can understand both the mathematical underpinnings of RLS and its conceptual alignment with emerging trends in machine learning, such as continual adaptation, dynamic model updates, and Bayesian interpretations.

15.6 Machine Learning at the Edge

Edge is a term that refers to a location, far from the cloud or a big data center, where we have a computing device (edge device) capable of running (edge) applications. Edge computing is the act of running workloads on these edge devices.

Machine learning at the edge (ML@Edge) is a concept that brings the capability of running ML models locally to edge devices. These ML models can then be invoked by the edge application. ML@Edge is important for many scenarios where raw data is collected from field sources far from the cloud.

Edge machine learning (Edge ML) is the process of running machine learning algorithms on computing devices at the periphery of a network to make decisions and predictions as close as possible to the originating source of data. It allows AI/ML-powered applications to be placed physically closer to data sources like sensors, cameras, and mobile devices, gathering insights faster and identifying patterns without relying on traditional cloud networks.

For real-time and online machine learning at the edge, RLS method is one of the most useful techniques available for real-world applications.

15.7 Summary

From both a technical and conceptual perspective, the recursive least-squares (RLS) method embodies core principles of machine learning, particularly in the domain of online learning. RLS performs iterative, data-driven updates to refine a predictive model and minimize errors over time. While traditionally taught in signal processing and adaptive control, RLS closely aligns with the philosophy of machine learning, which emphasizes learning from observed data to improve performance. As such, RLS can be regarded as a foundational technique in incremental or online learning.

Although RLS lacks the complexity of modern deep learning models, it remains a legitimate machine learning method because it addresses the essential task of parameter adjustment in response to new data. This incremental adaptation of model parameters aligns with the central goal of machine learning: continuously refining models as new data streams in. Therefore, RLS is not just an early precursor to modern machine learning, but also a valuable tool for practical machine learning tasks that prioritize efficiency and real-time performance.

RLS plays a pivotal role in streaming analytics, real-time decision-making, and adaptive control systems. Despite the growth of more complex models, such as deep neural networks and advanced optimization techniques like stochastic gradient descent, the fundamental principle of continual parameter updating remains central to both RLS and modern online machine learning algorithms.

As machine learning continues to evolve, the demand for real-time adaptation and solutions to handle streaming data will grow. In this context, RLS retains its relevance, especially in scenarios where linear models suffice, computational resources are constrained, or interpretability is crucial. Hybrid approaches that combine RLS with modern machine learning frameworks have significant potential, leveraging the strengths of classical adaptive methods alongside cutting-edge algorithms.

In summary, RLS and machine learning converge conceptually, with both methods emphasizing the iterative improvement of model parameters based on new data. RLS remains an essential building block for modern online and adaptive machine learning techniques, proving valuable for applications requiring:

- Real-time adaptation.
- Computational efficiency.
- Interpretability.

Looking forward, integrating RLS with deep learning and probabilistic methods offers promising avenues for further innovation. Fields such as adaptive model predictive control (MPC), Bayesian filtering, and reinforcement learning exemplify where RLS principles can complement modern AI-driven frameworks, enriching the versatility and capability of contemporary control systems.

Ultimately, RLS exemplifies the core principle of machine learning: continuous learning and improvement from data. While it may lack the complexity of deep neural networks, the legacy of RLS in online adaptation and real-time learning continues to make it an invaluable tool in the ever-evolving landscape of machine learning for modeling and control.

References

1. Dehaene J, Moonen M, Vandewalle J (1995) Continuous-time recursive least-squares estimation, adaptive neural networks and systolic arrays. IEEE Trans Circuits Syst I Fundam Theory Appl 42(2):116–119

2. Shin H, Kim Y (2016) Flight envelope protection of aircraft using adaptive neural network and online linearisation. Int J Syst Sci 47(4):868–885
3. Osorio-Arteaga F, Giraldo E (2024) Adaptive neural network identification for robust multivariable systems. Int J Appl Math 54(1):68–76
4. Engel Y, Mannor S, Meir R (2004) The kernel recursive least-squares algorithm. IEEE Trans Signal Process 52(8):2275–2285
5. Kárný M (2014) Approximate Bayesian recursive estimation. Inf Sci 285(C):100–111
6. Chatterjee T, Shaw AD, Friswell MI, Khodaparast HH (2023) Sparse Bayesian machine learning for the interpretable identification of nonlinear structural dynamics: towards the experimental data-driven discovery of a quasi zero stiffness device. Mech Syst Signal Process 205:110858
7. Powell WB (2022) Reinforcement learning and stochastic optimization: a unified framework for sequential decisions. Wiley, Hoboken, New Jersey, USA
8. Haykin S (2014) Adaptive filter theory. Pearson Education, London, UK
9. Sayed AH (2003) Fundamentals of adaptive filtering. Wiley, Hoboken, New Jersey, USA
10. Sayed AH (2008) Adaptive filters. Wiley, Hoboken, New Jersey, USA
11. Theodoridis S (2020) Machine learning: a Bayesian and optimization perspective. Elsevier, Amsterdam
12. Bottou L (2012) Stochastic gradient descent tricks. In: Montavon G (ed) Neural networks: tricks of the trade. Springer, Berlin, pp 421–436
13. Brandt MW, Santa-Clara P (2006) Dynamic portfolio selection by augmenting the asset space. J Financ 60(5):2187–2217
14. Kivinen J, Smola AJ, Williamson RC (2004) Online learning with kernels. IEEE Trans Signal Process 52(8):2165–2176
15. Sayed AH, Kailath T (2010) Recursive least-squares adaptive filters. In: Madisetti VK (ed) The digital signal processing handbook: digital signal processing fundamentals. CRC Press, Boca Raton, pp 1–40
16. Hastie T, Tibshirani R, Friedman J (2009) The elements of statistical learning: data mining, inference, and prediction. Springer Science, Berlin
17. Mitchell TM (1997) Machine learning. McGraw-Hill, Columbus, Ohio, USA
18. Kalman RE (1960) A new approach to linear filtering and prediction problems. Trans ASME J Basic Eng 82(1):35–45

Chapter 16
Concluding Remarks

The central theme of this book is the effective use of model-based control for mass–stiffness–damping systems (MKC systems), emphasizing both theoretical rigor and practical applicability. This chapter provides a cohesive summary of the modeling and control frameworks presented throughout the book, with an outlook on future developments in the field.

16.1 Models: Foundations of Effective Control

Modern control theory and practice are deeply rooted in the principle that all control is model-based, either explicitly or implicitly [1]. Whether grounded in first-principles physics or derived from empirical data, mathematical models are indispensable for understanding system dynamics and synthesizing effective controllers.

Control-oriented modeling serves two fundamental purposes:

1. *Descriptive Modeling (Plant Knowledge):* A mathematical model encapsulates the behavior of the target system. Effective control begins with an adequate understanding of the plant, enabling predictive analysis, stability assessment, and performance evaluation.
2. *Prescriptive Modeling (Reference Behavior):* A reference model defines desired closed-loop dynamics. When a controller aligns the system output with a well-behaved reference model, control objectives such as tracking, stability, and disturbance rejection can be systematically achieved.

Throughout this book, model-based strategies have been consistently employed. The early chapters introduce feedback linearization and dynamic decoupling for transforming nonlinear, cross-coupled mass–stiffness–damping systems into approximately linear, decoupled subsystems. Subsequently, outer-loop feedback and feed-forward controls are synthesized for these simplified structures, leveraging the

H. Zhu, *Model-Based Control of Mass–Stiffness–Damping Systems*, Advances in Industrial Control, https://doi.org/10.1007/978-3-031-97592-9_16

resulting integrator dynamics for ease of control design. Disturbance rejection and uncertainty attenuation are tackled using feedforward and enhanced control strategies—all grounded in a model-based framework.

Yet, model-based control systems are inherently sensitive to modeling uncertainties. To address this, this book introduces a class of uncertainty-reducing structures that enhance robustness by compensating for discrepancies between the modeled and actual plant dynamics. This augmentation is still model-driven, preserving the benefits of structure and predictability while extending applicability under non-ideal conditions.

The central message is clear: even when perfect models are unattainable, designing with model uncertainty in mind enables a robust, effective approach to real-world control. Techniques such as model-following control, feedforward control, adaptive control, and internal model control are shown to benefit from structural augmentation for uncertainty mitigation, illustrating a principled way forward in the face of imperfect knowledge.

16.2 Control Strategies: Balancing Sophistication and Simplicity

A major thrust of this book is the synthesis of high-performance control strategies that are both powerful and practical. With increasing system complexity and performance requirements, control refinement—the process of enhancing control performance without compromising tractability—has become a central concern.

This is achieved through a modular, multiloop architecture, where simple control building blocks are composed to meet higher-level specifications:

- Inner-loop linearization and decoupling isolate core dynamics.
- Outer-loop state feedback ensures desired dynamic response.
- Integral action eliminates steady-state error (Sect. 7.3.2).
- Feedforward compensation improves disturbance rejection (Chap. 9).
- Uncertainty-reduction schemes enhance robustness with minimal overhead (Chap. 10).

Each layer addresses a different aspect of performance while maintaining conceptual simplicity. The result is a layered control framework that is intuitive, flexible, and adaptable to evolving system requirements. In addition to structural augmentation, robust design can be used to fine-tune controller parameters—a natural complement to the multiloop designs introduced here [2, 3].

This book advocates for design parsimony—complexity only where justified. Techniques like sliding-mode control (SMC), traditionally used as standalone robust controllers, are repurposed here as complementary augmentation tools to enhance the disturbance rejection or tracking behavior of an existing feedback loop [4–7].

Such hybrid approaches demonstrate that advanced control need not be prohibitively complex. Instead, when layered thoughtfully, sophisticated control logic can be embedded into relatively simple and transparent architectures—bridging the gap between cutting-edge research and practical implementation.

16.3 Advanced Control: A Perspective on Future Directions

The evolution of control systems continues to be shaped by increased system complexity, uncertainty, and the demand for autonomy. As discussed throughout this book, classical model-based control remains foundational—but must be augmented to handle modern challenges.

Adaptive and intelligent control offer promising avenues:

- Adaptive control dynamically adjusts controller parameters to track changes in system behavior or the environment [8–10]. This capability is critical when plant models are incomplete or time-varying.
- Robust adaptive schemes [11] blend structural robustness with online learning [12–15], enabling performance preservation in the face of both structured and unstructured uncertainty.
- Intelligent control, including fuzzy logic and neural networks, introduces flexibility in handling nonlinearities, incomplete models, and human-like reasoning [16–19]. These techniques are especially valuable in systems with limited model information or rapidly changing dynamics.

Future control systems must also be fault-tolerant, capable of graceful degradation and active redundancy. Controllers operating in standby or complementary mode can take over when primary loops become ineffective, thereby enhancing reliability and performance without significant structural overhead.

Looking ahead, control engineers will face an environment where autonomy, adaptability, and computational efficiency are paramount. Control systems must operate across broader regions of uncertainty, satisfy stricter performance constraints, and do so cost-effectively. This necessitates:

- Methods for automated model refinement and learning.
- Hierarchical control architectures for large-scale systems.
- Integrated design frameworks combining physical modeling, machine learning, and real-time adaptation [20–22].

References

1. Fisher DG (1991) Process control: an overview and personal perspective. Can J Chem Eng 69(1):5–26

2. Abdallah C, Dawson D, Dorato P, Jamshidi M (1991) Survey of robust control for rigid robots. IEEE Control Syst Mag 11(2):24–30
3. Spong MW (1992) On the robust control of robot manipulators. IEEE Trans Autom Control 37(11):1782–1786
4. Utkin VI (1977) Variable structure systems with sliding modes. IEEE Trans Autom Control 22(2):212–222
5. Utkin VI (1992) Sliding modes in control and optimization. Springer-Verlag
6. Hong GS, Zhu HA, Teo CL, Poo AN (1993) Robust control of robotic manipulators with model-based precompensation and SMC postcompensation. Proc Inst Mech Eng Part I J Syst Control Eng 207(2):97–103
7. Zhu HA, Teo CL, Hong GS (1993) Robust quasi-SMC of mechanical manipulators with dual-model pre-compensation. In: Proceedings of the 1993 American control conference, vol 3, pp 2511–2515
8. Tomizuka M, Horowitz R, Anwar G, Jia YL (1988) Implementation of adaptive techniques for motion control of robotic manipulators. J Dyn Syst Meas Control 110(1):62–69
9. Narendra KS, Annaswamy AM (1989) Stable adaptive systems. Prentice-Hall, New Jersey, USA
10. Åström KJ, Wittenmark B (2008) Adaptive control. Dover Publications, New York, USA
11. Landau ID (1988) Future trends in adaptive control of robot manipulators. In: Proceedings of the 27th IEEE conference on decision and control, vol 2, pp 1604–1606
12. Peek MD, Antsaklis PJ (1990) Parameter learning for performance adaptation. IEEE Control Syst Mag 10(7):3–11
13. Moore KL (1993) Iterative learning control for deterministic systems. Springer, Berlin
14. Antsaklis PJ, Passino KM (1993) An introduction to intelligent and autonomous control. Kluwer Academic Publishers, Dordrecht
15. Antsaklis PJ, Passino KM, Wang SJ (1991) An introduction to autonomous control systems. IEEE Control Syst 11(4):5–13
16. Cela A, Hamam Y (1992) A fuzzy robust controller for trajectory tracking of robotic systems. In: Proceedings of the 1992 IEEE/RSJ International conference on intelligent robots and systems, pp 767–774
17. Passino KM, Yurkovich S (1998) Fuzzy control. Addison Wesley, Reading, USA
18. Miller WT, Sutton RS, Werbos PJ (1990) Neural networks for control. The MIT Press, Cambridge, USA
19. Hunt KJ, Sbarbaro D, Żbikowski R, Gawthrop PJ (1992) Neural networks for control systems—a survey. Automatica 28(6):1083–1112
20. Fu KS (1971) Learning control systems and intelligent control systems: an intersection of artificial intelligence and automatic control. IEEE Trans Autom Control 16(1):70–72
21. Åström KJ (1989) Toward intelligent control. IEEE Control Syst Mag 9(3):60–64
22. Åström KJ (1991) Where is the intelligence in intelligent control? IEEE Control Syst Mag 11(1):37–39

Afterword

This book has focused on developing practical, high-performance model-based controllers for mass–stiffness–damping systems (MKC systems). From foundational modeling to advanced structural enhancements, the methodologies presented emphasize a balance between theoretical rigor and engineering pragmatism.

Every effort has been made to ensure accessibility without sacrificing analytical clarity. It is hoped that the engineering-oriented presentation style used here will empower practitioners to apply and implement the methods across a wide range of applications.

While many advances have been consolidated in this volume, much work remains. The research trajectory outlined here is part of an ongoing story—one in which the tools of control theory continue to evolve, and where the intelligent integration of models, adaptation, and robustness will define the next generation of high-performance control systems.

H. Zhu, *Model-Based Control of Mass–Stiffness–Damping Systems*, Advances in Industrial Control, https://doi.org/10.1007/978-3-031-97592-9

Appendix A
Model-Based Linearization and Decoupling of MKC Systems

This appendix summarizes the foundational technique used throughout this book for model-based linearization and decoupling of mass–stiffness–damping systems (MKC systems). The method provides an essential inner-loop framework that transforms complex nonlinear, cross-coupled dynamics into a simplified, linear, and decoupled form suitable for advanced control design.

To linearize and decouple the general nonlinear dynamics of an n-degrees-of-freedom mass–stiffness–damping system described by:

$$\mathbf{D}(\boldsymbol{q})\ddot{q} + \boldsymbol{h}(\dot{q}, \boldsymbol{q}) = \boldsymbol{\tau}_C, \tag{A.1}$$

a model-based dynamic compensation is applied in the form:

$$\boldsymbol{\tau}_C := \hat{\mathbf{D}}(\boldsymbol{q})\boldsymbol{u} + \hat{h}(\dot{q}, \boldsymbol{q}), \tag{A.2}$$

where

- $\boldsymbol{q} \in \mathbb{R}^n$, $\dot{q}$, and $\ddot{q}$ are the vectors of displacement, velocity, and acceleration, respectively.
- $\mathbf{D}(\boldsymbol{q}) \in \mathbb{R}^{n\times n}$ is the positive-definite, configuration-dependent inertia matrix.
- $\boldsymbol{h}(\dot{q}, \boldsymbol{q}) \in \mathbb{R}^n$ captures Coriolis, centrifugal, and gravitational effects.
- $\boldsymbol{\tau}_C \in \mathbb{R}^n$ is the control input to the system.
- $\boldsymbol{u} \in \mathbb{R}^n$ is the virtual control input in the compensated linearized domain.
- Quantities with hats, $\widehat{\mathbf{D}}(\boldsymbol{q})$ and $\hat{h}(\dot{q}, \boldsymbol{q})$, denote model-based estimates.

This control structure is illustrated in Fig. A.1, which shows how the control loop effectively cancels the nonlinear terms by superimposing the estimated dynamics onto the real plant.

Assuming a sufficiently accurate model—i.e., $\widehat{\mathbf{D}}(\boldsymbol{q}) \approx \mathbf{D}(\boldsymbol{q})$ and $\hat{h}(\dot{q}, \boldsymbol{q}) \approx \boldsymbol{h}(\dot{q}, \boldsymbol{q})$—the nonlinear dynamics within the plant are cancelled by their model-based counterparts within the controller, yielding:

H. Zhu, *Model-Based Control of Mass–Stiffness–Damping Systems*, Advances in Industrial Control, https://doi.org/10.1007/978-3-031-97592-9

Fig. A.1 Model-based linearization and decoupling

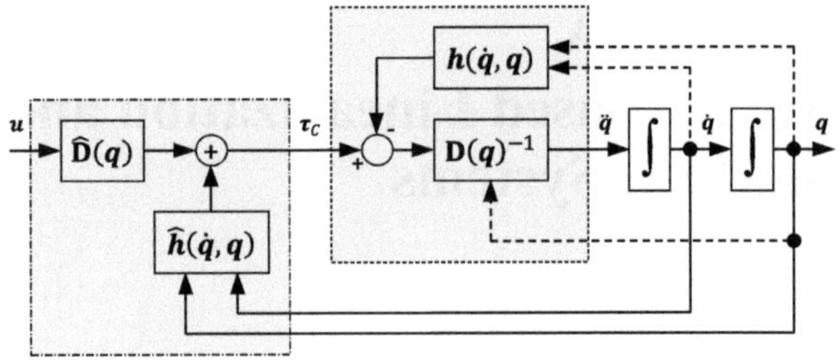

$$\ddot{q} = \boldsymbol{u}.$$

This results in an equivalent system of n decoupled second-order linear integrators:

$$\boldsymbol{Q}(s) = \left[\mathbf{I}s^2\right]^{-1}\boldsymbol{U}(s), \tag{A.3}$$

where $\boldsymbol{Q}(s)$ and $\boldsymbol{U}(s)$ are Laplace transforms of $\boldsymbol{q}(t)$ and $\boldsymbol{u}(t)$, respectively.

To define the desired closed-loop behavior, a PD control law in the outer loop is typically employed:

$$\boldsymbol{u} := \boldsymbol{r} - \mathbf{K}_v\dot{q} - \mathbf{K}_p\boldsymbol{q}, \tag{A.4}$$

where

- $\mathbf{K}_v$, $\mathbf{K}_p \in \mathbb{R}^{n\times n}$ are diagonal matrices defining damping and stiffness feedback gains, respectively.
- $\boldsymbol{r} \in \mathbb{R}^n$ is the reference input vector.

Substituting Eq. (A.4) into Eq. (A.3), the overall closed-loop dynamics become:

$$\boldsymbol{Q}(s) = \left[\mathbf{I}s^2 + \mathbf{K}_v s + \mathbf{K}_p\right]^{-1}\boldsymbol{R}(s). \tag{A.5}$$

This corresponds to a set of decoupled second-order systems whose poles can be arbitrarily assigned by tuning $\mathbf{K}_v$ and $\mathbf{K}_p$, offering precise control over transient and steady-state behavior.

For set-point regulation, where the objective is to move the system from an initial position $\boldsymbol{q}_i \in \mathbb{R}^n$ to a desired final position $\boldsymbol{q}_f \in \mathbb{R}^n$, the reference input $\boldsymbol{r}$ in the outer-loop controller can be chosen as:

$$\boldsymbol{r} := \mathbf{K}_p\left(\boldsymbol{q}_f - \boldsymbol{q}_i\right). \tag{A.6}$$

This ensures that the closed-loop system asymptotically converges to the specified final configuration.

For trajectory tracking, where the goal is to follow a continuously varying reference path $\boldsymbol{q}_d(t) \in \mathbb{R}^n$, the reference input $\boldsymbol{r}$ is constructed using model-based feedforward compensation:

$$\boldsymbol{r} := \ddot{q}_d(t) - \mathbf{K}_v\dot{q}_d(t) - \mathbf{K}_p\boldsymbol{q}_d(t). \tag{A.7}$$

Substituting this expression into the outer-loop PD controller and assuming ideal inner-loop dynamics (i.e., accurate model cancelation and no disturbances), the resulting closed-loop system satisfies:

$$\boldsymbol{Q}(s) = \boldsymbol{Q}_d(s) \quad \text{or} \quad \boldsymbol{q}(t) = \boldsymbol{q}_d(t). \tag{A.8}$$

This reflects perfect trajectory tracking, under the assumption of exact modeling and disturbance-free operation.

For ease of reference and practical implementation, the model-based linearization and decoupling procedure described above are summarized next as Algorithm A.1.

Algorithm A.1 Assume that the dynamics of the mass–stiffness–damping system, as described in Eq. (A.1), are perfectly modeled, with negligible actuator dynamics and no external disturbances. By applying the model-based compensating control input defined as:

$$\boldsymbol{\tau}_C := \widehat{\mathbf{D}}(\boldsymbol{q})u + \hat{h}(\dot{q}, \boldsymbol{q}),$$

the nonlinear, cross-coupled system dynamics are effectively cancelled. As a result, the inner-loop dynamics of the compensated system reduce to:

$$\ddot{q} = \boldsymbol{u}.$$

This transformation yields a set of decoupled second-order integrators, thereby linearizing and decoupling the original nonlinear, cross-coupled system dynamics.

This method establishes a powerful model-based framework for transforming nonlinear, cross-coupled multibody mass–stiffness–damping dynamics into simplified linear, decoupled forms, thereby enabling the design of high-performance control strategies. It is both conceptually foundational and practically indispensable in robotics, where it is widely recognized as the computed-torque control (CTC) approach.

While Algorithm A.1 is idealized and assumes perfect modeling, its structure serves as a basis for the more robust multiloop control strategies discussed throughout this book, such as feedforward disturbance rejection, adaptive control, and robust IMC enhancement.

In real-world applications where actuator bandwidth or model uncertainty is non-negligible, additional compensatory mechanisms (e.g., feedback linearization with robustness enhancement or adaptive schemes) are typically required.

Appendix B
Pole-Placement of Dynamic Systems

This appendix summarizes the pole-placement techniques for second- and third-order linear time-invariant (LTI) systems, as used throughout this book for designing controllers in mass–stiffness–damping systems (MKC systems). These methods are foundational in control system design, especially for systems requiring specific transient performance characteristics.

B.1 Pole-Placement for Second-Order Systems

Consider a second-order characteristic equation representing the closed-loop dynamics of a single-input, single-output (SISO) system:

$$s^2 + k_v s + k_p = 0. \tag{B.1}$$

Here, k_v and k_p are positive real constants representing feedback gains. For the system to be asymptotically stable, the following conditions must be satisfied:

$$k_v > 0 \quad \text{and} \quad k_p > 0. \tag{B.2}$$

For systematic pole-placement, the characteristic polynomial in Eq. (B.1) can be rewritten in terms of standard second-order system parameters as:

$$\begin{aligned} \Gamma_1(s) &= s^2 + k_v s + k_p \\ &= s^2 + 2\zeta\omega_n s + \omega_n^2 \\ &= \left[s + \omega_n\left(\zeta + \sqrt{\zeta^2 - 1}\right)\right]\left[s + \omega_n\left(\zeta - \sqrt{\zeta^2 - 1}\right)\right], \end{aligned} \tag{B.3}$$

where

H. Zhu, *Model-Based Control of Mass–Stiffness–Damping Systems*, Advances in Industrial Control, https://doi.org/10.1007/978-3-031-97592-9

- $\omega_n > 0$ is the undamped natural frequency.
- $\zeta \geq 0$ is the damping ratio.

The roots of Eq. (B.3) correspond to:

- A pair of complex conjugate poles for $0 < \zeta < 1$ (underdamped).
- A repeated real pole for $\zeta = 1$ (critically damped).
- Two distinct real poles for $\zeta > 1$ (overdamped).

In many motion control applications, for example in robotics, overdamped or critically damped behavior ($\zeta \geq 1$) is often preferred to avoid oscillatory response and to improve settling time.

From Eq. (B.3), it is evident that the feedback gains are directly related to the desired closed-loop pole locations. Specifically, the relationships are expressed as:

$$k_v = 2\zeta\omega_n \tag{B.4}$$

$$k_p = \omega_n^2. \tag{B.5}$$

Consequently, once the desired pole positions are selected—characterized by the damping ratio ζ and the natural frequency ω_n—these parameters can be systematically mapped to the corresponding velocity and position feedback gains. This direct mapping facilitates intuitive controller design, enabling engineers to achieve precise dynamic performance through appropriate gain selection.

B.2 Pole-Placement for Third-Order Systems

Now consider the third-order characteristic equation, often encountered in systems that include actuator dynamics or additional control loops:

$$s^3 + k_a s^2 + k_v s + k_p = 0. \tag{B.6}$$

To ensure asymptotic stability, the Routh–Hurwitz conditions for third-order systems impose the following constraints:

$$k_a > 0, \quad k_v > 0, \quad k_p > 0, \quad \text{and} \quad k_a k_v > k_p. \tag{B.7}$$

For design purposes, the characteristic equation in Eq. (B.6) is often constructed to yield one real pole and one complex conjugate pair. It can therefore be factored as:

$$\begin{aligned}\Gamma_2(s) &= s^3 + k_a s^2 + k_v s + k_p \\ &= (s+\lambda)\left(s^2 + 2\zeta\omega_n s + \omega_n^2\right)\end{aligned}$$

$$= (s + \lambda)\Big[s + \omega_n\Big(\zeta + \sqrt{\zeta^2 - 1}\Big)\Big]\Big[s + \omega_n\Big(\zeta - \sqrt{\zeta^2 - 1}\Big)\Big], \tag{B.8}$$

where

- $-\lambda < 0$ is a real pole.
- $(s^2 + 2\zeta\omega_n s + \omega_n^2)$ represents a pair of stable poles.
- $\zeta \geq 1$ is typically selected to ensure non-oscillatory response.

From Eq. (B.8), it is evident that the feedback gains are explicitly related to the desired closed-loop pole locations. These relationships can be clearly expressed as:

$$k_a = \lambda + 2\zeta\omega_n \tag{B.9}$$

$$k_v = \lambda \cdot 2\zeta\omega_n + \omega_n^2 \tag{B.10}$$

$$k_p = \lambda \cdot \omega_n^2. \tag{B.11}$$

These equations provide a direct and intuitive mapping between the selected closed-loop pole parameters—namely, the damping ratio ζ, the natural frequency ω_n, and the additional pole-placement parameter λ —and the corresponding acceleration, velocity, and position feedback gains. This direct relationship simplifies control system design, allowing engineers to easily translate desired dynamic performance criteria into appropriate controller gains.

B.3 Remarks and Applications

- The pole-placement method provides direct control over transient response characteristics such as settling time, overshoot, and rise time.
- The above relationships assume a unity-gain open-loop system. If the actual system involves scaled dynamics or incorporates actuator gains, these expressions must be appropriately adjusted to account for those additional factors.
- In practical applications, it is critical to ensure robustness to model uncertainties when choosing pole locations. Overly aggressive pole-placement may result in poor noise attenuation or actuator saturation.

Pole-placement remains one of the most widely used techniques in both academic and industrial control system designs. It is particularly effective when the system is controllable and full-state feedback is available or can be estimated via observers.

These pole-placement formulas are repeatedly used in this book for motion control and stabilization of mass–stiffness–damping systems. For systems with non-minimum-phase dynamics or parameter uncertainties, robust pole-placement strategies (e.g., using LQR or $\mathcal{H}^\infty$ methods) may offer enhanced performance guarantees.

Appendix C
Motion Tasks of Robotic Manipulators

This appendix compiles supporting materials related to trajectory planning, motion task specifications, and disturbance models employed in the numerical examples throughout this book. These materials underpin the simulation studies conducted for validating various control strategies applied to mass–stiffness–damping systems (MKC systems), particularly in the context of robotic manipulators.

C.1 Polynomial Trajectory for Smooth Motion Planning

For trajectory-tracking tasks in robotic manipulators, smooth motion generation is essential to ensure bounded control efforts and avoid high-frequency excitation of unmodeled dynamics. In this book, a 7th-order polynomial trajectory is adopted for motion interpolation between an initial position q_0 and a final position $\left(q_0 + q_f\right)$ over a finite time interval $t \in \left[0, t_f\right]$.

The trajectory is designed to ensure zero boundary conditions not only for position but also for velocity, acceleration, and jerk (first through third derivatives), leading to a minimally exciting yet dynamically smooth trajectory.

The desired trajectory is defined as:

$$q_d(t) = q_0 + \left[35\frac{t^4}{t_f^4} - 84\frac{t^5}{t_f^5} + 70\frac{t^6}{t_f^6} - 20\frac{t^7}{t_f^7}\right]q_f. \tag{C.1}$$

This trajectory satisfies the boundary conditions:

- Position: $q_d(0) = q_0$ and $q_d\left(t_f\right) = q_0 + q_f$.
- Velocity: $\dot{q}_d(0) = \dot{q}_d\left(t_f\right) = 0$.
- Acceleration: $\ddot{q}_d(0) = \ddot{q}_d\left(t_f\right) = 0$.
- Jerk: $\dddot{q}_d(0) = \dddot{q}_d\left(t_f\right) = 0$.

H. Zhu, *Model-Based Control of Mass–Stiffness–Damping Systems*, Advances in Industrial Control, https://doi.org/10.1007/978-3-031-97592-9

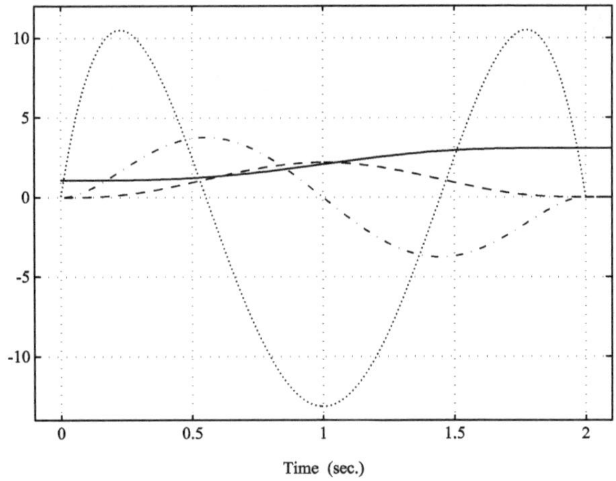

Fig. C.1 Trajectory and its derivatives

These smooth boundary conditions are crucial for avoiding abrupt actuator commands and for maintaining continuity in the dynamic response of the manipulator. For $q_0 = \pi/2$, $q_f = 2.0$, and $t_f = 2.0$, the reference trajectory $q_d(t)$ along with the profiles of its time derivatives is illustrated in Fig. C.1 {Solid line: $q_d(t)$. Dashed line: $\dot{q}_d(t)$. Dash-dot line: $\ddot{q}_d(t)$. Dotted line: $\dddot{q}_d(t)$} to provide intuition on its smoothness and suitability for actuator-friendly control inputs.

C.2 Physical Parameters and Motion Task Specifications

For the numerical simulations of both single- and two-link manipulators, the following physical parameters are used.

Single-Link Manipulator System:

- Link mass: $m_0 = 11.36$ kg.
- Link length: $l = 0.432$ m.
- Gear ratio: $N = 53.65$.
- Motor and drive parameters:
  - Back-EMF constant: $k_E = 0.19\,\text{V/rad/s}$.
  - Torque constant: $k_T = 0.2613\,\text{Nm/A}$.
  - Armature inductance: $L_a = 0.0048\ \Omega\text{s}$.
  - Armature resistance: $R_a = 1.6\,\Omega$.

Two-Link Manipulator System:

- Link mass: $m_1 = 15.91$ kg and $m_2 = 11.36$ kg.
- Link length: $l_1 = l_2 = 0.432$ m.
- Gear ratio: $N_1 = 107.81$ and $N_2 = 53.65$.

- Motor and drive parameters:
  - Back-EMF constant: $k_{E1} = k_{E2} = 0.19$ V/rad/s.
  - Torque constant: $k_{T1} = k_{T2} = 0.2613$ Nm/A.
  - Armature inductance: $L_{a1} = L_{a2} = 0.0048$ Ωs.
  - Armature resistance: $R_{a1} = R_{a2} = 1.6$ Ω.

  Payload applied at the end-effector for various case studies: $m_p = 5$ kg.

Motion Task for Two-Link Manipulator:
The following trajectory-tracking task is used to evaluate control performance:

- Initial condition:
  - Joint 1 (lower link): $q_{10} = 0.5$ rad (held constant).
  - Joint 2 (upper link): $q_{20} = \pi/2$ rad.
- Task requirement:
  - Joint 1 remains fixed, that is, $q_{1f} = 0.5$ rad.
  - Joint 2 moves counterclockwise by $q_{2f} = 2.0$ rad.
  - Motion duration: $t_f = 2.0$ s.
  - Trajectory defined using Eq. (C.1).

Other task configurations, including multi-joint motions, hybrid tracking, and regulation, are detailed within the relevant examples and experiments across Chaps. 6–14.

C.3 External Disturbances

External disturbances are introduced in the simulation examples to assess the robustness and disturbance-rejection capabilities of the discussed controllers. The following disturbance signals are considered:

- Impulse Disturbance:
  A short-duration disturbance of 50 Nm applied over $t \in [0.50, 0.51]$ s.
- Step Disturbance:
  A sustained disturbance of 10 Nm applied over $t \in [0.50, 1.00]$ s.

These disturbances are injected either at the actuator torque level or at specific joints, depending on the case study, and reflect common real-world perturbations such as sudden contact forces, frictional effects, or abrupt load changes.

The trajectory profiles and disturbance inputs summarized here form the experimental basis for validating a wide range of control algorithms presented in this book—including model-based feedback linearization, feedforward compensation, adaptive control, and robust IMC design. They serve as standard benchmarks to demonstrate controller effectiveness under both nominal and perturbed conditions.

Appendix D
A Reference Testbed

This appendix outlines the experimental testbed used to validate the various modeling and control algorithms introduced throughout this book. The setup centers on a single-link, vertically actuated, direct-drive robotic manipulator. Despite its apparent simplicity, the testbed presents a wide range of real-world challenges—such as nonlinearity, unmodeled dynamics, parameter variations, and disturbance effects—that are essential for rigorous control system evaluation.

The configuration of the test environment is illustrated schematically in Fig. D.1. A standard desktop PC serves as the real-time control platform, executing control algorithms in a software loop interfaced with the hardware system via high-speed data acquisition components.

The single-link manipulator is driven by a high-performance DC servomotor mounted at the base, which directly actuates the rigid link without gear reduction (direct-drive configuration). The motor assembly includes:

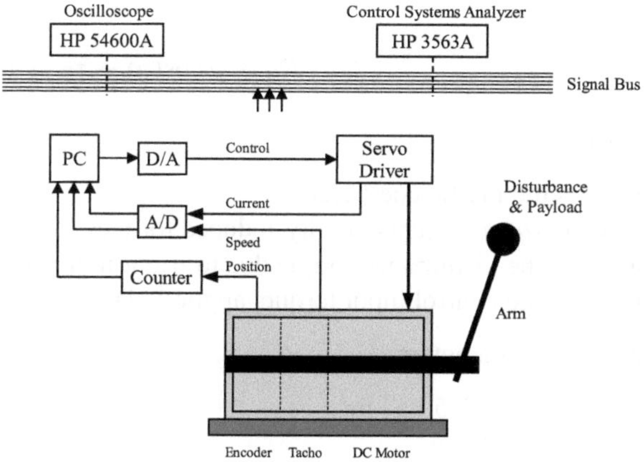

Fig. D.1 Experimental setup

H. Zhu, *Model-Based Control of Mass–Stiffness–Damping Systems*, Advances in Industrial Control, https://doi.org/10.1007/978-3-031-97592-9

- Tacho-generator (7 V/K rpm): Provides continuous analog voltage proportional to angular speed.
- Incremental Encoder (1000 pulses per revolution, ppr): Provides position feedback.

The control loop uses both position and velocity feedback for algorithm implementation. Current measurement is obtained via a built-in current sensor in the servo driver, delivering an analog signal proportional to the armature current—a critical signal for implementing torque-based control schemes, including: Algorithms 6.1, 9.1 and 9.2.

To enable real-time digital control, sensor signals and control commands are routed through the following signal interface components:

- ADDA-14 Analog–Digital/Digital–Analog Converter Card:
  - Input channels (14-bit ADC, < 42 μs conversion time).
  - Two output channels (14-bit DAC, 0.5 μs settling time).
  - Provides fast and precise analog–digital conversion for sensor signals and control outputs.
- Quadrature Decoder and 16-Bit Counter Interface:
  - Decodes encoder channels A and B to generate a $4\times$-interpolated count signal.
  - Converts 1000 ppr encoder to effective resolution of 4000 ppr, enhancing precision.
  - Counts are read via a digital I/O interface card (8255/8253 standard) into the PC control loop.

This position feedback enables accurate state estimation, making the testbed suitable for validating both classical and modern control strategies.

The rigid single-link moves in the vertical plane, where its dynamics are governed by:

$$\mathcal{D}\ddot{q} + \mathcal{G}\cos(q) + \tau_D = \tau_C$$

where

- $\mathcal{D}$ is the inertia coefficient.
- $\mathcal{G}\cos(q)$ represents gravity-induced torque.
- τ_D is the disturbance torque due to friction, unmodeled dynamics, etc.
- τ_C is the control input torque, applied via the motor.

Nominal Parameters:

- $\mathcal{D}_0 = 0.00765\ \text{kg}\cdot\text{m}^2$.
- $\mathcal{G}_0 = 0.52$ Nm.

Parameter Variations for Testing Robustness:

- Low Inertia/Payload: $\mathcal{D} = 0.000864$ and $\mathcal{G} = 0.080$.

- High Inertia/Payload: $\mathcal{D} = 0.028490$ and $\mathcal{G} = 1.053$.

These variations are used to test sensitivity and robustness of control laws under model mismatch and varying operating conditions.

Motor Electrical Parameters (Estimated):

- Back-EMF constant: $k_E = 0.1595$ V/rad/s.
- Torque constant: $k_T = 0.1524$ Nm/A.
- Armature inductance: $L_a = 0.0052$ Ωs.
- Armature resistance: $R_a = 2.2$ Ω.

Though the hardware setup is compact, it encompasses a broad spectrum of control challenges that mirror those found in large-scale or industrial applications:

- Parameter uncertainty (e.g., due to payload variation).
- Nonlinear dynamics (e.g., due to gravity and friction).
- Sensor noise and quantization errors (e.g., in position encoders).
- External disturbances, both impulsive and sustained.
- Unmodeled actuator dynamics, especially in transient regimes.

These characteristics make the testbed an ideal platform for validating model-based control strategies—such as feedforward control, model-following control, adaptive pole-placement, and internal model control—as developed throughout this book.

The reference testbed described in this appendix is more than a simple experimental fixture. It embodies a representative platform for evaluating modern control methods under real-world conditions. Its combination of rich dynamics, sensor integration, and flexible software interface allows for practical experimentation and validation of advanced theoretical developments, serving as a bridge between analytical modeling and industrial implementation.

Index

H. Zhu, *Model-Based Control of Mass–Stiffness–Damping Systems*, Advances in Industrial Control, https://doi.org/10.1007/978-3-031-97592-9

B

C

E

M

T

MIX
Papier aus verantwortungsvollen Quellen
Paper from responsible sources
FSC® C105338

FSC
www.fsc.org

If you have any concerns about our products,
you can contact us on
ProductSafety@springernature.com

In case Publisher is established outside the EU,
the EU authorized representative is:
Springer Nature Customer Service Center GmbH
Europaplatz 3, 69115 Heidelberg, Germany

Printed by Libri Plureos GmbH
in Hamburg, Germany